Bioremediation of Chlorinated Solvents

BIOREMEDIATION

The *Bioremediation* series contains collections of articles derived from many of the presentations made at the First, Second, and Third International In Situ and On-Site Bioreclamation Symposia, which were held in 1991, 1993, and 1995 in San Diego, California.

First International In Situ and On-Site Bioreclamation Symposium
1(1)   *On-Site Bioreclamation: Processes for Xenobiotic and Hydrocarbon Treatment*
1(2)   *In Situ Bioreclamation: Applications and Investigations for Hydrocarbon and Contaminated Site Remediation*

Second International In Situ and On-Site Bioreclamation Symposium
2(1)   *Bioremediation of Chlorinated and Polycyclic Aromatic Hydrocarbon Compounds*
2(2)   *Hydrocarbon Bioremediation*
2(3)   *Applied Biotechnology for Site Remediation*
2(4)   *Emerging Technology for Bioremediation of Metals*
2(5)   *Air Sparging for Site Bioremediation*

Third International In Situ and On-Site Bioreclamation Symposium
3(1)   *Intrinsic Bioremediation*
3(2)   *In Situ Aeration: Air Sparging, Bioventing, and Related Remediation Processes*
3(3)   *Bioaugmentation for Site Remediation*
3(4)   *Bioremediation of Chlorinated Solvents*
3(5)   *Monitoring and Verification of Bioremediation*
3(6)   *Applied Bioremediation of Petroleum Hydrocarbons*
3(7)   *Bioremediation of Recalcitrant Organics*
3(8)   *Microbial Processes for Bioremediation*
3(9)   *Biological Unit Processes for Hazardous Waste Treatment*
3(10) *Bioremediation of Inorganics*

Bioremediation Series Cumulative Indices: 1991-1995

For information about ordering books in the Bioremediation series, contact Battelle Press. Telephone: 800-451-3543 or 614-424-6393. Fax: 614-424-3819. Internet: sheldric@battelle.org.

Bioremediation of Chlorinated Solvents

Edited by

Robert E. Hinchee and Andrea Leeson
Battelle Memorial Institute

Lewis Semprini
Oregon State University

BATTELLE PRESS

Columbus • Richland

Library of Congress Cataloging-in-Publication Data

Hinchee, Robert E.
    Bioremediation of chlorinated solvents / edited by Robert E. Hinchee,
    Andrea Leeson, Lewis Semprini.
        p.   cm.
    Includes bibliographical references and index.
    ISBN 1-57477-005-5 (hc : acid-free paper)
    1. Solvents—Biodegradation—Congresses. 2. Organochlorine
    compounds—Biodegradation—Congresses. 3. Bioremediation—
    Congresses. I. Hinchee, Robert E. II. Leeson, Andrea. III. Semprini,
    Lewis.
    TD196.S54B56 1995
    628.5′2—dc20   95-32286
                                                                     CIP

Printed in the United States of America

Additional copies may be ordered through:
Battelle Press
505 King Avenue
Columbus, Ohio 43201, USA
1-614-424-6393 or 1-800-451-3543
Fax: 1-614-424-3819
Internet: sheldric@battelle.org

CONTENTS

FOREWORD

This book and its companion volumes (see overleaf) comprise a collection of papers derived from the Third International In Situ and On-Site Bioreclamation Symposium, held in San Diego, California, in April 1995. The 375 papers that appear in these volumes are those that were accepted after peer review. The editors believe that this collection is the most comprehensive and up-to-date work available in the field of bioremediation.

Significant advances have been made in bioremediation since the First and Second Symposia were held in 1991 and 1993. Bioremediation as a whole remains a rapidly advancing field, and new technologies continue to emerge. As the industry matures, the emphasis for some technologies shifts to application and refinement of proven methods, whereas the emphasis for emerging technologies moves from the laboratory to the field. For example, many technologies that can be applied to sites contaminated with petroleum hydrocarbons are now commercially available and have been applied to thousands of sites. In contrast, there are as yet no commercial technologies commonly used to remediate most recalcitrant compounds. The articles in these volumes report on field and laboratory research conducted both to develop promising new technologies and to improve existing technologies for remediation of a wide spectrum of compounds.

The articles in this volume discuss the use of aerobic and anaerobic biological degradation to dehalogenate sites contaminated with pesticides and chlorinated solvents such as trichloroethylene, tetrachloroethene, carbon tetrachloride, pentachlorophenols, and chlorinated benzenes.

Bench- and field-scale studies of the biological processes associated with in situ biodegradation of chlorinated solvent-contaminated soils and aquifers are described. Discussed are the use of microcosm studies and numerical simulation of dechlorination processes to support system design and the use of molecular- and cultural-based monitoring to manage system operation. Site characteristics (e.g., hydraulic properties, temperature, nitrogen availability) and their effect on the stability of the methanotrophic community are examined. Methods discussed include the use of air venting, alternative electron donors, biofilm reactors, surfactants, municipal digester sludge, iron enhancement, and sulfate reduction to improve conditions for the microbial consortia that effect biodegradation of chlorinated solvents.

The editors would like to recognize the substantial contribution of the peer reviewers who read and provided written comments to the authors of the draft articles that were considered for this volume. Thoughtful, insightful review is crucial for the production of a high-quality technical publication. The peer reviewers for this volume were:

Jens Aamand, *Geological Survey of Denmark*
Bruce Alleman, *Battelle Columbus*
Pedro J. Alvarez, *The University of Iowa*
John Armstrong, *The Traverse Group. Inc.*

Erik Arvin, *Technical University of Denmark*
Andrei Barkovskii, *University of Michigan*
Morton Barlaz, *North Carolina State University*
M.A. Becerra, *Mycotech Corporation*
Ned Black, *U.S. Environmental Protection Agency*
James S. Bonner, *Texas A & M University*
Dave A. Bower, *Waste Management, Inc.*
Angelo A. Bracco, *GE Corporate R&D Center*
Patrick V. Brady, *Sandia National Laboratories*
Edward Brown, *University of Northern Iowa*
Wil de Bruin, *Wageningen Agricultural University* (The Netherlands)
Verne T. Buehler, *Envirex, Inc.*
Gianni Chieruzzi, *Groundwater Technology, Inc.*
John Conrad, *Institute of Gas Technology*
Mary F. DeFlaun, *Envirogen, Inc.*
Greg Doyle, *IT Corporation*
James Duffy, *Occidental Chemical Corporation*
Chandra S. Dulam, *University of Connecticut*
Chris Du Plessis, *University of Natal* (Rep. of South Africa)
Soren Dyreborg, *Technical University of Denmark*
Elizabeth Edwards, *Beak Consultants, Ltd.* (Canada)
Udoudo Moses Ekanemesang, *Clark Atlanta University*
Roger L. Ely, *Oregon State University*
Charles T. Esler, *AGRA Earth & Environmental, Inc.*
Jim A. Field, *Wageningen Agricultural University* (The Netherlands)
Carl B. Fliermans, *Westinghouse Savannah River Co.*
Richard Gersberg, *San Diego State University*
John Glaser, *U.S. Environmental Protection Agency*
Roger B. Green, *Waste Management, Inc.*
Mark R. Harkness, *GE Corporate R&D Center*
Desma Hogg, *Woodward-Clyde Ltd.* (New Zealand)
J.C. Hughes, *University of Natal* (Rep. of South Africa)
Jonathan D. Istok, *Oregon State University*
Richard Johnson, *Alberta Environmental Centre*
Simeon Komisar, *University of Washington*
Suzanne E. Lantz, *SBP Technologies, Inc.*
Michael D. Lee, *DuPont Co.*
Jiyu Lei, *Biogénie inc.* (Canada)
Sarah Leihr, *North Carolina State University*
Charles Lovell, *University of South Carolina*
Thomas R. MacDonald, *Stanford University*
Andrzej Majcherczyk, *Universität Göttingen*
David W. Major, *Beak Consultants, Ltd.* (Canada)
Nigel Mark-Brown, *Woodward-Clyde Ltd.* (New Zealand)
Dean Martens, *University of California–Riverside*
Ilona McGhee, *University of Kent* (UK)

Mark C. Meckes, *U.S. Environmental Protection Agency*
William Mohn, *University of British Columbia*
Barry Molnaa, *Groundwater Technology, Inc.*
Peter Molton, *Battelle Pacific Northwest*
Pamela J. Morris, *Medical University of South Carolina*
Reynold Murray, *Clark Atlanta University*
Michael Niemet, *Oregon State University*
Minoru Nishimura, *Japan Research Institute, Ltd.*
John T. Novak, *Virginia Polytechnic & State University*
Donna Palmer, *Battelle Columbus*
Anthony V. Palumbo, *Oak Ridge National Laboratory*
Mary Peterson, *Battelle Pacific Northwest*
Tommy J. Phelps, *Oak Ridge National Laboratory*
Deborah J. Roberts, *University of Houston*
Amy G. Saberiyan, *AGRA Earth & Environmental, Inc.*
Rodney S. Skeen, *Battelle Pacific Northwest*
Robert J. Steffan, *Envirogen, Inc.*
Jaap J. van der Waarde, *Bioclear Environmental Biotechnology b.v.*
   (The Netherlands)
Timothy Vogel, *Rhône-Poulenc Industrialisation*
David White, *University of Tennessee*
Brian Wrenn, *University of Cincinnati*
Wei-Min Wu, *Michigan Biotechnology Institute*

The figure that appears on the cover of this volume was adapted from the article by Wu et al. (see page 49).

Finally, I want to recognize the key members of the production staff, who put forth significant effort in assembling this book and its companion volumes. Carol Young, the Symposium Administrator, was responsible for the administrative effort necessary to produce the ten volumes. She was assisted by Gina Melaragno, who tracked draft manuscripts through the review process and generated much of the correspondence with the authors, co-editors, and peer reviewers. Lynn Copley-Graves oversaw text editing and directed the layout of the book, compilation of the keyword indices, and production of the camera-ready copy. She was assisted by technical editors Bea Weaver and Ann Elliot. Loretta Bahn was responsible for text processing and worked many long hours incorporating editors' revisions, laying out the camera-ready pages and figures, and maintaining the keyword list. She was assisted by Sherry Galford and Cleta Richey; additional support was provided by Susan Vianna and her staff at Fishergate, Inc. Darlene Whyte and Mike Steve proofread the final copy. Judy Ward, Gina Melaragno, Bonnie Snodgrass, and Carol Young carried out final production tasks. Karl Nehring, who served as Symposium Administrator in 1991 and 1993, provided valuable insight and advice.

The symposium was sponsored by Battelle Memorial Institute with support from many organizations. The following organizations cosponsored or otherwise supported the Third Symposium.

Ajou University–College of Engineering (Korea)
American Petroleum Institute
Asian Institute of Technology (Thailand)
Biotreatment News
Castalia
ENEA (Italy)
Environment Canada
Environmental Protection
Gas Research Institute
Groundwater Technology, Inc.
Institut Français du Pétrole
Mitsubishi Corporation
OHM Remediation Services Corporation
Parsons Engineering Science, Inc.
RIVM–National Institute of Public Health and the Environment
    (The Netherlands)
The Japan Research Institute, Limited
Umweltbundesamt (Germany)
U.S. Air Force Armstrong Laboratory–Environics Directorate
U.S. Air Force Center for Environmental Excellence
U.S. Department of Energy Office of Technology Development
    (OTD)
U.S. Environmental Protection Agency
U.S. Naval Facilities Engineering Services Center
Western Region Hazardous Substance Research Center–
    Stanford and Oregon State Universities

Neither Battelle nor the cosponsoring or supporting organizations reviewed this book, and their support for the Symposium should not be construed as an endorsement of the book's content. I conducted the final review and selection of all papers published in this volume, making use of the essential input provided by the peer reviewers and other editors. I take responsibility for any errors or omissions in the final publication.

Rob Hinchee
June 1995

Aerobic Biodegradation of Trichloroethylene by Microorganisms that Degrade Aromatic Compounds

Chih-Jen Lu, Chu-Yin Chang, and Chi-Mei Lee

ABSTRACT

Aerobic biodegradation of trichloroethylene (TCE) at an initial concentration of 80 mg/L with and without the presence of an aromatic compound was conducted with a series of batch reactors. The target aromatic compounds were benzene, toluene, and catechol. The aromatics-acclimated microorganisms were used as the cell source for the batch study. The results indicated that the presence of an aromatic compound was required to initiate the aerobic biodegradation of TCE by the aromatic-utilizing microorganisms. The addition of benzene or toluene initiated the removal of TCE. However, TCE removal was not proportional to the initial concentration of the aromatic compounds. The presence of an aromatic compound at an initial concentration of 5 mg/L resulted in better TCE removal in comparison with that at 1 or 20 mg/L. TCE removal was still significant after the depletion of the aromatic compound, but at a lower rate. The presence of catechol, an intermediate of the biodegradation of an aromatic compound, did not initiate the biodegradation of TCE by the catechol-utilizing microorganisms.

INTRODUCTION

TCE is one of the most frequently reported contaminants in groundwaters contaminated by organic chemicals. Chlorinated aliphatics are suspected carcinogens and recalcitrant to biodegradation. The highly chlorinated aliphatics have been reported to be biotransformed to less-chlorinated aliphatics in an anaerobic environment through reductive dehalogenation, yet they are biologically recalcitrant under aerobic conditions (Bouwer et al. 1981; Vogel et al. 1987; Egli et al. 1988). However, biodegradation of chlorinated aliphatics by methane- or aromatic-utilizing bacteria has been shown (Fogel et al. 1986; Little et al. 1988; Nelson et al. 1988; Wackett & Gibson 1988). Oxygenase was a key enzyme for the aerobic biodegradation of TCE by aromatic-utilizing microorganisms (Wackett & Gibson 1988; Nelson et al. 1988; Zylstra et al. 1989). In this report, aerobic biodegradation of TCE was shown in the presence of various initial concentrations of an aromatic compound.

MATERIAL AND METHODS

TCE was used as the model compound for the volatile chlorinated aliphatics. Benzene, toluene, and catechol were selected as the aromatic compounds. Microorganisms were collected from the effluent of a petroleum wastewater treatment plant and then acclimated in a chemostat fed simultaneously with phenol, benzene, toluene, *o*-xylene, and catechol. The acclimated cultures were used as the source of aromatic-utilizing microorganisms for this study.

The effects of the presence of an aromatic compound on the aerobic biodegradation of TCE were examined using a series of batch reactors. Each reactor contained microorganisms and nutrient-containing solution. The reactor was completely filled to eliminate any headspace. The nutrient-containing solution had been well aerated before it was added into the batch reactors. TCE and the appropriate aromatic compound were then injected into the sealed bottles with a gastight syringe. The initial TCE concentration was 80 mg/L. A set of reactors that received nutrients and TCE only, but without microorganisms, was used as a control to measure the abiotic loss during the incubation period. Another set of bottles that received TCE, acetate (10 mg/L), and acetate-utilizing microorganisms was used to determine the effect of acetate-degrading microorganisms on the aerobic removal of TCE. The batch reactors were then inverted and shaken at 100 rpm in dark at room temperature. The batch reactors were sampled to analyze the remaining concentrations of the aromatic compound by high-pressure liquid chromatography (HPLC) and TCE by gas chromatography/electron capture detection (GC/ECD).

RESULTS AND DISCUSSION

Biodegradation of TCE in the Presence of Benzene

The data shown in Figure 1 illustrate the effect of benzene and acetate on the removal of TCE in batch reactors. TCE was not cometabolized by acetate-utilizing microorganisms in reactors fed with acetate. TCE removal by benzene-utilizing microorganisms was minimal in the absence of benzene. The microorganisms used in this study were harvested from a culture enriched on benzene. Benzene-utilizing microorganisms could not effectively cometabolize TCE without benzene, even if these microorganisms carried the degradative enzymes metabolizing benzene.

The data in Figure 1 show that the addition of benzene at 1 or 5 mg/L significantly enhanced the removal of TCE. After 11 days of incubation, 89% of

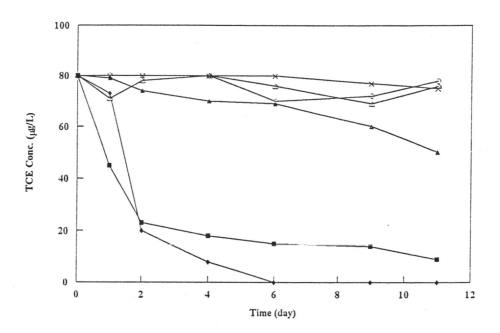

FIGURE 1. **Biodegradation of trichloroethylene in the presence of benzene (benzene conc.: O: blank, □: 0 mg/L, ■: 1 mg/L, ◆ : 5 mg/L, ▲: 20 mg/L, × : acetate-utilizing bacteria).**

TCE was removed with an initial benzene concentration of 1 mg/L. TCE was completely removed within 6 days at an initial benzene concentration of 5 mg/L. However, TCE removal decreased to only 30% when the initial benzene concentration was 20 mg/L. Benzene was completely removed within 1 day, when its initial concentration was 1 mg/L (Table 1). The TCE concentration decreased from 80 to 45 mg/L (44% removal) in 1 day in reactors initially containing 1 mg/L of benzene and further decreased to 8.8 mg/L, even though benzene was depleted.

A power function (concentration = a Time^b) was fitted to the data shown in Figure 1, and its derivation was used to estimate the TCE degradation rate (μg-TCE/L-day) (Figure 2). The results indicated that TCE was removed at a relatively lower rate after benzene had been completely removed. TCE was significantly cometabolized by benzene-utilizing microorganisms, when initial benzene concentration was 5 mg/L. As the only substrate, benzene was completely removed within 2 days, when its initial concentration was 5 mg/L (Table 1). TCE removal was 75% within 2 days and was completely removed after 6 days in the reactors that received benzene at 5 mg/L. The TCE degradation rate was much higher when benzene was still available. The results suggested that TCE cometabolism took place when benzene was present. However, Figure 1 also indicates that TCE removal was hindered if the initial

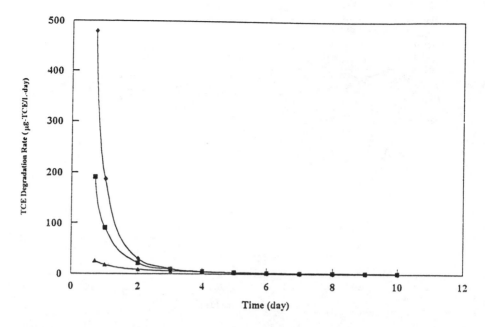

FIGURE 2. Trichloroethylene biodegradation rate (mg-TCE/day-L) in the presence of benzene (benzene initial conc.: ■: 1 mg/L, ◆: 5 mg/L, ▲: 20 mg/L).

TABLE 1. The remaining benzene and dissolved oxygen (DO) concentrations (mg/L) in reactors with both benzene and TCE.

	Initial Benzene Concentration					
	1 mg/L		5 mg/L		20 mg/L	
	Benzene	DO	Benzene	DO	Benzene	DO
1	<0.1	4.3	1.2	1.5	15	1.4
2	<0.1	4.2	<0.1	0.5	14	0.4
4	<0.1	4.2	<0.1	0.3	12	<0.3
6	<0.1	4.1	<0.1	<0.3	12	<0.3

benzene concentration was relatively high. When the initial benzene concentration was 20 mg/L, the TCE removal rate was lower compared to those for the initial benzene concentrations of 1 and 5 mg/L.

This result suggested that benzene may have competed with TCE for the degradative enzymes. Therefore, the TCE cometabolism rate was lower at the higher initial concentration of benzene, even if the higher concentration of benzene might initiate the higher production of biomass. The lower TCE removal rate may be due to the depletion of oxygen. The complete removal of benzene at the initial concentration of 20 mg/L theoretically required 30.8 mg/L of oxygen. However, the initial dissolved oxygen concentration was only 8.6 mg/L. The remaining dissolved oxygen concentration was less than 0.5 mg/L after 2 days of incubation. The lower dissolved oxygen concentration limited the microbial activity resulting in the lower TCE removal rates.

Figure 2 also shows that the TCE degradation rate decreased significantly after 2 days, when benzene had been completely removed. The TCE degradation rate decreased with decreases in the concentrations of the remaining benzene. However, cometabolism of TCE was still continuous even after benzene had been completely removed.

Biodegradation of TCE in the Presence of Toluene

The data in Figure 3 show the effect of toluene on the aerobic removal of TCE by culture enriched on toluene. TCE was not removed in the absence of toluene. The results indicated that toluene was required to initiate the cometabolism of TCE by the toluene-utilizing microorganisms. TCE removal was 20%, 86%, and 25% after 9 days of incubation at an initial toluene concentration of 1, 5, and 20 mg/L, respectively. These results suggested that toluene was

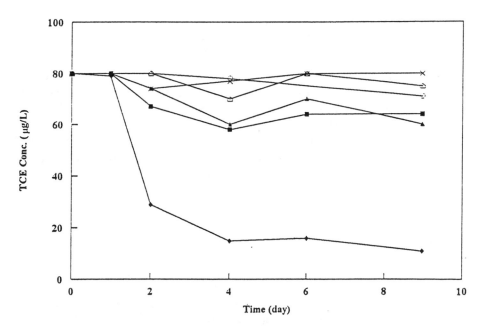

FIGURE 3. Biodegradation of trichloroethylene in the presence of toluene (toluene (toluene conc.: O: blank, □: 0 mg/L, ■: 1 mg/L, ◆ : 5 mg/L, ▲: 20 mg/L, × : acetate-utilizing bacteria).

required to initiate the cometabolism of TCE, but TCE removal was not proportional to the initial toluene concentrations. The lower TCE removal at the initial toluene concentration of 20 mg/L may result from the limited oxygen supply or from the less competitive to oxygenase for TCE compared to toluene. Figure 3 shows that TCE was kept intact in the reactors seeded with acetate-degrading microorganisms and fed only with acetate (10 mg/L). TCE was not cometabolized by acetate-utilizing microorganisms, even if the presence of acetate initiated the abundant production of biomass.

A power function was fitted to the data shown in Figure 3 to estimate the TCE degradation rate. The results were similar to those for the benzene reactors. When the toluene concentration was low, 1 mg/L, TCE was cometabolized at a lower rate. When toluene concentration was low, the capacity of the induced degradative enzymes was low and TCE was not significantly cometabolized. After 1 day of incubation 1 mg/L of toluene was completely removed and the TCE removal rate significantly decreased after this time. When the initial toluene concentration was 20 mg/L, toluene was more competitive for the affinity on the degradative enzymes and TCE was cometabolized at a lower rate. TCE cometabolism was most rapid at an initial toluene concentration of 5 mg/L. The results also indicated that TCE continued to be cometabolized at a lower rate after the depletion of toluene. The TCE degradation rate decreased with decreases in the concentrations of the remaining toluene. When toluene had been completely removed, TCE still continued to be cometabolized by the toluene-utilizing microorganisms, but at a lower rate.

Biodegradation of TCE in the Presence of Catechol

Catechol is an intermediate for the aerobic biodegradation of an aromatic compound. When catechol was used as the primary substrate to initiate the cometabolism of TCE, the results indicated that catechol did not initiate the removal of TCE. In this study, catechol was completely removed within 1 and 2 days, when its initial concentrations were 1 and 5 mg/L, respectively. The catechol-utilizing microorganisms did not cometabolize TCE in the presence and absence of catechol. Although catechol could be converted by catechol 1,2- or 2,3-dioxygenase, the results indicated that catechol was not an inducer to initiate the production of degradative enzymes cometabolizing TCE.

SUMMARY

The presence of an aromatic compound, such as benzene or toluene, initiated the cometabolism of TCE by aromatic-utilizing microorganisms. The biodegradation of TCE by aromatic-utilizing microorganisms was insignificant without the presence of an aromatic compound. The TCE removal efficiency was not proportional to the initial concentrations of aromatic compounds. The presence of benzene or toluene at an initial concentration of 5 mg/L resulted in the best TCE removal in comparison with that these compounds presented at

1 or 20 mg/L. The presence of an aromatic compound was required to initiate TCE removal. However, TCE removal was still observed but at a lower rate after the depletion of the aromatic compound. The presence of catechol, an intermediate for the biodegradation of aromatic compounds, did not initiate the cometabolism of TCE, even if catechol is an aromatic compound and its biodegradation is also catalyzed by a catechol dioxygenase.

ACKNOWLEDGMENT

This research was supported by National Research Council, Taiwan, ROC (NSC-81-0410-E-005-22).

REFERENCES

Bouwer, E. J., B. E. Rittmann, and P. L. McCarty. 1981. "Anaerobic Degradation of Halogenated 1- and 2-carbon Organic Compounds." *Environ. Sci. Technol.* 15 (5): 596-599.

Egli, C., T. Tschan, R. Scholtz, A. M. Cook, and T. Leisinger. 1988. "Transformation of Trichloromethane to Dichloromethane and Carbon Dioxide by *Acetobacterium woodii*." *Appl. Environ. Microbiol.* 54 (11): 2819-2824.

Fogel, T. M., A. R. Taddeo, and S. Fogel. 1986. "Biodegradation of Chlorinated Ethenes by a Methane-Utilizing Mixed Culture." *Appl. Environ. Microbiol.* 51 (4): 720-724.

Little, C. D., A. V. Palumbo, S. E. Herbes, M. E. Lidstrom, R. L. Tyndall, and P. J. Gilmer . 1988. Trichloroethylene Biodegradation by a Methane-Oxidizing Bacterium." *Appl. Environ. Microbiol.* 54 (4): 951-956.

Nelson, M. J. K., S. O. Montgomery, and P. H. Pritchard. 1988. "Trichloroethylene Metabolism by Microorganisms that Degrade Aromatic Compounds." *Appl. Environ. Microbiol.* 54 (2): 604-606.

Vogel, T. M., C. S. Criddle, and P. L. McCarty. 1987. "Transformations of Halogenated Aliphatic Compounds." *Environ. Sci. Technol.* 21 (8): 722-736.

Wackett, L. P. and D. T. Gibson. 1988. "Degradation of Trichloroethylene by Toluene Dioxygenase in Whole-cell Studies with *Pseudomonas putida* F1." *Appl. Environ. Microbiol.* 54 (7): 1703-1708.

Zylstra, G. J., L. P. Wackett, and D. T. Gibson. 1989. "Trichloroethylene Degradation by *Escherichia coli* Containing the Cloned *Pseudomonas putida* F1 Toluene Dioxygenase Genes." *Appl. Environ. Microbiol.* 55 (12): 3162-3166.

Comparison of Alternative Electron Donors to Sustain PCE Anaerobic Reductive Dechlorination

Donna E. Fennell, Michael A. Stover,
Stephen H. Zinder, and James M. Gossett

ABSTRACT ━━━━━━━━━━━━━━━━━━━━━━━━━━━━

Anaerobic reductive dechlorination of tetrachloroethene (PCE) to ethene (ETH) appears to use hydrogen as the direct electron donor (DiStefano et al. 1992). Hydrogen addition may be problematic for large-scale treatment systems. Adding an electron donor which is fermented to hydrogen may be more practical. Competition for substrate or reduction equivalents by methanogens should be minimized. Studies were performed with methanol, ethanol, lactic acid, and butyric acid to determine their suitability for maintaining reductive dechlorination by an anaerobic mixed culture. Electron donors were examined in semicontinuously operated serum bottles with a nominal PCE concentration of 110 μmol/L (neglecting partitioning to the gas space) and a 2:1 ratio of electron donor to PCE on an equivalent basis. The patterns of electron donor degradation, hydrogen formation, dechlorination, and methanogenesis were determined for each substrate. Dechlorination was sustained better with butyric acid, lactic acid, or ethanol than with methanol. Amendment with methanol stimulated methanogenesis and resulted in less complete dechlorination. Amendment with lactic acid or ethanol resulted in a hydrogen peak of 10^{-3} atm in 2 to 3 h. Butyric acid gave the most promising results. It was degraded about 10 times more slowly than ethanol or lactic acid, and its degradation constantly maintained 10^{-4} atm of hydrogen during the test. The data suggest that if hydrogen is supplied at low concentrations and low rates, dechlorination is favored over methanogenesis.

INTRODUCTION

Microbially mediated anaerobic reductive dechlorination of the common groundwater contaminants PCE and trichloroethene (TCE) has been investigated as a potential remediation tool since the early 1980s (Bouwer and McCarty 1982, 1983, 1985; Bouwer et al. 1981; Parsons et al. 1985; Vogel et al.

1987; Vogel and McCarty 1985). PCE and TCE are reductively dechlorinated to dichloroethene isomers (DCEs), vinyl chloride (VC), and ETH (Freedman and Gossett 1989), and in some cultures ethene is converted to ethane (Holliger et al. 1993). Many substrates such as glucose, sucrose, acetate, lactic acid, and methanol have been shown to support dechlorination both in microbial cultures and in sediment microcosms (Bouwer and McCarty 1983; Freedman and Gossett 1989; Gibson and Sewell 1992; Holliger et al. 1993). The dechlorinating organism(s) present in the mixed culture used during this study apparently use(s) hydrogen, which is evolved during degradation or conversion of more complex substrates, to carry out dechlorination (DiStefano et al. 1992). The dechlorinating organisms are not methanogens, but compete with hydrogenotrophic methanogens for available hydrogen. Attempts to sustain the dechlorinating mixed culture on hydrogen alone have been unsuccessful thus far without the addition of a complex nutrient source such as sludge supernatant (DiStefano et al. 1992; Tandoi et al. 1994). When PCE concentrations that were inhibitory to methanogens were used, methanol was an effective electron donor for dechlorination (DiStefano et al. 1992). The methanol was converted to acetate in these cultures, and the dechlorinating organisms presumably scavenged hydrogen from acetogens. When methanogenesis was not inhibited, competition for methanol or reducing equivalents by methanogens was a very important sink for electron donor.

The objective of this study was to compare the suitability of the electron donors lactic acid, ethanol, butyric acid, and methanol for maintaining dechlorination in a mixed anaerobic culture at noninhibitory PCE concentrations. For each substrate, the completeness of dechlorination during long-term operation and the patterns of hydrogen production, dechlorination, and methanogenesis during short-term tests were examined in semicontinuously operated serum bottles. The results were examined to determine if the pattern of hydrogen production influenced the favorability of dechlorination over methanogenesis.

EXPERIMENTAL PROCEDURES

Microbial Culture

The source culture used during these experiments was enriched with methanol, PCE, and yeast extract, operated with a nominal hydraulic retention time (HRT) of 40 days at 35°C and is described in detail elsewhere (DiStefano et al. 1992). For these experiments, aliquots of 100 mL were removed from the source culture and placed in 160-mL serum bottles. The bottles were capped with gray-butyl, Teflon™-lined septa (Wheaton) and crimped with aluminum caps. Following anaerobic transfer of the culture to the serum bottles, the bottles were maintained with the source culture protocol (addition of 55 μmol PCE, 156 μmol methanol, and 2 mg of prefermented yeast extract on Day −4 and Day −2) for 4 days to ensure that the transfer was successful. After 4 days the experiment was begun (defined as Day 0).

Cultures were grown in a basal salts medium, which has been used to develop and work with this culture. The solution was adapted (Freedman and Gossett 1989) from one used for culturing methanogens (Zeikus 1977). Prefermented yeast extract consisted of an anoxic mixture of 5 g yeast extract and 90 mL distilled water to which 10 mL of the source culture was added. The mixture was allowed to ferment for 10 days prior to use to remove readily available electron donor. At each feeding, 50 μL of this solution was added as a source of trace nutrients.

Experimental Protocol

During the electron donor experiment, duplicate bottles were amended every second day with 11 μmol PCE and either methanol (29.3 μmol), ethanol (44 μmol), lactic acid (44 μmol), butyric acid (44 μmol), or no electron donor. The electron donor was added at a 2:1 ratio to the PCE fed on an equivalent basis (176 μeq electron donor to 88 μeq PCE). Control bottles received PCE but no electron donor. All bottles were fed prefermented yeast extract as a trace-nutrient source.

On Day 0 and on every second day thereafter during long-term operation, the headspace of each bottle was sampled for dechlorination products, hydrogen, and methane, and then feeds were added. Every fourth day, after headspace samples were removed, an anoxic purge and a 10-mL exchange of culture and basal medium were made which resulted in a nominal HRT of 40 days. When liquid was removed, samples were taken for measurement of pH (and fatty acids for some bottles). After the basal medium exchange and anoxic purge, the bottles were fed.

The bottle sets were maintained with their respective protocols for 42 to 52 days for a long-term determination of the suitability of the various substrates as electron donors for PCE dechlorination. Time-intensive studies were performed initially within the first three weeks of set-up and finally during the last few days of operation on the same bottles to give a more detailed picture of culture performance, during which the disappearance of electron donor and the formation of dechlorination products, hydrogen, and methane were followed.

To initiate the time-intensive studies, the bottles were purged, 10 mL of culture was removed, and 9.0 mL of a PCE-saturated basal medium (1,200 μM PCE) and 1 mL of regular basal medium were added. Methanol and butyric acid were added to the bottles in neat form, whereas lactate and ethanol were added as 20% stock solutions in distilled water. Prefermented yeast extract was not added during the time-intensive studies to allow more accurate determination of the fate of reduction equivalents. Headspaces were analyzed periodically for 48 h and six to eight 0.25-mL liquid samples were removed during the test to monitor electron donor degradation.

Analytical Methods

A gas chromatography system incorporating packed columns, two flame-ionization detectors, a reduction-gas detector, a hot-wire detector, and switching valves to divert streams of carrier gas to specific detectors was used to quantify PCE, TCE, VC, ETH, CH_4, and H_2 in 0.1 mL gas samples. The method has been described elsewhere (Freedman and Gossett 1989).

A Perkin-Elmer Corporation Model 8500 gas chromatograph with a 0.53-mm Nukol® capillary column and flame ionization detector was used for analysis of butyric acid. The nitrogen carrier gas flowrate was 10 mL/min (80 PSI), and the flame was maintained with hydrogen (40 PSI, 40 mL/min) and air (40 PSI, 450 mL/min). The injector temperature was 200°C and the detector temperature was 250°C. The column was held at 110°C for 4 min then ramped to 150°C at 15°C per min and held for 30 s. Culture samples of 0.25 or 0.5 mL were filtered and preserved to pH 1 to 2 with 5 or 10 μL of 6 N H_2SO_4. Then 6N NaOH was added to bring the sample pH to 4 prior to analysis.

Lactic acid was determined with a Hewlett Packard 1090 high-performance liquid chromatograph (HPLC) with a 300-mm HPX-87H ion-exchange column (Bio-Rad Laboratories). The mobile phase was 0.013N H_2SO_4, and the temperature was 65°C. Culture samples of 0.25 or 0.5 mL were filtered and preserved with 5 or 10 μL of 6N H_2SO_4.

Ethanol and methanol were determined by employing an enzymatic/spectrophotometric method (Herzberg and Rogerson 1985). Samples were prepared in the same way as for HPLC.

RESULTS

Long-Term Operation

Initially in all of the bottles, most of the added PCE was converted to ETH. As the test progressed, dechlorination became less complete to varying degrees in all bottles. The decline was most pronounced in the methanol-amended bottles. In these methanol-amended bottles near the end of the study, at the end of a 2-day period, up to 20% of the PCE added remained and 70 to 80% of the reduction equivalents added as methanol were channeled to methanogenesis (Figure 1). In contrast, at the end of a 2-day period in a butyric-acid-amended bottle, PCE was completely absent, having been largely converted to ETH with only 10 to 35% remaining as VC. Of the reduction equivalents added as butyric acid, 42% was used for dechlorination, while the remainder was used for methane production. Ethanol, lactic acid, and butyric acid all sustained more complete dechlorination than methanol; however, all of the bottles exhibited some loss of dechlorination efficiency over time (data not shown). It was later shown with butyric-acid-amended cultures that loss of efficiency and eventual failure after 70 days under the same operational conditions as are described here were a result of nutrient limitation.

Final Time-Intensive Test

Time-intensive tests revealed more detail about how the electron donors were converted to hydrogen during 48-h periods following feeding. Final time-intensive tests were considered closer to "steady conditions" and will be

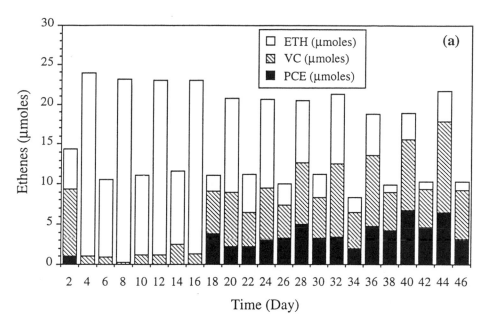

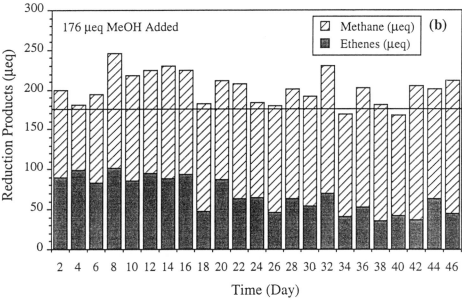

FIGURE 1. Results of dechlorination and reduction product distribution during long-term operation of a methanol-amended culture.

reported here. Graphs of dechlorination, methane production, and hydrogen levels during the final time-intensive test for an ethanol-amended bottle are shown in Figure 2. During the final time-intensive test, ethanol-amended bottles produced a pool of 2,500 and 3,000 nmol of hydrogen (10^{-3} atm) after

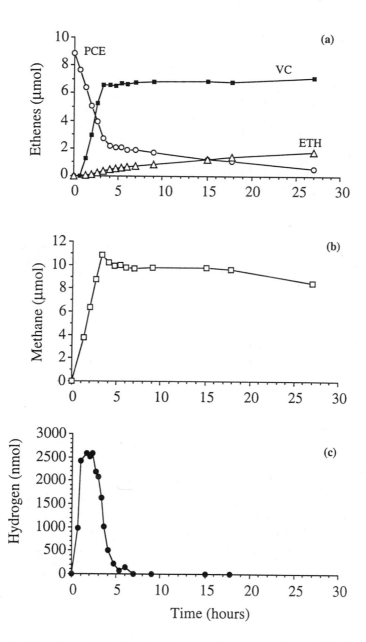

FIGURE 2. Results of dechlorination, methanogenesis, and hydrogen production during the final time-intensive study in an ethanol-amended bottle.

only 2 h and hydrogen was then quickly used primarily for methanogenesis. Dechlorination continued slowly in the ethanol-amended bottles, but PCE remained after 27 h. A similar pattern of hydrogen formation was observed in lactic-acid-amended bottles; however, PCE dechlorination was even less complete than in the ethanol-amended bottles. The butyric-acid-amended bottles, in contrast, exhibited no hydrogen peak, only a steady pool of 200 to 250 nmol per bottle (10^{-4} atm) over a 48-h period (data not shown). PCE was completely converted to VC and ETH after 15 h in the butyric-acid-amended bottles. Addition of methanol, which is not converted directly to hydrogen, resulted in a hydrogen pool of less than 100 nmol per bottle (10^{-4} atm) during the 48 h after feeding. However, most of the reduction equivalents added were channeled directly to methanogenesis by methanol-using methanogens, and dechlorination of PCE was far from complete.

DISCUSSION

The source culture normally received PCE and electron-donor loadings that were five times those used in the experiment reported here. The results from this experiment, therefore, cannot be said to come from steady conditions since the bottles were only operated for 42 to 52 days—only about one HRT. However, the experiment can be used to compare the outcome of dechlorination when different electron donors are used.

The decline in the completeness of dechlorination in all of the bottles may have been, in part, a result of a population shift following the change in electron donor type and loading; however, it may also have been a result of nutrient limitation. A nutrient limitation was observed later in butyric-acid-enriched cultures, which were operated with the same protocol as the butyric-acid-amended bottles reported here. This nutrient limitation was overcome by adding a vitamin solution (Balch et al. 1979).

Because the methanogens were inhibited by the higher PCE loading rate in the source culture, but were free to grow in the bottles that received a lower PCE loading, they began to compete for hydrogen (or for the supplied substrate itself, in the case of the methanol-amended bottles). This explains the shift of reduction equivalents from dechlorination to methanogenesis during long-term operation. Thus, the nonmethanogenic substrates maintained dechlorination better than methanol. The electron donor that resulted in a lower hydrogen partial pressure, butyric acid, sustained dechlorination the best. More of the reducing equivalents released from its degradation continued to be channeled to dechlorination. The results suggest that non-methanogenic substrates that produce a low and steady supply of hydrogen will result in more complete, and better sustained, dechlorination of PCE and lesser chlorinated ethenes than those that produce a higher amount of hydrogen in a short amount of time. This information suggests that the dechlorinating organisms and hydrogenotrophic methanogens have different affinities for hydrogen.

REFERENCES

Balch, W.E., G.E. Fox, L.J. Magrum, C.R. Woese, and R.S. Wolfe. 1979. "Methanogens: Reevaluation of a unique biological group." *Microbiology Reviews* 43: 260-296.

Bouwer, E. J., and P. L. McCarty. 1982. "Removal of trace chlorinated organic compounds by activated carbon and fixed-film bacteria." *Environmental Science and Technology* 16 (12): 836-843.

Bouwer, E. J., and P. L. McCarty. 1983. "Transformations of 1- and 2-carbon halogenated aliphatic organic compounds under methanogenic conditions." *Applied and Environmental Microbiology* 45 (4): 1286-1294.

Bouwer, E. J., and P. L. McCarty. 1985. "Utilization rates of trace halogenated organic compounds in acetate-grown biofilms." *Biotechnology and Bioengineering* XXVII 1564-1571.

Bouwer, E. J., B. E. Rittmann, and P. L. McCarty. 1981. "Anaerobic degradation of halogenated 1- and 2-carbon organic compounds." *Environmental Science & Technology* 15 (5): 596-599.

DiStefano, T. D., J. M. Gossett, and S. H. Zinder. 1992. "Hydrogen as an electron donor for the dechlorination of tetrachloroethene by an anaerobic mixed culture." *Applied and Environmental Microbiology* 58: 3622-3629.

Freedman, D. L., and J. M. Gossett. 1989. "Biological reductive dechlorination of tetrachloroethylene and trichloroethylene to ethylene under methanogenic conditions." *Applied and Environmental Microbiology* 55 (9): 2144-2151.

Gibson, S. A., and G. W. Sewell. 1992. "Stimulation of reductive dechlorination of tetrachloroethene in anaerobic aquifer microcosms by addition of short-chain organic acids or alcohols." *Applied and Environmental Microbiology* 58 (4): 1392-1393.

Herzberg, G. R., and M. Rogerson. 1985. "Use of alcohol oxidase to measure the methanol produced during the hydrolysis of D- and L-methyl-3-hydroxybutyric acid." *Analytical Biochemistry* 149: 354-357.

Holliger, C., G. Schraa, A. J. M. Stams, and A. J. B. Zehnder. 1993. "A highly purified enrichment culture couples the reductive dechlorination of tetrachloroethene to growth." *Applied and Environmental Microbiology* 59 (6): 2991-2997.

Parsons, F., G. B. Lage, and R. Rice. 1985. "Biotransformation of chlorinated organic solvents in static microcosms." *Environmental Toxicology and Chemistry* 4: 739-742.

Tandoi, V., T. D. DiStefano, P. A. Bowser, J. M. Gossett, and S. H. Zinder. 1994. "Reductive dehalogenation of chlorinated ethenes and halogenated ethanes by a high-rate anaerobic enrichment culture." *Environmental Science and Technology* 28 (5): 973-979.

Vogel, T. M., C. S. Criddle, and P. L. McCarty. 1987. "Transformations of halogenated aliphatic compounds." *Environmental Science and Technology* 21 (8): 722-736.

Vogel, T. M., and P. L. McCarty. 1985. "Biotransformation of tetrachloroethylene to trichloroethylene, dichloroethylene, vinyl chloride, and carbon dioxide under methanogenic conditions." *Applied and Environmental Microbiology* 49 (5): 1080-1083.

Zeikus, J. G. 1977. "The biology of methanogenic bacteria." *Bacteriology Reviews* 41 (2): 514-541.

Anaerobic Biodegradation of Chlorinated Solvents: Comparative Laboratory Study of Aquifer Microcosms

J. Martin Odom, JoAnn Tabinowski,
Michael D. Lee, and Babu Z. Fathepure

ABSTRACT

Our strategy for exploring the key variables involved in in situ anaerobic dechlorination is a comparative multisite analysis of the relationship between dechlorination activity, nutritional amendments, and environmental factors. This strategy involves the following steps: (1) characterize a number of geographically distinct chlorinated solvent plumes in terms of groundwater chemistry, pollutants, and overall environmental conditions; and (2) examine microbiological response patterns to various nutritional conditions across the range of distinct sites. The results show that (1) microbial dechlorinating activity may not always be present or be nutritionally enhanced even within abundant and diverse microbial populations; (2) dechlorination is often partial or solvent-specific; and (3) dechlorinating cultures were most reliably produced under heterotrophic, fermentative conditions.

INTRODUCTION

Chlorinated solvents have been found in groundwaters and soils adjacent to or underlying many industrial sites. Costs of conventional remediation technologies are high, mostly due to the persistence of these chemicals in the soil-water matrix. This persistence necessitates long-term approaches (i.e., pump and treat) that really only contain the plume instead of destroying the contamination in place. An alternative approach is the in situ biodegradation of these solvents via the ability of indigenous anaerobic bacteria to carry out anaerobic dehalogenations (Beeman et al. 1994). The power of this approach is that degradation of the solvents could be achieved in a relatively short time and result in the formation of innocuous end-products.

The chemical mechanism for this microbial process is known as reductive dehalogenation (Krone and Thauer 1989). This process has been described for a number of crude, as well as pure, cultures. However, the microbial process is poorly understood, and it is not possible to specifically enhance or manipulate this process as a remediation technology at this time. Existing literature suggests that anaerobic dehalogenation may not be associated with any one taxonomic or physiological grouping, and it is not clear whether the process confers a significant growth advantage to the organism (Mohn and Tiedje 1992).

Reductive dehalogenation has been described for chlorinated methanes and chlorinated ethylenes (Krone and Thauer 1989, Wackett et al. 1992, DiStefano et al. 1991) and consists of a series of sequential hydrogen additions and halogen eliminations that can result in the complete dehalogenation of carbon tetrachloride (CT), or perchloroethylene (PCE), to methane and ethylene, respectively, as shown below.

$$PCE => TCE => DCE =>VC => ethylene$$

where PCE is perchloroethylene; TCE is trichloroethylene; DCE is dichloroethylene; and VC is vinyl chloride.

$$CT => CF => MC => CM => methane$$

where CT is carbon tetrachloride; CF is chloroform; MC is methylene chloride; and CM is chloromethane.

Understanding of the microbiology of this process is very poor at this time. It is equally unclear how to specifically stimulate this process. Therefore, a "trial and error" approach for each site is the approach often taken. Anaerobic dehalogenation may occur under fermentative, sulfate-reducing or methanogenic conditions (Bouwer 1994). Fermentative bacteria break down complex substrates such as yeast extract to simpler organic acids and alcohols that may serve as substrates for the sulfate-reducers or methanogens (Atlas and Bartha 1993). The sulfate-reducers reduce sulfate to sulfide, whereas the methanogens reduce carbon dioxide to methane. These processes may be viewed as terminal anaerobic respirations that are vital for mineralization of organic matter under anaerobic conditions. Dechlorination may occur under different nutritional conditions depending on the site. In this report we nutritionally stimulated a variety of aquifer sediments under fermentative, methanogenic, or sulfate-reducing conditions in an attempt to extract correlations between dechlorination and the nutritional amendment.

EXPERIMENTAL PROCEDURES AND MATERIALS

Aquifer Sampling

Aquifer core material was obtained by split spoon sampling at depths ranging from 10 to 80 ft (3 to 24 m), depending on the depth of the particular aquifer. The cores were taken and maintained in sterile stainless steels cylinders or placed in sterile glass vials. Where possible, immediately upon acquisition, the cylinders were placed in anaerobic GasPak jars (BBL) with elimination of oxygen by using an H_2/CO_2-generating GasPak pouch (BBL). The glass vials were purged with nitrogen. The cores were then shipped to the laboratory at ambient temperatures under anaerobic conditions. Typically, multiple cores were taken from each site and then blended under an anaerobic glove bag (Coy Laboratory Products Inc., Ann Arbor, Michigan) atmosphere (10% H_2, 5% CO_2, 85% N_2) in the laboratory. Groundwaters were obtained and shipped from the site to the laboratory in plastic bottles at 4°C. For some studies, groundwater was sterilized in the laboratory by direct filtration through 0.45-μ filters (Gelman). Some groundwaters required centrifugation prior to filtration to remove suspended matter.

Laboratory Microcosms

Laboratory microcosms were prepared in either 500-mL or 250-mL Wheaton bottles (Wheaton, Millville, New Jersey) within an anaerobic chamber. Duplicate microcosms were prepared for each nutritional condition. Three nutritional conditions were tested in the absence of added sulfate: 0.05% yeast extract, 5 mM sodium benzoate, and 15 mM sodium acetate. Three identical nutritional conditions were tested with the inclusion of 20 mM sodium sulfate. Soil:groundwater ratios were typically 1:10 vol:vol. If multiple cores were obtained form a site, these were blended to provide a single composite inoculum for the microcosms. The microcosms were filled to the top such that there was little or no headspace, and then were stoppered with Teflon™-lined disks and crimp-sealed with aluminum seals (Wheaton, Millville, New Jersey). The groundwaters were always amended with a redox indicator, resazurin (200 μL of 0.1% per liter stock solution), 2 mM ammonium chloride, and 2 mM potassium phosphate. The resazurin addition permitted visualization of low-potential anaerobic conditions by the change in color from pink to colorless. Each microcosm was spiked with up to 120 μM PCE or TCE. The spikes were prepared via direct injection of the solvent, or as a solution in water or methanol. The microcosms were incubated on their sides, in the dark, at ambient room temperatures or approximately 22°C for up to 180 days.

Samples for analyses of chlorinated solvents, sulfide, ethylene, and methane were obtained by two methods that yield equivalent results over a 30-day time course. The first method involved syringe puncture of the Teflon™ septa with a 27.5-gauge needle to obtain 200 to 1,000 μL of liquid sample. For chlorinated solvent analysis, this sample was immediately mixed 1:0.5 (vol:vol) with methanol in an autosampler vial and stored at –80°C for later analysis by GC/MS according to SW 846 EPA method 8240. Sulfide analysis was performed by the methylene blue method of Siegal (Siegal 1965). Ethylene and methane were determined by GC/TCD analysis on an HP 5880a equipped with a PoraPak Q column at ambient temperatures with argon as carrier gas. Acridine orange counts were performed by diluting a sample into 10 mL of water containing 10 μL of 0.1% acridine orange and then passing the entire 10 mL through an acrodisc filter (0.2 μM) to collect the bacteria.

Field Measurements

Field measurements of redox potential, dissolved oxygen, and pH generally were made in situ using a Hydrolab H20 Gprobe and meter. Total organic carbon (TOC) was measured using a TOC analyzer.

RESULTS

Table 1 shows the geochemical diversity of the aquifers that were sampled and values of various parameters as measured in situ. Redox values were, for the most part, quite positive, indicative of aerobic conditions. The one exception was site E where a low redox value was observed. Generally, only very low levels of metabolic products such as methane or sulfide were detected, which is in line with fairly low TOC values.

Table 2 shows that there is good correlation between detection of *c*DCE in the ground and its generation in laboratory microcosms. One exception to this was site F. This site consistently yielded negative dechlorination results in the lab, even when the microcosm pH values were between 6 and 7 or well above the very acidic pH of 3 to 3.8, which was found in the ground. VC detection or generation is less consistent where it is either found in the ground but not in the laboratory microcosms or vice versa. Only trace levels of ethylene were detected in laboratory incubations when it was present at all.

TABLE 1. Characteristics and geochemistry of different aquifers.

Site	Aquifer Type	pH	Redox Potential mV	DO[a] mg/L	TOC[b] mg/L	Sulfide mg/L	Methane μg/L
A	Sandy	5.8 to 7.6	NA	NA	1 to 12	0	0
B	Saprolite/ Weathered Rock	5.4 to 6.3	140 to 355	0.06 to 0.34	19	1.3	520
C	Silt/Sand	5.4 to 6.2	−17 to 360	0.4 to 2.5	1 to 8	NA	0 to 80
D	Sandy	6 to 7.5	0 to −50	0	10 to 30	NA	20
E	Fill over Fractured Bedrock	5.5 to 8.5	−210	0.5	2.1	2.6	34
F	Clay/Sand	3 to 3.8	380	0.3	1 to 5	0	0

(a) DO: Dissolved Oxygen
(b) TOC: Total Organic Carbon

TABLE 2. Dechlorination products found in situ or in laboratory microcosms.

Site	cDCE		VC		Ethylene	
	In Situ[a]	Lab[b]	In Situ	Lab	In Situ	Lab
A	+	+	–	–	–	–
B	+	+	+	+	+	–
C	+	+	+	–	+	+/–
D	+	+	–	+	–	+/–
E	+	+	+	+	+	+/–
F	+	–	–	–	–	–

(a) Dechlorination product found in situ at the site.
(b) Dechlorination product found in laboratory microcosms (see Table 3).

Several organic substrates were tested with and without additional sulfate (20 mM) to examine the response of indigenous bacteria for dechlorination of TCE (Table 3). Other defined carbon sources were tested, including formate, methanol, lactate, sucrose, and cellobiose. Dechlorination rates with these substrates were generally very low (not shown). The highest rates of dechlorination were always achieved with complex substrates, such as yeast extract with or without added sulfate. However, acetate was a very effective donor for dechlorination in four out of six cases. Acetate was equally effective as yeast extract in sites D and E. Sulfate additions generally had little effect on the rate of dechlorination.

A detailed comparison was made between results from microcosm experiments on site E (the most active dechlorinating activity observed) and site F (a site where no dechlorination could be demonstrated in the laboratory, as shown in Table 4). Cell growth and sulfide respiration were used as general parameters to normalize the data and show that bacteria were in fact present in F microcosms. Acridine orange cell counts show that indeed cell growth was comparable in both microcosms E and F and that the most growth was observed with yeast extract and the least growth observed with benzoate. Benzoate supported significant levels of dechlorination in site E microcosms but supported only very low levels of sulfide formation. Acetate metabolism stimulated sulfate reduction in microcosms from both sites, as well as significant levels of growth.

DISCUSSION

Aquifer sediments and groundwaters from chemically and geographically distinct areas were tested for the ability of indigenous microflora to carry out

TABLE 3. Dechlorination rates in nutritionally stimulated microcosms.

Carbon/Energy Source	cDCE produced nmol/day/mL					
	Site					
	A	B	C	D	E	F
Yeast Extract	1.0	2.6	0.32	10	20	0
Yeast extract + Sulfate	3.8	0.6	0.36	10	30	0
Benzoate	0.06	nd	nd	7	4	0
Benzoate + Sulfate	0	nd	nd	4	2.4	0
Acetate	0	0.38	0.30	12	11	0
Acetate + Sulfate	0	0.22	0.04	12	35	0
No Addition	0	0.012	0	0	10	0

nd: Not determined

TABLE 4. Comparative microbiology of dechlorinating versus nondechlorinating sites. Site E exhibited the highest dechlorination rates whereas site F exhibited no dechlorination.

Carbon/Energy Source	Bacteria Cells/mL × 10^7		Sulfide nmol/Day/10^7 Cells		cDCE nmol/Day/10^7 Cells	
Site	E	F	E	F	E	F
Yeast Extract	28	130	1	1.7	20	0
Yeast extract + Sulfate	32	90	2.0	1.7	30	0
Benzoate	10	1	0.2	0	4	0
Benzoate + Sulfate	24	2	0.2	0	2.4	0
Acetate	20	32	3.6	6.2	11.0	0
Acetate + Sulfate	10	5	4.4	0.2	3.5	0
No Addition	7	1	0	0	1.0	0

dechlorination under distinct types of nutritional stimulation. A variety of organic compounds were tested as carbon and electron sources for growth and dechlorination. The rationale for the substrates chosen derives from the concept of the anaerobic food chain, with complex polymeric organics being fermented to organic acids and alcohols such as lactate or ethanol, which subsequently are fermented to acetate, hydrogen, and carbon dioxide. The final fermentation products may be substrates for methanogenic bacteria, as well as certain sulfate-reducing bacteria. It would be expected that distinct microbial populations would be generated during growth on yeast extract versus, for example, acetate. Thus, dechlorination was compared, most broadly, for growth on yeast extract at one extreme versus acetate at the other extreme of the food chain. Benzoate may be viewed as intermediary in complexity and would be expected to produce a microbial population distinct from either yeast extract or acetate. Other organics such as lactate, methanol, formate, and certain sugars were not as broadly reliable at producing dechlorination as yeast extract.

The multisite analysis for the presence of indigenous dechlorinating bacteria and the ability to nutritionally stimulate this activity resulted in the following broad conclusions.

1. Indigenous bacterial species can partially dechlorinate chlorinated ethylenes in most of the aquifers tested. Site F was an exception. This site contained comparable numbers of active bacteria to dechlorinating sites; however, dechlorination was clearly absent in the site F microcosm experiments.
2. The dechlorination observed was almost always partial, in that ethylene was only a trace level product in most incubations. The most consistent dechlorination data were obtained by monitoring *c*DCE formation. At several of the sites, either the subsequent dechlorination of *c*DCE and VC was too slow, or the resulting products did not accumulate sufficiently for detection.
3. Dechlorination could not be correlated 100% of the time with sulfate reduction. For example, high rates of sulfate respiration were observed with acetate in site F microcosms, but no dechlorination was detected.

REFERENCES

Atlas, R. M., and R. Bartha. 1993. *Microbial Ecology Fundamentals and Applications*. The Benjamin/Cummings Publishing Company Inc. Redwood City, CA.

Beeman, R. E., J. E. Howell, S. H. Shoemaker, E. A. Salazar, and J. R. Buttram. 1994. "A Field Evaluation of In Situ Microbial Reductive Dehalogenation by the Biotransformation of Chlorinated Ethylenes." In R. E. Hinchee, A. Leeson, L. Semprini, and S. K. Ong (Eds.), *Bioremediation of Chlorinated and Polycyclic Aromatic Hydrocarbon Compounds*, pp. 14-27. Lewis Publishers, Boca Raton, FL.

Bouwer, E. J. 1994. "Bioremediation of Chlorinated Solvents using Alternate Electron Acceptors." In R. D. Norris, R. E. Hinchee, R. E. Brown, P. L. McCarty, L. Semprini, J. T. Wilson, D. H. Kampbell, M. Reinhard, E. J. Bouwer, R. C. Borden, T. M. Vogel, J. M. Thomas and C. H. Ward (Eds.), *Handbook of Bioremediation*. pp. 149-175. Lewis Publishers, Boca Raton, FL.

DiStefano, T. D., J. M. Gossett, and S. H. Zinder. 1991. "Reductive Dehalogenation of High Concentrations of Tetrachloroethene to Ethene by an Anaerobic Enrichment Culture in the Absence of Methanogenesis." *Applied and Environmental Microbiology* 57:2287-2292.

Krone, U. E., and R. K. Thauer. 1989. "Reductive Dehalogenation of Chlorinated C1-Hydrocarbons Mediated by Corrinoids." *Biochemistry* 28:4908-4914.

Mohn, W. W., and J. M. Tiedje. 1992. "Microbial Reductive Dehalogenation." *Microbiological Reviews* 56:482-507.

Siegal, L. M. 1965. "A Microdetermination Method for Sulfide." *Analytical Biochemistry* 11:126-142.

Wackett, L. P., M. S. P. Logan, F. A. Blocki, and C. Bao-li. 1992. "A Mechanistic Perspective on Bacterial Metabolism of Chlorinated Methanes." *Biodegradation* 3:19-36.

Influence of Hydraulic Aquifer Properties on Reductive Dechlorination of Tetrachloroethene

Olaf Cirpka

ABSTRACT

Stimulation of in situ reductive dechlorination of chlorinated contaminants is like any in situ bioremediation approach dependent on hydraulic properties of the aquifer to be cleaned up. Model results are shown demonstrating how spatial variability of hydraulic conductivity in a moderate heterogeneous aquifer is leading to a highly spatial variable transformation pattern in the subsurface. These model calculations are part of a prestudy for experiments in a well-defined anaerobic artificial aquifer system.

INTRODUCTION

In situ reductive dechlorination has been addressed as an alternative approach for the remediation of sites contaminated by chlorinated ethenes (Vogel et al. 1987; DiStefano et al. 1991, 1992; de Bruin et al. 1992; Holliger et al. 1993; Beeman et al. 1994). Stimulation is done by injecting suitable electron donors and additional compounds, if necessary. The injection of electron donors leads to a simultaneous stimulation of competitive biotransformation processes such as acetogenesis and methanogenesis. As a consequence, these competitive processes have to be taken into account for the design of a bioremediation system for reductive dechlorination.

In situ reductive dechlorination is limited by certain factors, including the high sensitivity of the microbial system, soil properties, slow sorption kinetics, and low hydraulic conductivity areas. The focus of this study is on the influence of hydraulic aquifer properties, namely hydraulic conductivity, on microbial activity.

MODEL ASSUMPTIONS

A finite-element model has been developed to simulate the interaction of transformation processes and advective-dispersive transport in groundwater systems. Details of the numerical method are described in Cirpka and Helmig (1994).

The dechlorination model is based on the assumption, that reductive dechlorination of chlorinated ethenes is catalyzed by specific microorganisms, which are not identical to the unspecific methanogenic or acetogenic biomass being present in anaerobic systems (DiStefano et al. 1991, 1992). This assumption does not stand in contrast to the fact, that first observations of reductive dechlorination were made in methanogenic systems (Vogel et al. 1987), as in these studies the dechlorinating organisms were not identified.

In natural mixed-culture systems, some of the dechlorination steps might be catalyzed by several microorganisms, whereas certain microbes might be capable to catalyze at least two reduction steps. Especially for the first two dechlorination steps, several pure-culture systems could be established (e.g., Holliger et al. 1993; Scholz-Muramatsu et al. 1995). To the author's knowledge, until now no organism has been isolated that is capable of dechlorinating PCE completely to ethene. As it seems uncertain how many organisms are involved in complete reductive dechlorination, the model system was simplified to one specific kind of biomass X_i for each reduction step i. Note that X_1, X_2, X_3, and X_4 are only model biomasses that need not reflect specific organisms.

The work of Scholz-Muramatsu et al. (1995) indicates that at least the microorganism *Dehalospirillum multivorans* is able to grow specifically on reductive dechlorination of chlorinated ethenes. As a consequence, the model assumes that the specific biomasses grow exclusively on reductive dechlorination. In this regard, microbial growth and transformation rates are expressed by classical double Monod terms taking the specific chlorinated ethene CHC_i as electron acceptor for each specific growth rate $k_{gr,i}$:

$$k_{gr,i} = \mu_{max,i} \cdot \frac{[CHC_i]}{[CHC_i] + K_m^{CHC_i}} \cdot \frac{[E_{don}]}{[E_{don}] + K_m^{E_{don}}} \cdot X_i \tag{1}$$

where $\mu_{max,i}$ is the maximum growth rate of biomass X_i. For the first dechlorination step, CHC_i is tetrachloroethene, for the second trichloroethene, for the third cis-dichloroethene and for the fourth vinyl chloride. Only dissolved compounds are assumed to be bioavailable. The electron donor is not specified. A wide variety of electron donors, such as glucose, formate, methanol, hydrogen, lactate, propionate, pyruvate, and ethanol, have been used successfully in laboratory experiments. DiStefano et al. (1992) give a possible explanation for this variety. The model takes only one kind of electron donor into account.

Including transformation processes and mass transfer to the advection-dispersion equation leads to the following set of equations:

$$\frac{\partial [E_{don}]}{\partial t} = -\sum_{i=1}^{4} \frac{k_{gr,i}}{Y_i} - k_{comp}[E_{don}] + div\left(\underline{v}[E_{don}] - \underline{\underline{D}}\,grad[E_{don}]\right)$$
$$- \alpha_{sorb}\left([E_{don}] - \frac{[E_{don}^{sorb}]}{K_d^{E_{don}}}\right) \tag{2}$$

$$\frac{\partial [CHC_i]}{\partial t} = -\frac{k_{gr,i}}{Y_i} + \frac{k_{gr,i-1}}{Y_{i-1}} + div\left(\underline{v}[CHC_i] - \underline{\underline{D}}\,grad[CHC_i]\right)$$

$$- \alpha_{sorb}\left([CHC_i] - \frac{[CHC_i^{sorb}]}{K_d^i}\right) \tag{3}$$

$$\frac{\partial [CHC_i^{sorb}]}{\partial t} = \alpha_{sorb}\left([CHC_i] - \frac{[CHC_i^{sorb}]}{K_d^{CHC_i}}\right) \tag{4}$$

$$\frac{\partial [E_{don}^{sorb}]}{\partial t} = \alpha_{sorb}\left([E_{don}] - \frac{[E_{don}^{sorb}]}{K_d^{E_{don}}}\right) \tag{5}$$

$$\frac{\partial X_i}{\partial t} = k_{gr,i} - k_{dec}^i \cdot X_i \tag{6}$$

It is assumed that competitive transformation of the electron donor E_{don} is catalyzed by an unspecific, unidentified biomass that is different from the reductive dechlorinators. In a simplified approach, these transformations are expressed by a first-order decay of the electron donor, assuming a decay constant k_{comp}, which is set to 10^{-6}/s in the model calculation.

Y_i is the yield coefficient for reduction step i. Note that per mole dechlorinated hydrocarbon, one mole electron donor is oxidized. In cases where this assumption does not hold, stoichiometric coefficients have to be included. The effective velocity is $\underline{v}$, $\underline{\underline{D}}$ the dispersion tensor, and α_{sorb} the mass transfer coefficient for sorption, which is set to 1/day for all compounds. Concentrations of sorbed compounds are marked with the superscript sorb, K_d^i is the partitioning coefficient for compound i. For reasons of simplification, sorbed concentrations are related to the pore volume. The decay rate of biomass X_i is k_{dec}^i. The terms including $k_{gr,0}$ and $k_{gr,5}$, appearing in the application of Equation (3) to CHC_1 (PCE) and CHC_5 (ethene), are to be canceled.

The chemical and microbiological parameters chosen for calculating the examples are shown in Table 1. K_d values are calculated by the f_{OC}-K_{OC}-concept:

$$K_d = f_{OC} \cdot K_{OC} \cdot \rho_{soil} \cdot \frac{1-n_e}{n_e} \tag{7}$$

with the organic carbon content f_{OC}, the partitioning coefficient between organic carbon and water K_{OC}, the porosity n_e, and the soil density ρ. Monod constants, maximum growth rates, and yield coefficients are in the range of parameters measured for the first two dechlorination steps in a pure culture system by

Scholz-Muramatsu et al. (1995). Maximum growth rates related to the third and fourth reduction step are extrapolated to lower values.

In the model calculation, an initial PCE concentration of 200 μmol/L is assumed over the entire domain, being in equilibrium with the sorbed phase. TCE, DCE, VC, ethene, and electron donor concentrations are assumed to be zero. The input concentration of the electron donor is set to 4,000 μmol/L. No chlorinated ethenes are injected. As initial biomass concentration, 0.01 mg/L is assumed for all reduction steps.

NUMERICAL MODEL RESULTS

One-Dimensional Test Case

The 1D system is a 100-m-long confined aquifer, hydraulic conductivity is 10^{-3} m/s, and hydraulic gradient 1%, leading to a Darcy velocity of 10^{-5} m/s and an effective velocity v of $3.3 \cdot 10^5$ m/s. Longitudinal dispersivity is 0.01 m leading to a dispersion coefficient D of $3.3 \cdot 10^{-7}$ m^2/s. Figure 1 shows concentration profiles of dissolved compounds and biomasses at 30 days after start of injection.

The total mass flux of the electron donor is consumed by microbial processes, partly by reductive dechlorination and partly by the competitive process, limiting the required transformation. Only in a small reaction zone are both the chlorinated ethenes and the electron donor available to the microbes. In this zone biomass concentrations reach their maximum. As growth rates are decreasing with increasing reduction steps, dechlorination is not complete. The low transformation rates from VC to ethene lead to enrichment of VC. The VC has, compared

TABLE 1. Chemical and microbiological parameters for the model calculations.

Compound-related parameters						
	E_{don}	PCE	TCE	*cis*-DCE	VC	ETH
K_m^i [μmol/L]	100	100	100	100	100	100
K_{oc}^i [L/kg]	2.2	364	126	59	8.2	6.0
Biomass-related parameters						
	Step 1	Step 2	Step 3	Step 4		
$\mu_{max,i}$ [1/s]	$7.7 \cdot 10^{-5}$	$5.0 \cdot 10^{-5}$	$2.5 \cdot 10^{-5}$	$1.0 \cdot 10^{-5}$		
Y_i [mg/μmol]	0.001	0.001	0.001	0.001		
k_{dec} [1/s]	$1.0 \cdot 10^{-6}$	$1.0 \cdot 10^{-6}$	$1.0 \cdot 10^{-6}$	$1.0 \cdot 10^{-6}$		
Soil properties						
ρ = 2.6 kg/L	f_{oc} =0.2%	n_e = 0.3	α_{sorb} = 1/d = $1.157 \cdot 10^{-5}$/s			

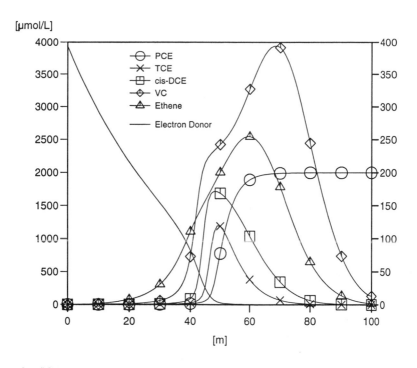

(a)

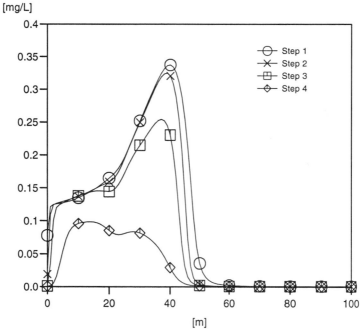

(b)

FIGURE 1. One-dimensional model results. Concentrations of dissolved compounds (a) and biomass concentrations (b) 30 days after start of injection.

to the other chlorinated ethenes, the lowest K_{OC} and is therefore retarded the least. As a consequence, the peak VC concentration is located in an area where no electron donor is present and, hence, no further dechlorination is possible. This behavior would change if additionally a phase of entrapped gas (e.g., due to methane production) was introduced into the model. The high volatility of VC would lead to enhanced retardation and consequently to more effective mixing with the electron donor.

Ethene and VC are both hardly retarded: the retardation of ethene is 1.073, and the retardation factor of VC is 1.097. Nevertheless the ethene peak is located closer by to the injection point than the VC peak. This can be explained by the slow stimulation of the specific biomass, dechlorinating VC. The VC and the electron have to be available for a certain duration, so that the initially low biomass can grow up. During this time the VC peak is moving further, while the biomass is immobile. As a consequence, the concentration of biomass 4 reaches its maximum in an area where the VC concentration is already decreasing. As a reasonable production of ethene requires a high specific biomass, the ethene peak occurs behind the VC peak. Note that the production of ethene has almost finished at the time shown in Figure 1, as VC and the electron donor are hardly mixed.

Two-Dimensional Test Case

The 2D system is a confined aquifer, 100 m in length and 20 m in width. The geometric mean of hydraulic conductivity is 10^{-3} m/s, standard deviation of logarithmic hydraulic conductivity σ_{lnK} is 1.0, which is a typical value for sandy aquifers (Gelhar 1976). Geostatistical distribution of hydraulic conductivity can be described by a Gaussian semivariogram with a longitudinal correlation length of 5 m and an anisotropy factor of 2. The distribution was generated with the software package GSLIB (Deutsch and Journel 1992).

The hydraulic head is fixed at the left and right boundary (head difference 1 m), the upper and lower boundaries are impermeable. Longitudinal dispersivity is 0.1 m and transverse dispersivity 0.01 m. Darcy velocities vary from $7.5 \cdot 10^{-7}$ m/s to $9.2 \cdot 10^{-5}$ m/s. Figures 2 and 3 show model results 30 days after start of injection.

It is obvious that the dissolved compounds are moving faster in the high-permeability areas leading to irregular concentration waves. Preferential areas of biomass growth are interfaces between low-permeability and high-permeability areas where the electron donor is delivered through the high-permeability area and the contaminant is slowly leached out of the low-permeability area.

DISCUSSION

For the chosen set of parameters, initial and boundary conditions, the delivery of the electron donor is the most limiting factor in both examples. Note that the assumed maximum growth rates and the assumed ratio of reductive dechlorination to competitive substrate consumption is still very optimistic.

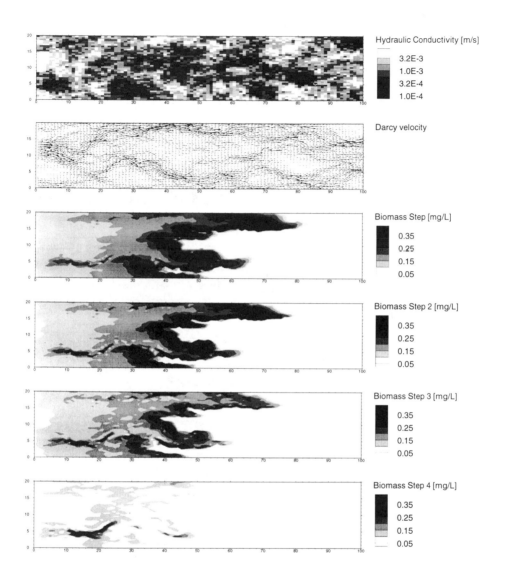

FIGURE 2. Two-dimensional model results. Distribution of hydraulic conductivity and Darcy velocities. Distribution of biomass concentrations for the four reduction steps 30 days after start of injection.

For the chosen standard deviation of logarithmic hydraulic conductivity of 1.0, spatial variability seems not to be a key factor limiting biodegradation. Anyhow, finite element modeling leads to artificial mixing, so that the differences between low-permeability areas and high-permeability areas might be underestimated.

Concentration profiles in the 2D test case are along stream lines quite similar to the 1D test case. As transverse mixing is secondary to longitudinal mixing, heterogeneity on a larger scale might be described by set of noninteracting stream

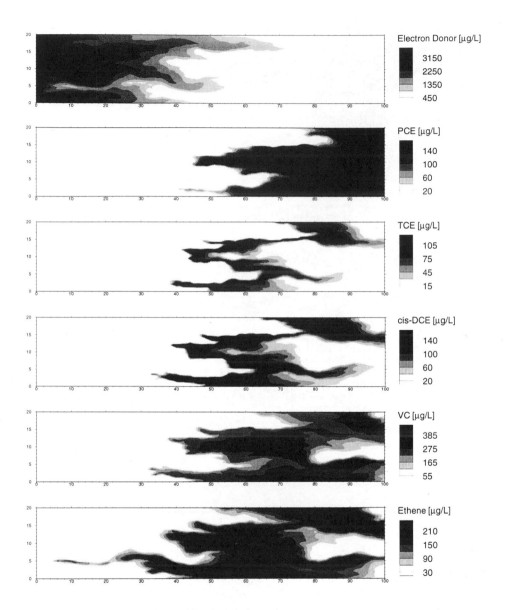

FIGURE 3. Two-dimensional model results. Distribution of dissolved compounds 30 days after start of injection.

pipes, which can be characterized by a certain distribution of travel times. Such a model approach would decrease the computational effort and overcome some numerical difficulties of multidimensional modeling.

The interactions of microbial populations in anaerobic mixed culture systems are not yet well enough understood. Kinetic parameters are scarce and difficult to evaluate. The large variety of electron donors that can be used for reductive dechlorination might indicate that some of the competitive transformations are to be seen as primary steps for the electron donor use. Inhibitory effects by metabolites or cocontaminants lead to further complications. Transport of microbes, change of hydraulic properties due to microbial activity, and occurrence of nonaqueous-phase liquids are not yet included in the model.

Hence, the presented model calculations are not able to predict the efficiency of in situ reductive dechlorination as bioremediation technology. On the other hand, the complexity of process interaction and, additionally, the high uncertainty of site characterization make field studies difficult to interpret. Therefore, a large-scale artificial aquifer (10 m × 0.2 m × 0.7 m) has been constructed at the University of Stuttgart. In this domain, the interaction of flow and transport properties in a heterogeneous aquifer with microbial activity is to be investigated. Microbiological experiments will start in autumn 1995 and will be compared to model results and scaled-up laboratory experiments.

Notation

α_{sorb} Mass transfer coefficient for sorption [1/s]
$[CHC_1]$ PCE concentration in aqueous phase [µmol/L]
$[CHC_2]$ TCE concentration in aqueous phase [µmol/L]
$[CHC_3]$ *cis*-DCE concentration in aqueous phase [µmol/L]
$[CHC_4]$ VC concentration in aqueous phase [µmol/L]
$[CHC_5]$ Ethene concentration in aqueous phase [µmol/L]
$\underline{\underline{D}}$ Dispersion coefficient [m²/s]
$[E_{don}]$ Electron donor concentration in aqueous phase [µmol/L]
f_{OC} Mass fraction of organic carbon in soil [–]
K_d^i Partitioning coefficient [–]
k_{dec}^i Decay rate for biomass i [1/s]
$k_{gr,i}$ Growth rate for biomass i [mg/s]
K_m^i Monod constant for compound i [µmol/L]
K_{OC} Partitioning coefficient between organic carbon and water [L/kg]
$\mu_{max,i}$ Maximum growth rate for biomass i [1/s]
n_e Effective porosity [–]
ρ_{soil} Soil density [kg/L]
$\underline{v}$ Effective velocity [m/s]
X_1 Biomass catalyzing reduction of PCE to TCE (related to pore volume) [mg/L]
X_2 Biomass catalyzing reduction of TCE to *cis*-DCE (related to pore volume) [mg/L]
X_3 Biomass catalyzing reduction of *cis*-DCE to VC (related to pore volume) [mg/L]
X_4 Biomass catalyzing reduction of VC to ethene (related to pore volume) [mg/L]
Y_i Yield coefficient for CHC_i in reduction step i [mg/µmol]
^{sorb} Concentration in the sorbed phase (related to pore volume) [µmol/L]

REFERENCES

Beeman, R. E., J. E. Howell, S. H. Shoemaker, E. A. Salazar, and J. R. Buttram. 1994. "A field evaluation of in situ microbial reductive dehalogenation by the biotransformation of chlorinated ethenes." In R. E. Hinchee et al. (Eds.), *Bioremediation of Chlorinated and Polycyclic Aromatic Hydrocarbon Compounds*, pp. 14-36. Lewis Publishers, Boca Raton, FL.

de Bruin, W. P., M.J.J. Kotterman, M. A. Posthumus, G. Schraa, and A.J.B. Zehnder. 1992. "Complete biological reductive transformation of tetrachloroethene to ethene." *Appl. Environ. Microbiol. 58*(6): 1996-2000.

Cirpka, O., and R. Helmig. 1994. "Numerical simulation of contaminant transport and biodegradation in porous and fractured-porous media." In A. Peters et al. (Eds.), *Computational Methods in Water Resources X*, pp. 605-612. Kluwer Academic Publishers, Dordrecht, The Netherlands.

Deutsch, C. V., and A. G. Journel. 1992. *GSLIB: Geostatistical Software Library and User's Guide.* Oxford University Press, New York, NY.

DiStefano, T. D., J. M. Gossett, and S. H. Zinder. 1991. "Reductive dechlorination of high concentrations of tetrachloroethene to ethene by an anaerobic enrichment culture in the absence of methanogenesis." *Appl. Environ. Microbiol. 57*(8): 2287-2292.

DiStefano, T. D., J. M. Gossett, and S. H. Zinder. 1992. "Hydrogen as an electron donor for dechlorination of tetrachloroethene by an anaerobic mixed culture." *Appl. Environ. Microbiol. 58*(11): 3622-3629.

Gelhar, L. 1976. "Effects of hydraulic conductivity variations on groundwater flows." In International Association of Hydraulic Research (Ed.), *Proc. of 2nd Int. Symp. on Stochastic Hydraulics*, Lund, Sweden.

Holliger, C., G. Schraa, A.J.M. Stams, and A.J.B. Zehnder. 1993. "A highly purified enrichment culture couples the reductive dechlorination of tetrachloroethene to growth." *Appl. Environ. Microbiol. 59*(9): 2991-2997.

Scholz-Muramatsu, H., A. Neumann, M. Meßmer, E. Moore, and G. Diekert. 1995. "Isolation and characterization of *Dehalospirillum multivorans.*" *Arch. Microbiol. 163*(1): 48-56.

Vogel, T. M., C. S. Criddle, and P. L. McCarty. 1987. "Transformation of halogenated aliphatic compounds." *Environ. Sci. Technol. 21*(8): 722-736.

Chemical-Biological Catalysis for In Situ Anaerobic Dehalogenation of Chlorinated Solvents

J. Martin Odom, Eva Nagel, and JoAnn Tabinowski

ABSTRACT

Laboratory studies, using bacterial enrichments from geographically diverse aquifers, show that certain transition metal catalysts can effect partial dechlorination of chlorinated methanes when these catalysts are added to aquifer enrichments. The mechanism for this activity appears to involve direct catalysis by the transition metal complexes. Catalytic enhancement involves electron transfer reactions between the extracellular cytochrome c_3 of sulfate-reducing bacteria and the exogenously added transition metal complex. This reaction is shown to occur with purified proteins, with pure cultures of sulfate-reducers, and in crude sulfate-reducing aquifer enrichments. The results suggest a defined process for aquifer remediation involving: (1) generation of sulfate-reducing populations within the aquifer that do not necessarily need to possess dechlorination activity, and (2) addition of a transition metal complex that accepts electrons from the extracellular cytochrome of the sulfate-reducer and catalyzes the reductive dehalogenation.

INTRODUCTION

Many naturally occurring transition metal complexes and even transition metals themselves are known to catalyze reductive dehalogenations of chlorinated solvents (Wackett et al. 1992). These reactions may be partially or wholly responsible for microbiologically catalyzed reductive dehalogenations observed in cultures of many anaerobic bacteria. However, application of bacterial anaerobic dehalogenation as a remediation technology for in situ aquifer contamination is limited by our understanding of how to specifically enhance this activity.

Low potential reductants are required as electron donors for transition metal catalysis (Krone and Thauer 1989). Utilization of direct chemical reductants for in situ dehalogenation is impractical due to the toxicities of the chemicals and the stoichiometric amounts needed. Anaerobic bacteria usually generate low potential reductants as part of their normal physiological process; i.e. hydrogen metabolism (E_0' –400 mV, pyruvate oxidation occurs at E_0' –600 mV) (Thauer et al. 1977). Reduction of cobalt- or nickel-centered tetrapyrroles necessitates reductants of at least –400 mV (Krone and Thauer 1989). Therefore, it is feasible that electron transfer processes in anaerobic bacteria could drive reduction of exogenous tetrapyrrole if there is a conduit for extracellular electron transfer to the tetrapyrrole.

The sulfate-reducing bacteria are likely candidates for driving an extracellular redox process in situ for two reasons: (1) the high concentrations of a low potential (E_0' –300 to –400 mV) cytochrome in the periplasmic space of these bacteria, and (2) these organisms are ubiquitous in nature and can be enriched for in almost any aquifer environment.

Our approach to the practical application of anaerobic dehalogenation for aquifer remediation is to combine the catalytic activity of exogenously added transition metal complexes with the indigenous reducing power provided by enriched sulfate-reducing populations.

EXPERIMENTAL PROCEDURES AND MATERIALS

Growth of Bacteria

Pure strains of sulfate-reducing bacteria were obtained from the laboratory of H.D. Peck Jr., University of Georgia, Athens, Georgia. *Escherichia coli* ATCC 4157; *Paracoccus denitrificans* ATCC 13543; *Clostridium beijerinckii* ATCC 858; and *Clostridium pasteurianum* ATCC 6013 were all obtained from the American Type Culture Collection, Rockville, Maryland. All sulfate-reducing bacteria and sulfate-reducing enrichments (other than aquifer slurries) were cultivated in a basal mineral BTZ-3 medium (Odom et al. 1991) amended with 40 mM L-sodium lactate (Sigma Chemical Co. St. Louis, Missouri) and 20 mM sodium sulfate. Bacterial cultures were maintained as 10-mL cultures at 30°C without shaking.

Dechlorination Assays

For experiments using purified enzymes and cofactors, the following components were used: 20 μL partially purified *Desulfovibrio gigas* hydrogenase, purified by the periplasmic extraction (Van der Westen et al. 1980); 4 nmol purified *D. gigas* cytochrome c_3, purified according to LeGall and Forget (1979); riboflavin 100 nmol (Sigma Chemical Co.); vitamin B_{12} (cyanocobalamin) 100 nmol (Sigma Chemical Co.); HEPES buffer pH 7.5 50 mM for a final volume of 2 mL.

For experiments with whole cells, crude enrichments, or aliquots of aquifer slurries, the following components were used: 200 μL of a 10× concentrated cell suspension, enrichment or aquifer slurry (10^9 to 10^{10} cells/mL) in 50 mM HEPES buffer pH 7.5; 100 nmol riboflavin; 100 nmol cyanocobalamin (unless specified otherwise); and HEPES buffer pH 7.5, 50 mM for a final volume of 2 mL.

Reactions were started with the addition of 1 μL of carbon tetrachloride. All reactions were carried out in a 10-mL total volume, tinted Wheaton bottle under 100% hydrogen gas. Teflon™-coated gray butyl rubber stoppers were used in all experiments.

GC Analysis for Hydrogen or Dechlorination Products

For detection and quantification of chloroform, methylene chloride, and methane, 100-μL injections of reaction mixture headspace gases were injected into an HP 5890 gas chromatograph equipped with an HP-5 column (5M × 0.32 mm × 1.05-μM film thickness), with helium carrier. Hydrogen was detected by thermal conductivity on an HP 5880a gas chromatograph, equipped with a Porapak Q column at ambient temperature with an argon carrier gas.

Determination of Protein and Sulfide

Total protein was determined by the Coomassie Blue dye-binding procedure (Bradford 1976). Sulfide was determined by the microassay procedure involving formation of methylene blue (Siegal 1965).

RESULTS

The formation of chloroform, with hydrogen as the sole source of electrons, unequivocally demonstrates that electron transfer from hydrogen to carbon tetrachloride has occurred as shown in Figure 1. Chloroform formation required the presence of both riboflavin and vitamin B_{12}. Omission of either of these components resulted in loss of dechlorination activity. For a 20-h time course, only chloroform formation is observed, whereas with extended time courses small amounts of methylene chloride, methane, and formic acid also are observed (typically less than 10% of total carbon tetrachloride present).

Whole cell incubations, under 100% hydrogen, in the presence of riboflavin and vitamin B_{12} also supported vitamin B_{12} catalyzed dechlorination of carbon tetrachloride, but only when sulfate-reducing bacteria were utilized (Table 1). Other microbial types (*E. coli, P. denitrificans, C. beijerinckii*, and *C. pasteurianum*) did not produce dechlorination, under the conditions of the assay, in the presence or absence of riboflavin and vitamin B_{12}. Crude sulfate-reducing enrichments from an anaerobic digestor and one from an aquifer also

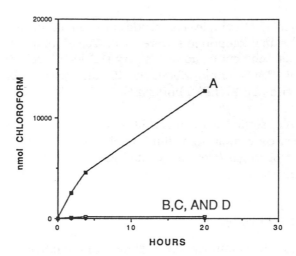

FIGURE 1. Chloroform production from purified proteins. All incubations contained hydrogenase and 4 nmol cytochrome c_3. The following incubations contained additionally: (A) 100 nmol riboflavin and 100 nmol vitamin B_{12}; (B) 100 nmol riboflavin; (C) 100 nmol vitamin B_{12}; (D) no addition.

TABLE 1. Dechlorination by pure cultures and enrichments. Dechlorinating activity of pure bacterial cultures and sulfate-reducing enrichments in the presence and absence of 50 μM riboflavin and 50 μM vitamin B_{12}.

Culture	Cells alone[a]	Cells + Riboflavin/B_{12}[a]
Desulfovibrio		
gigas	1	45
desulfuricans (Norway)	0	1
desulfuricans (27774)	0	1
salexigens	0	32
desulfuricans G100A	0	4
Anaerobic digestor enrichment	0	0.4
Aquifer enrichment	0.004	0.1
Bacteria lacking cytochrome c3		
E. coli	0	0
P. denitrificans	0	0
C. beijerinckii	0	0
C. pasteurianum	0	0

(a) nmol chloroform/min/mg protein.

displayed large increases in dechlorination in the presence of riboflavin and vitamin B_{12}. Only *D. gigas* carried out significant dechlorination in the absence of the added reagents. Note that all sulfate-reducing cultures responded to the added reagents, whereas none of the organisms lacking a periplasmic, low-potential cytochrome c_3 produced dechlorinating activity. The whole cell data support the role for cytochrome c_3 as a key element in electron transfer from the cell to vitamin B_{12}.

The influence of vitamin B_{12} concentration on the dechlorination rate in the presence of excess sulfate (20 mM) is shown in Figure 2. In the absence of vitamin B_{12} and riboflavin, there is a very low but detectable rate of carbon tetrachloride dechlorination with a ratio (dechlorination/sulfate respiration) of only 0.008. This indicates that less than 1% of the total electron flux in the cell is directed toward dechlorination. However, in the presence of 20 μM vitamin B_{12} and 50 μM riboflavin, that ratio increases to approximately 20%. We estimate that there is nearly a 30-fold increase in the efficiency of dechlorination by *D. gigas* in the presence of the riboflavin/B_{12} combination.

A mixed, defined culture, composed of *D. gigas* and *C. beijerinckii* growing by glucose fermentation with interspecies hydrogen transfer to the sulfate-reducer, was placed under a 100% argon headspace. The incubation was made 50 μM in riboflavin and vitamin B_{12} each. The culture was amended with 5 mM glucose, and hydrogen formation followed over time as shown in Figure 3. Carbon tetrachloride was then injected. At that point, hydrogen evolution ceased and chloroform production commenced. Hydrogen gas was reutilized as chloroform was produced for approximately 5 days. Somewhat more hydrogen was consumed than can be accounted for by chloroform formation.

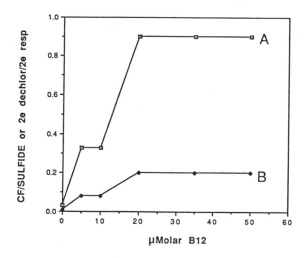

FIGURE 2. **Dechlorination versus sulfate respiration. A. Ratio of chloroform production (CF) to sulfide formation (mol/mol). B. Ratio of electron pairs (2e) appearing as chloroform to those appearing as sulfide. Experiments performed on a resting cell suspension of *D. gigas*.**

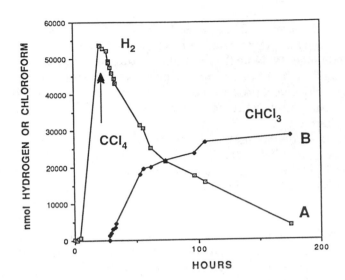

FIGURE 3. **Dechlorination by a mixed culture. Production of hydrogen (A)
and chloroform (B) from a mixed culture of *C. beijerinckii* and *D. gigas*.
Resting cell suspension of each organism was mixed 1:1 and amended
with 5 mM glucose. When chloroform production ceased, 30,000 nmol of
chloroform were produced and 40,000 nmol of hydrogen were consumed
The amount of carbon tetrachloride consumed was not determined; however, an excess carbon tetrachloride was still present when chloroform
production ceased.**

There was still an excess of carbon tetrachloride in the incubation when
chloroform production ceased indicating, that either the catalyst had become
inactivated or the electron flow from the bacteria had ceased. No additional
dechlorination products other than chloroform were detected. The data suggest that chloroform is the predominant product and that only a small fraction,
if any, of the chloroform is further dehalogenated to dichloromethane.

Crude aquifer slurries (10/90 vol/vol of sediment/groundwater) were incubated in the presence of 5 mM glucose and 20 mM sodium sulfate over a period
of 20 days until the bottles turned black due to sulfide formation from the
growth of sulfate-reducing bacteria. Supernatant fractions of these incubations
were concentrated 10-fold and assayed as 2-mL incubations for hydrogen
transformations and dechlorination in the presence and absence of carbon
tetrachloride, riboflavin, and vitamin B_{12}, as shown in Figure 4. In the absence
of carbon tetrachloride, hydrogen evolution occurs continuously and no
chloroform is produced. In the presence of carbon tetrachloride (1 μL), but in
the absence of any riboflavin or vitamin B_{12}, hydrogen formation ceases
abruptly after about 6 hours and chloroform begins to appear after 20 hours. In
a third incubation that contained the carbon tetrachloride, riboflavin, and
vitamin B_{12}, hydrogen is produced but is rapidly reconsumed concomitant
with a rapid burst of chloroform formation.

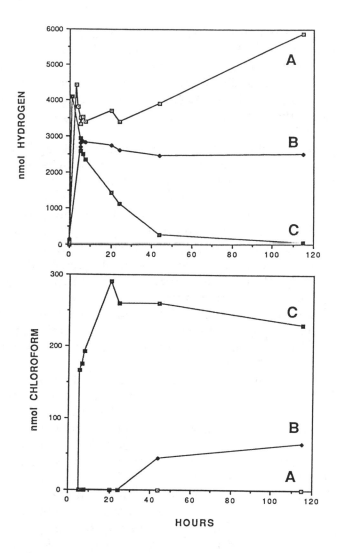

FIGURE 4. Dechlorination by an aquifer enrichment. Hydrogen (top panel) and chloroform (bottom panel) production from a subsample of pre-enriched aquifer slurry enriched on glucose and sulfate. The subsample was incubated with glucose and sulfate and additional carbon tetrachloride. Incubation (A) cells alone, no amendments; (B) carbon tetrachloride added but no riboflavin or vitamin B_{12}; (C) carbon tetrachloride, 50 μM riboflavin, 50 μM vitamin B_{12} added.

DISCUSSION

Sulfate-reducing bacteria are a diverse and ubiquitous group of strict anaerobes that can be found in almost any anaerobic environment. The ability to dechlorinate chlorinated solvents is not, however, a prevalent trait among these bacteria, although dechlorination is often found in association with "sulfate-reducing conditions" (Mohn and Tiedje 1992). It has been known for some time that some biologically occurring transition metal cofactors such as protoporphyrin IX, vitamin B_{12}, or other corrinoids can carry out a catalytic reductive dechlorination reaction in the presence of a strong reducing agent (Krone and Thauer 1989).

We have shown that it is possible to electrochemically couple dechlorination of carbon tetrachloride by transition metal cofactors with bacterial electron transfers in the *Desulfovibrio* species such that the combination carries out dechlorination using up to 20% of the reducing equivalents intended for sulfate respiration. Dehalogenation of carbon tetrachloride to chloroform was used here as a model system to demonstrate the concept of electrochemical coupling. The process we describe is most likely an extracellular catalytic enhancement. This is different from a nutritional enhancement that might be observed upon adding vitamin B_{12} to an organism having a significant requirement for this vitamin for a respiratory process.

The requirement for riboflavin reflects poor electrochemical coupling between vitamin B_{12} and the cytochrome c_3. There is, however, a very low level of electron transfer in the absence of riboflavin. The riboflavin is apparently acting as a redox mediator in the process. A number of other redox mediators have been investigated, but none are as effective and nontoxic as riboflavin.

A role for hydrogen as the ultimate source of electrons is implicated in the experiments shown. This is in line with the physiological role for cytochrome c_3 as a hydrogenase cofactor in *Desulfovibrio* species. Other physiological reductants, such as pyruvate, lactate, and ethanol, have been investigated in our laboratory (not shown), and only pyruvate was found to be an effective donor for vitamin B_{12}-catalyzed dechlorination. This is consistent with the known midpoint redox potential values for oxidation of these compounds.

In nature, hydrogen is a key intermediate in anaerobic food chains and it may be an effective reductant even at very low partial pressures due to the very high affinity of bacterial hydrogenases (LeGall and Forget 1979) for hydrogen. This is particularly true for the hydrogenases from the sulfate-reducing bacteria which have an exceptional affinity for hydrogen (Km values below 1 μM are known). Thus we could anticipate that an aquifer, under nutritional stimulation, may generate adequate levels of hydrogen to drive dechlorination via the extracellular process described here and that hydrogen formation would be sustained as long as fermentable substrates are supplied. The stability of the catalyst itself, over a long time course, and the ability of the catalyst to permeate the aquifer matrix are important factors that could limit application of this concept.

The practice of this chemical-biological approach would entail two steps: (1) generation of a sulfate-reducing aquifer enrichment in situ, and (2) addition of the catalytic/redox mediator combination to effectively scavenge hydrogen and carry out rapid dechlorination. The advantages of this approach are multiple. First, dechlorination is a site-specific phenomenon that is poorly understood at the microbial level. The artificial process described here eliminates the requirement for dechlorinating bacteria. Second, the rate and, more importantly, the efficiency of dechlorination may be dramatically enhanced. Incomplete dechlorination and solvent specificity are limitations of the process and are currently being addressed in a structure-function study of tetrapyrrole compounds.

REFERENCES

Bradford, M. 1976. "A Rapid and Sensitive Method for Quantitation of Micrograms Quantities of Protein Utilizing the Principle of Protein Dye-Binding." *Analytical Biochemistry* 72:248-252.

Krone, U. E. and R. K. Thauer. 1989. "Reductive Dehalogenation of Chlorinated C-1 Hydrocarbons Mediated by Corrinoids." *Biochemistry* 28: 4908-4914.

LeGall, J., and N. Forget. 1979. "Purification of the Electron-Transfer Components from the Sulfate-Reducing Bacteria." *Methods in Enzymology LIII* pt D. 613-634.

Mohn, W. W. and J. M. Tiedje. 1992. "Microbial Reductive Dehalogenation." *Microbiological Reviews* 56:482-507.

Odom, J. M., K. Jessie, E. Knodel, and M. Emptage. 1991. "Immunological Cross-Reactivities of Adenosine 5' Phosphosulfate Reductases From Sulfate-Reducing and Sulfide-Oxidizing Bacteria." *Applied and Environmental Microbiology* 57:727-733.

Siegal, L. M. 1965. "A Microdetermination Method for Sulfide." *Analytical Biochemistry* 11:126-142.

Thauer, R.K., K. Jungerman, and K. Decker. 1977. "Energy Conservation in Chemotrophic Anaerobic Bacteria." *Bacteriological Reviews* 41: 100-180.

Van der Westen, H.M., S.G. Mayhew, and C. Veeger. 1980. "Separation of Hydrogenase from Intact Cells of *Desulfovibrio vulgaris.*" *FEBS Letters* 86:122-126.

Wackett, L.P., M.S.P. Logan, F.A. Blocki, and C. Bao-li. 1992. "A Mechanistic Perspective on Bacterial Metabolism of Chlorinated Methanes." *Biodegradation* 3:19-36.

Dechlorination of PCE and TCE to Ethene Using an Anaerobic Microbial Consortium

Wei-Min Wu, Jeffery Nye, Robert F. Hickey, Mahendra K. Jain, and J. Gregory Zeikus

ABSTRACT

An anaerobic microbial consortium capable of dechlorinating chlorinated ethenes to ethylene has been developed as anaerobic granules in a laboratory-scale upflow anaerobic reactor (34 L) under ambient temperature conditions. Dechlorination of tetrachloroethene (PCE), trichloroethene (TCE), and dichloroethenes (DCEs) to ethene occurred under a wide temperature range (10 to 30°C) tested. TCE, *cis*-1,2-DCE, and vinyl chloride (VC) are sequential intermediates of PCE dechlorination. The consortium also dechlorinated *trans*-1,2-DCE and 1,1-DCE to ethene. Various substrates for methane production can support dechlorination by this consortium. The feasibility of application of the microbial consortium as a microbial inoculum for treatment of groundwater and saturated soils was examined in bench-scale systems. Preliminary results indicate that the consortium can effectively dechlorinate PCE and TCE to ethene in saturated soils.

INTRODUCTION

PCE and TCE are widely used as industrial solvents and in dry cleaning fluids and hence are common contaminants in groundwater and soils. PCE is not degraded by microorganisms under aerobic conditions (Vogel et al. 1987). If an indigenous, anaerobic microbial population, capable of performing complete dechlorination, is present in a contaminated site and a proper substrate/electron donor is available, PCE and TCE will be dechlorinated eventually to ethene. However, if a site has a population that cannot completely dechlorinate PCE and TCE, their metabolic intermediates, the DCEs and VC, will be persistent. No dechlorination will be observed at sites having no dechlorinating organisms. Many contaminated sites appear to lack a significant dechlorinating population or have populations able only to partially dechlorinate. Delivery of dechlorinating organisms to the subsurface is necessary for effective biotreatment of these sites.

We have developed anaerobic dechlorinating microbial granules in anaerobic upflow reactors (Wu et al. 1993). These granules are a self-immobilized microbial consortium consisting of anaerobic dechlorinating organisms, methanogens, aceto-genes, and hydrolytic-fermentative bacteria. The advantage of growing a dechlor-inating consortium in granules is that the granules can be mass-produced in reactors and easily utilized for both on-site and in situ biotreatment of PCE, TCE, and other chlorinated ethenes.

EXPERIMENTAL PROCEDURES AND MATERIALS

Growth of PCE/TCE-Dechlorinating Granules in an Anaerobic Reactor

PCE/TCE dechlorinating granules developed in a 2.65 L volume laboratory anaerobic reactor (Wu et al. 1993) were used as an inoculum for scale-up. The medium used for growth of the dechlorinating consortium contained the follow-ing components (per liter of tap water): KH_2PO_4, 0.05 g; $FeSO_4 \cdot 7H_2O$, 0.028 g; urea, 0.05 g; cane molasses, 1 g; propionic acid, 2 g; butyric acid, 1.5 g; and a trace element solution, 0.5 mL (Wu et al. 1993). The reactor was a water-jacketed glass column with a diameter of 15.2 cm and height of 2.0 m (volume 34 L). The details of the reactor system were essentially the same as reported for the 2.65-L reactor (Wu et al. 1993). The 34-L reactor was operated at 20 to 22°C with contin-uous feed of the medium. Recycling treated effluent from the top of the reactor to the reactor inlet (recycle ratio of approximately 40 to 1 of the influent flow rate) was used to maintain the required hydraulic flux rate. A solution comprised of a mixture of PCE-TCE-methanol was fed to the reactor inlet using a syringe pump. The ratio of PCE, TCE, and methanol in this solution was 1:1:8 (w/w/w). The influent PCE and TCE concentration was calculated based on the mass of PCE and TCE fed and the flowrate of the medium.

Batch Dechlorination Tests

The tests for the characterization of the PCE/TCE-dechlorination pathway and the temperature and substrate effects were performed in 158-mL serum bottles sealed with Teflon™-lined butyl rubber stoppers. Liquid medium (100 mL) and 0.3 mL of granules were used; the headspace of the bottles was vacuum-flushed with a N_2-CO_2 (95:5, vol/vol) gas mixture to remove volatile compounds (chlori-nated ethene, ethene, and methane). The chlorinated ethenes were added neat using a 10-µL microsyringe except for VC, which was added as a gas. All bottles were incubated upside-down in a shaker operated at 100 strokes/min. The tests were conducted at 20 to 22°C except for the assay designed to study the effect of temperature on dechlorination. This assay was conducted at 4, 15, 20, and 30°C, respectively. Cane molasses was used as the primary substrate for all batch tests except for the substrate effect test. Two replicates were employed in all experiments.

Dechlorination in Saturated Soils

The feasibility of using the granular consortium as an inoculum was tested using three different sandy soils. Soil No. 1 contained a trace of organic matter (0.25% VS), Soil No. 2 contained 10.3% VS (mainly plant residues) which were relatively biodegradable), and Soil No. 3 contained 32% of VS (mainly humus) which were biodegradable to only a limited extent. Soil (80 g) was added to each 158-mL serum bottle and then saturated by adding tap water (buffered with 20 mM of sodium phosphate, pH 7.0) into the bottle. The bottle was flushed with the N_2-CO_2 (95%:5%) gas mixture to remove oxygen and supplemented with Na_2S solution to obtain a starting concentration of 0.1 mM. A suspension of disrupted dechlorinating granules (0.5 ml, 0.5 g (volatile suspended solids) VSS/L) was added to each bottle as inoculum. In the bottles containing Soil No. 1, cane molasses solution (50 µL of 100 mg/mL) was supplemented as the primary substrate. Soils No. 2 and No. 3 did not receive additional substrate. TCE (0.5 µL) was added to each bottle to achieve a TCE content of approximately 9 mg/kg soil. Duplicate was employed for each soil. The bottles were then incubated upside-down under static conditions at ambient temperature (20 to 22°C). The headspace gas composition was monitored periodically. When chlorinated ethenes were consumed and ethene production stopped, the headspace of each bottle was purged with the N_2-CO_2 (95%:5%) gas mixture to remove ethene and methane and TCE was again added to the bottle. Cane molasses also was periodically supplemented to the bottles containing Soil No. 1. This procedure was repeated over a 170-day period. In addition, the bottles containing respective soils and without added dechlorinating consortium were used as controls and incubated at the same condition.

Analytical Methods

Methane, methanol, and volatile fatty acids (VFAs) were determined using gas chromatography (GC) (Wu et al. 1993). Ethane, ethene, and chlorinated ethenes in the gas phase were analyzed with a Hewlett-Packard 5890A GC equipped with a flame ionization detector (FID). Separation was performed by using a metal column containing Carbopack B/1%SP-100 (Supelco, Bellefonte, PA) at 40°C for ethane, ethene, and vinyl chloride and 190°C for DCEs, TCE, and PCE. For determination of dissolved ethene and chlorinated ethenes in the reactor, liquid samples (5 mL) were injected into a Teflon™-covered rubber-capped test vial (total volume 12.5 mL), and the vial was heated in a 70°C water bath to release dissolved compounds to the headspace. The dissolved ethene concentration was calculated by using the ethene concentration in the headspace and the volume ratio between headspace and liquid sample. A Varian GC system (Varian Analytical Instruments, Sunnyvale, CA), consisting of a Varian model 3500 GC equipped with a FID and Vocol™ capillary column (30 m, 0.53 µm ID, 3.0 µm film, Supelco, Bellefonte, PA) with helium as a carrier, and a Varian Genesis 705 Tecmar headspace autosampler were used for the analysis of dissolved VC, DCEs, TCE, and PCE.

RESULTS AND DISCUSSION

Performance of the Anaerobic Reactor

After five months of operation, a full granule bed (15 to 16 L) with biomass concentration of 45 to 50 g VSS/L had been developed in the reactor under steady state conditions (hydraulic retention time 1.0 day, pH 7.0). The influent PCE (16,000 µg/L) and TCE (16,000 µg/L) were completely converted to ethene; chlorinated ethene concentrations in the effluent were below detection limits (PCE, TCE, DCEs, and VC < 5 µg/L). This indicates that dechlorinating granules can be mass-produced in an anaerobic reactor, which is of significance in applications of the dechlorinating cultures for both on-site and in situ bioremediation.

Dechlorination Pattern of the Consortium

A time course of PCE dechlorination by the dechlorinating consortium in liquid medium is presented in Figure 1. TCE, *cis*-1,2-DCE and VC accumulated as intermediates at significant levels. Trace levels of *trans*-1,2-DCE, and 1,1-DCE were detected during the time course assay, and ethene was the primary end product. When dechlorination of TCE was assayed, *cis*-1,2 DCE, and VC were observed to be the predominant intermediates while trace levels of *trans*-1,2-DCE and 1,1-DCE also were observed.

The initial dechlorination rates of different chlorinated ethenes by the anaerobic consortium were estimated using PCE, TCE, *cis*-DCE, *trans*-DCE, 1,1-DCE, or

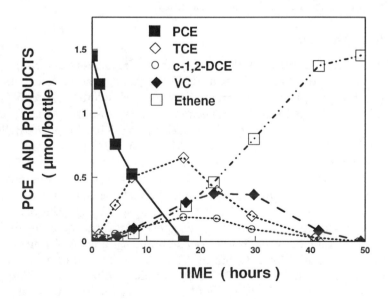

FIGURE 1. PCE dechlorination by PCE/TCE dechlorinating granules at ambient temperature.

VC as the sole added chlorinated ethene in sequentially controlled assays. Dechlorination rates for these six chlorinated ethenes from most to least rapid were: PCE > TCE > *cis*-1,2-DCE > 1,1-DCE > VC > *trans*-1,2-DCE. Based on the time course experiment for PCE and TCE dechlorination and initial dechlorination rates of *cis*-1,2-DCE, *trans*-1,2-DCE, and 1,1-DCE, the predominant intermediate DCE during PCE and TCE dechlorination was *cis*-1,2-DCE. This is based on the observation that *cis*-1,2-DCE accumulated to the highest concentrations of the DCEs while exhibiting a higher dechlorination rate than the other two isomers. The dechlorination pathway for PCE is presented in Figure 2. TCE, *cis*-1,2-DCE, and VC are the major intermediates observed during PCE dechlorination. This pathway also indicates that this dechlorinating consortium is capable of dechlorinating all chlorinated ethenes, i.e., PCE, TCE, DCEs, and VC.

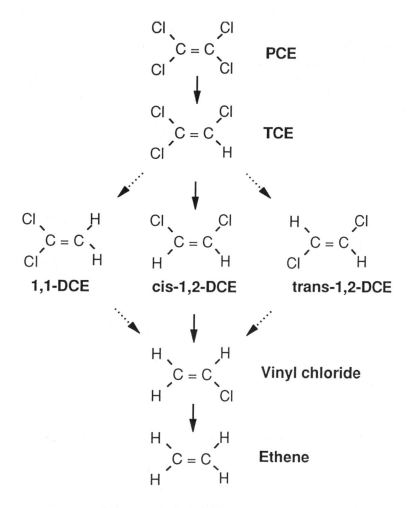

FIGURE 2. Proposed PCE dechlorination pathway by PCE/TCE-dechlorinating granules.

Factors Influencing Dechlorination

Various substrates have been reported as suitable electron donors for PCE-dechlorination in the literature (Freedman and Gossett 1989; Gibson and Sewell 1992; de Bruin et al. 1992; DiStefano et al. 1992). The primary substrates tested in this study that supported PCE and TCE dechlorination included hydrogen, formate, acetate, propionate, butyrate, lactate, benzoate, methanol, ethanol, glucose, and cane molasses. Rapid dechlorination was observed when hydrogen, formate, glucose, and molasses were used. Dechlorination rates initially were slow but then became rapid when propionate, butyrate, lactate, benzoate, or ethanol was used. Dechlorination was slow with acetate and methanol as substrate. This is different from results presented for a PCE and methanol-enrichment culture (DiStefano et al. 1992; Tandol et al. 1994). The granules had a poor dechlorination rate with methanol as the primary substrate. It appears that any substrate from which hydrogen is produced during degradation will support dechlorination by the granules. The poor dechlorination rate with methanol may be attributed to differences in the populations of the different anaerobic consortia used in the respective studies. TCE dechlorination was observed to occur between 4 and 30°C (Figure 3). The rate of dechlorination increased as the temperature increased. The results indicate that the granules can dechlorinate TCE at the typical temperatures (10 to 18°C) encountered at most sites. The higher temperature results in more rapid dechlorination.

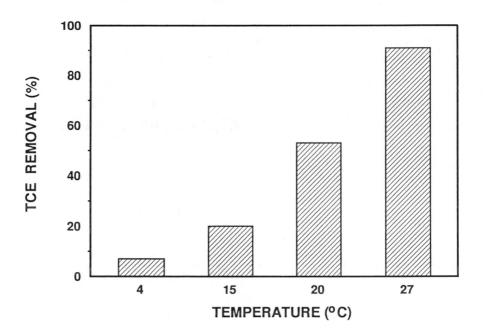

FIGURE 3. Effect of temperature on TCE removal by dechlorinating granules.

Dechlorination of TCE in Saturated Soils

Dechlorination of TCE to ethene was observed in all three saturated soils (Figure 4). Neither production of dechlorinated intermediates (DCEs, VC, and ethene) nor methane production was observed in the headspace in the serum bottles containing soils which were not amended with added dechlorinating

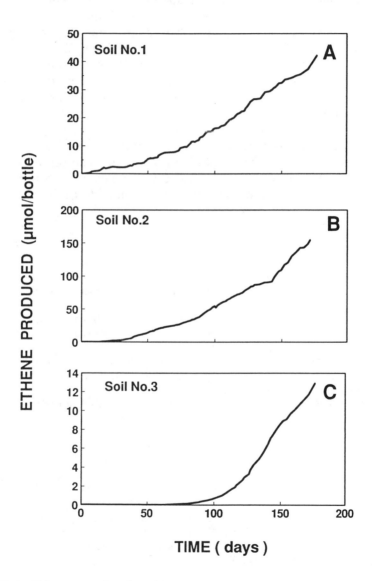

FIGURE 4. Ethene production from TCE in three different saturated soils inoculated with anaerobic dechlorinating consortium. Cane molasses was supplemented to Soil No. 1 as primary substrate (A). No substrate was added to Soils No. 2 and 3 (B and C).

consortium after 170-day incubation at 20 to 22°C. This indicated that the original soils used did not have detectable dechlorinating activities and methanogenic activity. In Soil No. 1, addition of substrate (cane molasses) was essential for the dechlorination (Figure 4A). In Soils No. 2 and No. 3, no additional substrate was required since soil organic matter supported anaerobic activity and dechlorination. After 170 days incubation, the average ethene produced from TCE in Soils No. 1, 2, and 3 was 42, 170, and 13 µmol, respectively, while average methane production was 820, 1080, and 5.4 µmol, respectively. Therefore, the molar ratio of methane to ethene produced was 20, 7.1, and 0.41 for Soils No. 1, 2, and 3, respectively. The results indicate that the dechlorinating consortium in the granules can use either soluble organic matter (such as cane molasses) or soil organic material as a source of electron donors for dechlorination. In addition, dechlorination seemed to out-compete methanogenesis when poorly biodegradable materials were used as substrate.

Bioaugmentation is necessary for the sites that lack dechlorinating organisms. These results suggest that the dechlorinating granules can serve as inoculum for soil/aquifer bioremediation. The need to supplement substrate/electron donors will depend on the organic content of soil and aquifer material.

REFERENCES

de Bruin, W. P., M.J.J. Kotterman, M. A. Posthumus, G. Schraa, and A.J.B. Zehnder. 1992. "Complete biological reductive transformation of tetrachloroethene to ethane." *Applied and Environmental Microbiology* 58(6):1996-2000.

DiStefano, T. D., J. M. Gossett, and S. H. Zinder. 1992. "Hydrogen as an electron donor for dechlorination of tetrachloroethene by an anaerobic mixed culture." *Applied Environmental Microbiology* 58(11):3622-3629.

Freedman, D. L. and J. M. Gossett. 1989. "Biological reductive dechlorination of tetrachloroethylene and trichloroethylene to ethylene under methanogenic conditions." *Applied Environmental Microbiology* 55(9):2144-2151.

Gibson, S. A. and G. W. Sewell. 1992. "Stimulation of reductive dechlorination of tetrachloroethene in anaerobic aquifer microcosms by addition of short-chain organic acids or alcohols." *Applied Environmental Microbiology* 58(4):1392-1393(1992).

Tandol, V., T. D. DiStefano, P. A. Bowser, J. M. Gossett, and S. H. Zinder. 1994. "Reductive dehalogenation of chlorinated ethenes and halogenated ethanes by a high-rate anaerobic enrichment culture." *Environmental Science and Technology* 28(5):973-979.

Vogel, T. M., C. S. Criddle, and P. L. McCarty. 1987. "Transformations of halogenated aliphatic compounds." *Environmental Science and Technology* 21(8):722-736.

Wu, W.-M., J. Nye, R. F. Hickey, and L. Bhatnagar. 1993. "Anaerobic granules developed for reductive dechlorination of chlorophenols and chlorinated ethylene." In *Proceedings of the 48th Industrial Waste Conference*, May 10-12, 1993, Purdue University, pp. 483-493. Lewis Publishers, Boca Raton, FL.

Effect of Temperature on Perchloroethylene Dechlorination by a Methanogenic Consortium

Jianwei Gao, Rodney S. Skeen, and Brian S. Hooker

ABSTRACT ━━━━━━━━━━━━━━━━━━━━━━

The effect of temperature on the kinetics of growth, substrate metabolism, and perchloroethylene (PCE) dechlorination by a methanogenic consortium is reported. In all cases, a simple kinetic model accurately reflected experimental data. Values for the substrate and methane yield coefficients, and the maximum specific growth rate are consistent at each temperature. Also, the substrate, methane, and PCE dechlorination yield coefficients show little temperature sensitivity. In contrast, the maximum specific growth rate is temperature dependent.

INTRODUCTION

Biological treatment has become one of the most appealing technical solutions for contaminated soils and groundwater. In situ bioremediation for chlorinated solvents is one such technology that has gained popularity. Microorganisms rapidly transform chlorinated solvents into nontoxic compounds. In situ biological processes also can circumvent the mass-transport issues that limit the effectiveness of pump-and-treat systems. This advantage results from contaminants being destroyed in place rather than extracted (Skeen et al. 1993).

Most research efforts on bioremediation for chlorinated solvents have focused on exploiting aerobic microbial metabolisms to achieve contaminant transformation. As a result, much is known about the biochemistry of these systems (McCarty and Semprini 1994). In contrast, little is known about the environmental conditions necessary to initiate and sustain anaerobic dechlorination (Bouwer 1994). Hence, laboratory screening tests using microcosms with various additives have served as the basic tool to evaluate in situ bioremediation potential (Petersen et al. 1994; Gibson and Sewell 1992; Sewell and Gibson 1991). The primary goals of these tests are to determine the potential for in situ contaminant destruction and to develop kinetic information to help design an in situ remediation process. This paper reports the effects of temperature on the kinetics of growth, substrate metabolism, and PCE dechlorination by a methanogenic consortium.

MATERIALS AND METHODS

Sediments were obtained from the Yakima River delta in southeastern Washington State and cultivated anaerobically with methanol. A seed culture for these experiments was prepared by combining 20 g of sediment and 40 mL of culture media with 3 g/L of methanol in a 100-mL serum bottle. The culture was continuously maintained by periodically exchanging 20-mL culture solution with fresh media and methanol. The media used both in the seed culture and in subsequent experiments contained (per liter of deionized water): 270 mg KH_2PO_4, 350 mg K_2HPO_4, 530 mg NH_4Cl, 75 mg $CaCl_2 \cdot 2H_2O$, 100 mg $MgCl_2 \cdot 6H_2O$, 20 mg $FeCl_2 \cdot 4H_2O$, 1,200 mg $NaHCO_3$, 250 mg $Na_2S \cdot 9H_2O$, 10 mg $C_6H_7NO_6Na_2$, 5.0 mg $MnCl_2 \cdot 4H_2O$, 0.5 mg H_3BO_3, 0.5 mg $ZnCl_2$, 0.5 mg $CoCl_2 \cdot 6H_2O$, 0.5 mg $NiSO_4 \cdot 6H_2O$, 0.3 mg $CuCl_2 \cdot 2H_2O$, 0.1 mg $NaMoO_4 \cdot 2H_2O$, and 1.0 mg resazurin. One liter of culture medium also contained the following vitamins: 0.02 mg biotin, 0.02 mg folicin, 0.1 mg vitamin B_6, 0.1 mg riboflavin, 0.1 mg thiamine (HCl), 0.05 mg pantothenic acid, 0.05 mg nicotinamide, 0.1 mg vitamin B_{12}, 0.05 mg PAPA, and 0.06 mg lipoic acid.

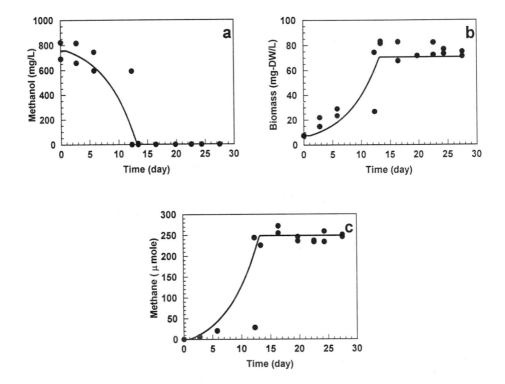

FIGURE 1. Example measured and predicted results of a growth experiment at 17°C; (a) methanol concentration, (b) biomass concentration, and (c) methane in the culture.

Growth tests were conducted at 17 and 30°C to test kinetic expressions for anaerobic metabolism of methanol. In these experiments, 10 mL culture media with approximately 1 g/L methanol was added to 28 mL balsh tubes. Each tube also received 0.5 mL inoculum from a 7 day old subculture of the original microbial stock. The subculture was prepared by adding 10 mL of the seed culture to a 100 mL serum bottle that contained 50 mL media with 3 g/L methanol. At every sample time, duplicate tubes were sacrificed and analyzed for the concentrations of methane, hydrogen, protein, methanol, and acetate.

Perchloroethylene (PCE) dechlorination tests were conducted at 17 and 30°C using the fed-batch reactor system reported previously (Petersen et al. 1994). The system was charged with sufficient PCE-saturated water and biomass to achieve a nominal aqueous concentration of 1 mg/L and 20 mg dry weight per liter (mg-DW/L), respectively.

Aqueous samples were periodically removed from the reactor and analyzed for the concentration of protein, methanol, acetate, and chlorinated organics (Petersen et al. 1994). Simultaneously, headspace samples were removed and analyzed for methane and hydrogen. Methane and hydrogen were quantified by

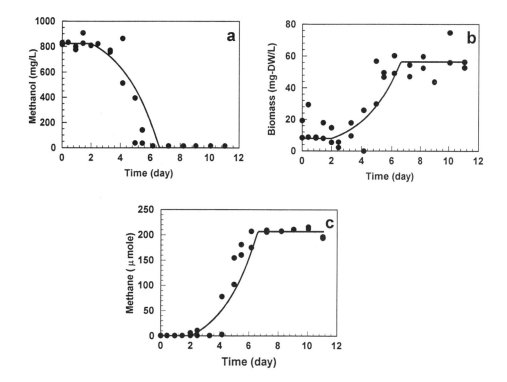

FIGURE 2. Example measured and predicted results of a growth experiment at 30°C; (a) methanol concentration, (b) biomass concentration, and (c) methane in the culture.

gas chromatography (GC) using a thermal conductivity detector and a 4.5-m Carboxen 1000 packed column having a 3.2-mm ID. Chlorinated organics such as PCE, trichloroethylene (TCE), dichloroethylene (DCE), and vinyl chloride (VC) were assayed by a GC using an electron capture detector and a 30-m DB-624 column (J&W Scientific, Fulsom, California) having a 0.53-mm ID. Methanol concentration was analyzed by GC using a flame ionization detector with a 30-m DB-Wax column (J&W Scientific, Fulsom, CA) having a 0.53-mm ID. Acetate concentrations were determined from a filtered sample using a Dionex 4000i (Dionex, Sunnyvale, California) ion chromatograph with a Dionex PAX 100 anion exchange column (Dionex, Sunnyvale, California). Aqueous solutions were analyzed for protein with the Pierce Micro BCA Protein Assay Kit. Cell concentrations were converted to mg-DW/L assuming biomass is composed of 51% protein on a dry weight basis.

RESULTS AND DISCUSSION

Equations 1 through 4 were used to describe growth, substrate consumption, methane production, and PCE dechlorination to TCE by the methanogenic consortium.

$$\frac{d[X]}{dt} = \frac{\mu_{max}[X][S]}{([S] + K_s)} \tag{8}$$

$$\frac{d[S]}{dt} = -Y_{SX}\frac{\mu_{max}[X][S]}{([S] + K_s)} \tag{9}$$

$$\frac{dM}{dt} = Y_{MX}\frac{\mu_{max}[X][S]}{([S] + K_s)}V_l \tag{10}$$

$$\frac{d[TCE]}{dt} = Y_{PCE}\,Y_{MX}\frac{\mu_{max}[X][S]}{([S] + K_s)} \tag{11}$$

In these equations [X] represents the aqueous biomass concentration (mg-DW/L); [S] the methanol concentration (mole/L); M the total moles of methane in the reactor; [PCE] and [TCE] the aqueous concentrations of PCE and TCE (mg/L), respectively; and V_l the liquid volume in the system (L). Only PCE conversion to TCE was described since this was the sole dechlorination step observed during the limited duration of the experiments. A K_s value of 1.7×10^{-4} mole-methanol/L reported by Gupta et al. (1994) was used in all cases. The form of Equation 4 was chosen based on the observation that PCE dechlorination by methanogenic organisms is strongly linked to methane formation (Rasmussen et al. 1994,

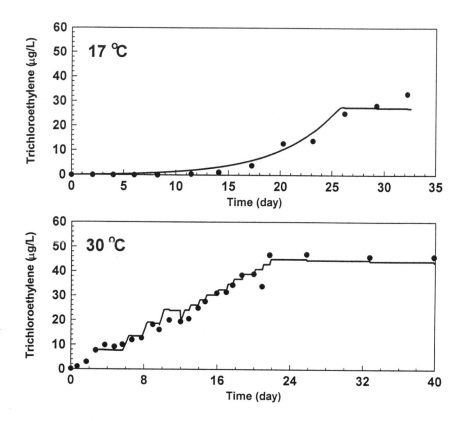

FIGURE 3. Measured and predicted trichloroethylene concentration for dechlorination tests at 17 and 30°C.

Fathepure and Boyd 1988). Dechlorination was assumed independent of PCE concentration since the levels used in these tests were always much higher than published values of the half-saturation coefficient (Tandoi et al. 1994).

In all cases, the model accurately reflected the experimental data. This is demonstrated in Figures 1 and 2, which show the results from growth tests conducted at 17 and 30°C, respectively. In addition, Figure 3 displays experimental data and model response curves for TCE production at both temperatures. The 30°C data represent an experiment in which multiple additions of methanol were made up to day 22.

Table 1 summarizes the kinetic coefficients that best described the experimental data for growth tests and dechlorination tests at 17 and 30°C. These values were determined using the SimuSolv® program (Dow Chemical Company, Midland, Michigan) to calculate optimal model parameter values based on input experimental data sets. Values for μ_{max}, Y_{SX}, Y_{MX} appear to be fairly consistent at each temperature. Also, there is no statistically significant temperature sensitivity in the substrate, methane, and PCE dechlorination (Y_{PCE}) yield coefficients.

TABLE 1. Summary of rate data at 17 and 30°C.

Temperature (°C)	Test Type	μ_{max} (day^{-1})	Y_{SX} (mole/g-DW)	Y_{MX} (mole/g-DW)	Y_{PCE} (mole/mole)
30	Dechlorination	0.43	0.49	0.27	3.3×10^{-6}
30	Dechlorination	0.56	0.39	0.26	2.0×10^{-6}
30	Growth	0.43	0.54	0.41	na
30°C Averages:		0.47 ± 0.08	0.47 ± 0.08	0.31 ± 0.08	$(2.7\pm0.9)\times10^{-6}$
17	Dechlorination	0.12	0.32	0.35	1.3×10^{-6}
17	Dechlorination	0.19	0.38	0.34	8.0×10^{-6}
17	Growth	0.19	0.37	0.25	na
17	Growth	0.16	0.37	0.37	na
17°C Averages:		0.17 ± 0.03	0.36 ± 0.03	0.33 ± 0.05	$(4.7\pm4.7)\times10^{-6}$

na = not applicable or not measured.

In contrast, the maximum growth rate is temperature dependent with an average of 0.48 day^{-1} at 30°C and 0.18 day^{-1} at 17°C.

The fact that Y_{PCE} is insensitive to temperature suggests that the dechlorination reaction is not mechanistically different at the two temperatures. In addition these results suggest that it is possible to apply anaerobic dechlorination kinetics measured at higher temperatures to lower temperature situations if the maximum specific growth rate is known at both temperatures. Conducting tests to measure dechlorination rates at higher temperatures reduces the cost of treatability analysis by reducing experiment duration.

ACKNOWLEDGMENTS

This work was supported by the U.S. Department of Energy, Office of Technology Development, under Contract DE-AC06-76RLO 1830, as part of the In Situ Remediation Integrated Program. Pacific Northwest Laboratory is operated by Battelle Memorial Institute for the U.S. Department of Energy under contract DE-AC06-76RLO-1830.

REFERENCES

Bouwer, E. J. 1994. "Bioremediation of Chlorinated Solvents Using Alternate Electron Acceptors." In Norris, Hinchee, Brown, McCarty, Semprini, Wilson, Kampbell, Reinhard, Bouwer, Borden, Vogel, Thomas, and Ward (Eds.), *Handbook of Bioremediation*. pp. 149-175. Lewis Publishers, Ann Arbor, MI.

Fathepure, B. Z., and S. A. Boyd. 1988. "Reductive Dechlorination of Perchloroethylene and the Role of Methanogens." *FEMS Microbiology Letters 49*:149-156.

Gibson, S. A., and G. W. Sewell. 1992. "Stimulation of Reductive Dechlorination of Tetrachloroethylene in Anaerobic Aquifer Microcosms by Addition of Short-Chain Organic Acids or Alcohols." *Appl. Environ. Micro. 58*:1392-1393.

Gupta, A., J.R.V. Flora, G. D. Sayles, and M. T. Suidan. 1994. "Methanogenesis and Sulfate Reduction in Chemostats — II. Model Development and Verification." *Wat. Res. 28*: 795-803.

Hooker, B. S., R. S. Skeen, and J. N. Petersen. 1994. "Biological Destruction of CCl_4 Part II: Kinetic Modelling." *Biotechnol. Bioeng. 44*:211-218.

McCarty, P. L., and L. Semprini. 1994. "Groundwater Treatment for Chlorinated Solvents." In Norris, Hinchee, Brown, McCarty, Semprini, Wilson, Kampbell, Reinhard, Bouwer, Borden, Vogel, Thomas, and Ward (Eds.), *Handbook of Bioremediation*. pp. 87-116. Lewis Publishers, Ann Arbor, MI.

Petersen, J. N., R. S. Skeen, K. M. Amos, and B. S. Hooker. 1994. "Biological Destruction of CCl_4 Part I: Experimental Design and Data." *Biotechnol. Bioeng. 43*:521 528.

Rasmussen, G., S. J. Komisar, and J. F. Ferguson. 1994. "Transformation of Tetrachloroethene to Ethene in Mixed Methanogenic Cultures: Effect of Electron Donor, Biomass Levels, and Inhibitors." In R. E. Hinchee, A. Leeson, L. Semprini, and S. K. Ong (Eds.), *Bioremediation of Chlorinated and Polycyclic Aromatic Hydrocarbon Compounds*. pp. 307-313. Lewis Publishers, Ann Arbor, MI.

Sewell, G. W., and S. A. Gibson. 1991. "Stimulation of Reductive Dechlorination of Tetrachloroethylene in Anaerobic Aquifer Microcosms by Addition of Toluene." *Environ. Sci. Technol. 25*:982-984.

Skeen, R. S., S. P. Luttrell, T. M. Brouns, B. S. Hooker, and J. N. Petersen. 1993. "In-situ Bioremediation of Hanford Groundwater." *Remediation 3*:353-367.

Tandoi, V., T. D. DiStefano, P. A. Bowser, J. M. Gossett, and S. H. Zinder. 1994. "Reductive Dehalogenation of Chlorinated Ethenes and Halogenated Ethanes by a High-rate Anaerobic Enrichment Culture." *Environ. Sci. Technol. 28*:973-979.

Transformation of Tetrachloromethane by *Shewanella putrefaciens* MR-1

Erik A. Petrovskis, Timothy M. Vogel, Daad A. Saffarini, Kenneth H. Nealson, and Peter Adriaens

ABSTRACT

The dissimilatory metal-reducing facultative anaerobe *Shewanella putrefaciens* MR-1 has been shown to transform chlorinated methanes by biotic and abiotic processes under Fe(III)-respiring conditions. MR-1 produces magnetite and other Fe(II)-containing compounds that dehalogenate tetrachloromethane (CT) and trichloromethane (CF) abiotically. The biological CT transformation activity has been localized utilizing metabolic inhibitors and respiratory mutants. Cytochrome and iron-sulfur complex inhibitors did not block CT transformation. Terminal reductase mutants showed wild-type CT transformation activity. However, menaquinone-deficient mutants lost 90% of the CT transformation activity. This represents the first direct evidence of respiratory chain involvement in reductive dehalogenation.

INTRODUCTION

Investigations of pollutant transformations by pure cultures may enhance our understanding of in situ natural attenuation processes in the environment. Biotic and abiotic pollutant transformation can occur in Fe(III)-reducing environments. Dissimilatory iron-reducing bacteria transform pollutants directly or generate reductants, such as Fe(II), for abiotic transformation reactions. For example, oxidation of aromatic hydrocarbons can be coupled to Fe(III) reduction by *Geobacter metallireducens* (Lovley et al. 1989). Abiotic reduction of 4-chloronitrobenzene and tetrachloromethane on Fe(II)-mineral surfaces has also been demonstrated (Heijman et al. 1993; Kriegman-King & Reinhard 1994). In addition, tetrachloromethane (CT) can be dechlorinated by *Shewanella putrefaciens* (Picardal et al. 1993; Petrovskis et al. 1994).

S. putrefaciens is a facultative anaerobe with the unique ability to use a wide range of terminal electron acceptors for growth, including Fe(III) oxides (Nealson & Saffarini 1994). In this study, the biotic and abiotic transformations of chlorinated aliphatic compounds by *S. putrefaciens* MR-1 were investigated. Utilizing metabolic inhibitors and respiratory mutants, we evaluated the hypothesis that an anaerobic respiratory chain component is responsible for CT transformation in MR-1.

MATERIALS AND METHODS

MR-1 was isolated from anaerobic sediment of Oneida Lake, New York (Myers & Nealson 1988). Growth conditions, dechlorination assays, and analytical procedures have been previously described (Petrovskis et al. 1994). Briefly, MR-1 was anaerobically grown in a defined media, washed, and resuspended in 10 mM HEPES buffer, pH 7.5. CT was added and headspace concentrations of CT, $CHCl_3$ (CF), and CH_2Cl_2 (DCM) were determined by gas chromatography. For experiments presented in Table 1, MR-1 cell suspensions were incubated with metabolic inhibitors for 45 to 60 minutes before addition of CT. MR-1 mutants (Table 2), deficient in the ability to use one or more terminal electron acceptors, were isolated after ICR-191 treatment or after transposon (Tn5) mutagenesis, as described earlier (Saffarini et al. 1994). Menaquinone-deficit mutants of MR-1 were provided by Dr. Charles Myers (Myers & Myers 1993) and were grown with TMAO as a terminal electron acceptor.

TABLE 1. Effects of metabolic inhibitors on CT transformation.[a]

Inhibitor	Site of Action	Concentration/ Carrier Solvent	Relative CT Trans- formation Activity[b]	
Rotenone	NADH dehydrogenase	100 µM/acetone	0.91	0.67
Quinacrine	flavins	1 mM/water	1.34	1.15
p-CMPS	Fe-S complexes	500 µM/ethanol	1.03	1.03
HQNO	cyt *b* to cyt *c*	10 µM/ethanol	1.01	1.09
Antimycin A	cyt *b* to cyt *c*	100 µM/ethanol	0.98	0.99
NaN_3	cytochrome oxidase	1 mM/water	0.95	1.05
KCN	cytochrome oxidase	1 mM/water	0.98	1.01
CCCP	uncoupler	100 µM/ethanol	0.92	1.09

(a) MR-1 cells were grown under nitrate-respiring conditions with lactate, resuspended in LM media (Petrovskis et al. 1994) with 10 mM sodium lactate, and incubated with inhibitor for 60 minutes at 30°C before addition of CT.
(b) Activity relative to cells treated with solvent used for inhibitor. Duplicate experiments are shown.

TABLE 2. Screening MR-1 respiratory mutants.[a]

Phenotype	Number screened	CT transformation
No Mn(IV) reduction	8	Yes
No Fe(III) reduction	7	Yes
No fumarate reduction	7	Yes
No anaerobic growth[b]	3	Yes
No menaquinone[c]	2	No

(a) Cells were grown under nitrate-respiring conditions when possible (Petrovskis et al. 1994). Otherwise, the cells were grown anaerobically in LB broth, aerobically in a defined media (b), or under TMAO-respiring conditions (c).

RESULTS AND DISCUSSION

Effect of Fe(III) on CT Transformation

Earlier work has shown that MR-1 transforms CT to chloroform (CF, 24%), cell-bound material (50%), minor amounts of CO_2 (7%), and unidentified non-volatile products (4%) under anaerobic conditions (Figure 1A, Petrovskis et al. 1994). MR-1 did not transform other chlorinated aliphatic compounds under nitrate- or fumarate-respiring conditions. In the presence of 8 mM Fe(III) as iron citrate, the transformation of CT was enhanced three-fold, as shown in Figure 1B. Moreover, CF was dechlorinated to dichloromethane (DCM). In both cases, abiotic controls containing Fe(III) did not show CF or DCM production. In an independent experiment, CF transformation to DCM was confirmed in MR-1 cell suspensions incubated with Fe(III) (data not shown).

Although the form of Fe(II) was not determined, a fine, black precipitate was observed in MR-1 cell suspensions after growth using Fe(III) as a terminal electron acceptor. This strain is known to reduce Fe(III) to magnetite (Fe_3O_4) and to siderite ($FeCO_3$) (Kostka, personal communication). Reductive dechlorination of CT has been previously observed in solutions containing Fe(II) (Doong & Wu 1992). The reduction of chlorinated aliphatic compounds by magnetite has not been reported. However, magnetite produced by a dissimilatory Fe(III)-reducing enrichment culture and by *G. metallireducens* was responsible for abiotic reduction of 4-chloronitrobenzene (Heijman et al. 1993). The transformation of CT and CF by MR-1 under dissimilatory Fe(III)-reducing conditions suggests the potential for abiotic and biotic mechanisms for pollutant transformation in the environment. The relative significance of the processes is unknown.

Effect of Metabolic Inhibitors on CT Transformation

Synthesis of anaerobic respiratory chain components may be necessary for significant CT transformation activity in MR-1 (Petrovskis et al. 1994). Metabolic

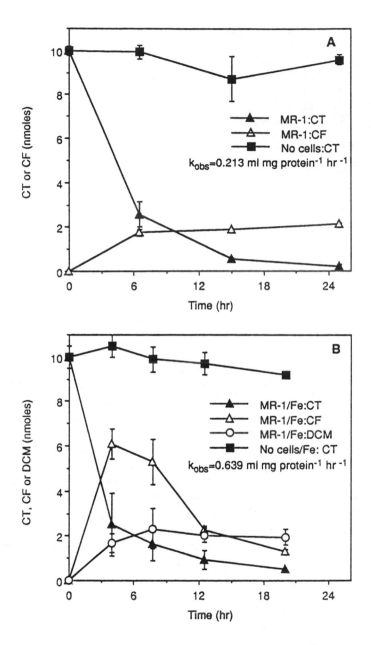

FIGURE 1. CT and CF production by MR-1 resting cells after growth under
Fe(III)-reducing conditions. Tetrachloromethane (CT) dechlorination and
trichloromethane (CF) production by *S. putrefaciens* MR-1 resting cells
after growth under Fe(III)-reducing conditions. CT transformation assays
were done in the absence (A) and presence (B) of 8 mM Fe(III) citrate.
Total protein concentration was 0.83 mg/mL (A) and 0.34 mg/mL (B). Error
bars represent ± one standard deviation of triplicate samples.

inhibitors have been used to determine components of anaerobic respiratory chains in *S. putrefaciens*. For MR-1, based on inhibition by *p*-chloromercuriphenyl-sulfonate (*p*-CMPS), HQNO, and antimycin A, iron-sulfur groups and cyto-chromes are involved in Fe(III) reduction (Myers & Nealson 1990; Myers & Myers 1993). Cytochromes are also involved in fumarate reduction (Myers & Nealson 1990). However, in *S. putrefaciens* 200, cytochrome inhibitors HQNO and CO blocked "constitutive" Fe(III) reduction, but had less or no effect on Fe(III) reduc-tion induced under microaerophilic conditions (Arnold et al. 1986). The effects of metabolic inhibitors on CT transformation were examined (Table 1).

These experiments were repeated in 10 mM HEPES (pH 7.5), as well as with increased concentrations of the inhibitors and no differences were observed from the results in Table 1 (data not shown). The cytochrome inhibitors, HQNO, and antimycin A had no effect on CT transformation in MR-1. Similar results were obtained for the iron-sulfur complex inhibitor, *p*-CMPS, and for cytochrome oxi-dase inhibitors, NaN_3 and KCN. Rotenone showed a slight inhibition of CT trans-formation activity. Lactate and hydrogen were the electron donors used in these experiments. Little effect would be expected, since electrons from lactate oxidation ($E_0' = -190$ mV) would enter the electron transport chain beyond the site of rotenone inhibition ($E_0' = -320$ mV). Interestingly, quinacrine, an inhibitor that blocks electron flow from flavins in dehydrogenases, enhanced CT transformation activity. Quinacrine may reduce CT chemically or, by blocking electron transport, may provide more electrons to be used for CT reduction upstream of the block.

It was anticipated that one of the compounds known to inhibit Fe(III) or fumarate reduction in MR-1 would have reduced CT transformation activity. If CT transformation was inhibited, then the respiratory chain component involved in this activity may be at the blocked site or further downstream in the electron transport pathway. These results suggest that, if a respiratory chain component is involved in CT transformation, then it is unlikely to be an iron-sulfur complex or a cytochrome. Alternatively, if the electron transfer rate to CT is significantly less than to a respiratory chain component or to a terminal electron acceptor, metabolic inhibitor effects may be obscured. The inhibitors are not 100% effective at blocking electron transport. With incomplete inhibition, electrons may still be transferred to CT at or after the blocked site. Electron transfer rates to CT are difficult to determine, as pseudo first-order CT trans-formation rates are based on nontoxic CT concentrations, which are much lower than concentrations of terminal electron acceptors. CO has been shown to inhibit CT transformation in *S. putrefaciens* 200, suggesting cytochrome involvement (Picardal et al. 1993). Whether this cytochrome is involved in anaerobic respira-tion is unknown, as CO had no effect on Fe(III) reduction induced under micro-aerophilic conditions (Arnold et al. 1986).

Screening of Respiratory Mutants
for CT Transformation Activity

MR-1 mutants were screened for CT transformation activity to determine whether respiratory chain components are involved in this activity. These mutants have lost the ability to utilize one or more anaerobic terminal electron acceptors for growth. Table 2 summarizes these experiments. Terminal reductase mutants, which have lost the ability to utilize Fe(III), Mn(IV) or fumarate for growth, transformed CT at rates equivalent to or greater than MR-1 wild-type. Some of these mutants were pigmentless, and were apparently unable to synthesize heme cytochromes. Two pleiotropic regulatory mutants, that were unable to grow anaerobically, transformed CT after aerobic growth. However, two mutants that failed to synthesize menaquinones (MK), and thereby lost the ability to reduce nitrate, Fe(III) or fumarate, also lost 90% of CT transformation activity. These mutants provide the first unequivocal evidence of the involvement of a respiratory chain component in reductive transformation of a chlorinated aliphatic compound. CT transformation is mediated by MK or a component with a more positive redox potential. Approximately 10% of the CT was dehalogenated stoichiometrically to CF by the MK-deficient mutants. This activity could involve a cytochrome. Similar dechlorination activity has been observed in MR-1 soluble fractions (data not shown).

Bacterial cofactors containing transition metals, such as cytochromes, have been shown to dehalogenate chlorinated methanes in in vitro systems (Assaf-Anid et al. 1994; Castro et al. 1994; Wackett et al. 1992). Extrapolation of these results to cells must be done with caution. Results of metabolic inhibitor studies and screening of respiratory mutants show that MK or a cofactor reduced by MK, but not a terminal reductase, is involved in the major CT transformation activity in MR-1. Low potential c-type cytochromes have been purified from periplasmic fractions of *S. putrefaciens* (Morris et al. 1990). These cytochromes could fit the results of the metabolic inhibitor studies, with CT transformation occurring at the reducing end of the electron transport chain. However, only 14% of the CT transformation activity is localized to the periplasmic fraction, with 78% to the membrane fraction and 8% to the cytoplasmic fraction (data not shown). Also, the midpoint redox potentials of these cytochromes (Em7 = −320 to −180 mV) suggest that MK ($E_0' = -74$ mV) is not a suitable electron donor. Quinones have been shown to mediate reduction of nitroaromatic compounds (Schwarzenbach et al. 1990). Based on the results from this study, the transformation of chlorinated compounds by menaquinones is suggested.

REFERENCES

Arnold, R.G., T.J. DiChristina, and M.R. Hoffman. 1986. "Inhibitor studies of dissimilative Fe(III) reduction by *Pseudomonas* sp. strain 200 ("*Pseudomonas ferriductans*")." *Appl. Environ. Microbiol.* 52:281-289.

Assaf-Anid, N., K.F. Hayes, and T.M. Vogel. 1994. "Reductive dechlorination of carbon tetrachloride by cobalamin(II) in the presence of dithiothreitol: mechanistic study, effect of redox potential and pH." *Environ. Sci. Technol.* 28:246-252.

Castro, C.E., M.C. Helvenston, and N.O. Belser. 1994. "Biodehalogenation, reductive dehalogenation by *Methanobacterium thermoautotrophicum.* Comparison with nickel(I) octaethylisobacteriochlorin anion, an F-430 model." *Environ. Toxicol. Chem.* 13:429-433.

Doong, R.-A., and S.-C. Wu. 1992. "Reductive dechlorination of chlorinated hydrocarbons in aqueous solutions containing ferrous and sulfide ions." *Chemosphere* 24:1063-1075.

Heijman, C.G., C. Holliger, M.A. Glaus, R.P. Schwarzenbach, and J. Zeyer. 1993. "Abiotic reduction of 4-chloronitrobenzene to 4-chloroaniline in a dissimilatory iron-reducing enrichment culture." *Appl. Environ. Microbiol. 59*:4350-4353.

Kriegman-King, M.R., and M. Reinhard. 1994. "Transformation of carbon tetrachloride by pyrite in aqueous solution." *Environ. Sci. Technol. 28*:692-700.

Lovley, D.R., M.J. Baedecker, D.J. Lonergan, I.M. Cozzarelli, E.J.P. Phillips, and D.I. Siegel. 1989. "Oxidation of aromatic contaminants coupled to microbial iron reduction." *Nature* 339:297-300.

Morris, C.J., D.M. Gibson, and F.B. Ward. 1990. "Influence of respiratory substrate on the cytochrome content of *Shewanella putrefaciens*. " *FEMS Microbiol. Lett. 69*:259-262.

Myers, C.R., and J.M. Myers. 1993. "Ferric reductase is associated with the membranes of anaerobically grown *Shewanella putrefaciens* MR-1." *FEMS Microbiol. Lett.* 108:15-22.

Myers, C.R., and K.H. Nealson. 1988. "Bacterial manganese reduction and growth with manganese oxide as the sole electron acceptor." *Science* 240:1319-1321.

Myers, C.R., and K.H. Nealson. 1990. "Respiration-linked proton translocation coupled to anaerobic reduction of manganese(IV) and iron(III) in *Shewanella putrefaciens* MR-1." *J. Bacteriol.* 172:6232-6238.

Nealson, K.H., and D. Saffarini. 1994. "Iron and manganese in anaerobic respiration: environmental significance, physiology, and regulation." *Ann. Rev. Microbiol.* In press.

Petrovskis, E.A., T.M. Vogel, and P. Adriaens. 1994. "Effects of electron acceptors and donors on transformation of tetrachloromethane by Shewanella putrefaciens MR-1." *FEMS Microbiol. Lett.* 1211:357-364.

Picardal, F.W., R.G. Arnold, H. Couch, A.M. Little, and M.E. Smith. 1993. "Involvement of cytochromes in the anaerobic biotransformation of tetrachloromethane by *Shewanella putrefaciens* 200." *Appl. Environ. Micobiol. 59*:3763-3770.

Saffarini, D.A., T.J. DiChristina, D. Bermudes, and K.H. Nealson. 1994. "Anaerobic respiration of *Shewanella putrefaciens* requires both chromosomal and plasmid-borne genes." *FEMS Microbiol. Lett.* 119:271-278.

Schwarzenbach, R.P., R. Stierli, K. Lanz, and J. Zeyer. 1990. "Quinone and iron porphyrin mediated reduction of nitroaromatic compounds in homogeneous aqueous solution." *Environ. Sci. Technol. 24*:1566-1574.

Wackett, L.P., M.S.P. Logan, F.A. Blocki, and C. Bao-li. 1992. "A mechanistic perspective on bacterial metabolism of chlorinated methanes." *Biodegradation* 3:19-36.

Biofactor-Mediated Transformation of Carbon Tetrachloride by Diverse Cell Types

Gregory M. Tatara, Michael J. Dybas, and Craig S. Criddle

ABSTRACT

Pseudomonas sp. strain KC transforms carbon tetrachloride (CT) to carbon dioxide, formate, and nonvolatile product(s), without the production of chloroform. Transformation of CT by strain KC requires (1) a small (<500 dalton) secreted factor(s) produced by strain KC, and (2) cells capable of regenerating the secreted factor(s). To determine whether specific cell types are required for regeneration, partially purified supernatant was combined with cells that are incapable of CT transformation or transform it slowly. Rapid CT transformation was obtained when the supernatant was combined with other pseudomonads (*P. stutzeri*; *P. fluorescens*); *Escherichia coli* K-12; *Bacillus subtilis*; a consortium (SC-1) derived from CT-contaminated groundwater at Schoolcraft, Michigan; and a consortium (HC-14) derived from CT-contaminated aquifer material at Hanford, Washington. The secreted factor(s) was stable indefinitely at −20°C after lyophilization to powder, and it was transported without retardation through columns packed with aquifer material from the Schoolcraft site.

INTRODUCTION

Under denitrifying conditions, transformation of carbon tetrachloride (CT) typically results in the slow accumulation of chloroform (Criddle et al. 1990a; Egli et al. 1988; Semprini et al. 1992), a compound that is persistent and potentially harmful to human health (Sittig 1985). CT remediation strategies that avoid chloroform production are desirable. *Pseudomonas* sp. strain KC rapidly transforms CT to carbon dioxide (Criddle et al. 1990b; Lewis & Crawford 1993; Tatara et al. 1993), formate (Dybas et al. 1994), and unidentified nonvolatile product(s), without the production of chloroform (Criddle et al. 1990b; Lewis & Crawford 1993; Tatara et al. 1993). Thus, this transformation offers pathway and kinetic advantages. As shown by Dybas et al. (1994), strain KC produces a small (<500 dalton) secreted factor(s) that is required for CT transformation. This factor(s) is regenerated in the presence of viable whole cells of *Pseudomonas* sp. strain KC.

We hypothesized that the secreted factor(s) might be regenerated by bacteria other than *Pseudomonas* sp. strain KC. In this paper, we provide evidence that a diverse range of microorganisms regenerate CT transformation activity, including organisms that are indigenous to CT-contaminated aquifers. We also evaluate the stability of the factor(s), and we demonstrate its transport through aquifer material.

EXPERIMENTAL PROCEDURES AND MATERIALS

Chemicals

Carbon tetrachloride (CT, 99% purity) was obtained from Aldrich Chemical Co., Milwaukee, Wisconsin. All chemicals used were ACS reagent grade (Aldrich or Sigma Chemical Co.). All water used in reagent preparation was deionized to 18 Mohm resistance or greater.

Media and Growth Conditions

Medium D was prepared as described by Criddle et al. (1990b). The resulting medium was autoclaved at 121°C for 20 minutes and transferred to an anaerobic glove box (Coy Laboratories, Ann Arbor, Michigan) for degassing. A recipe for Hanford simulated groundwater (SGW) was provided by R. Skeen (Battelle Pacific Northwest Laboratory) and contained per liter of deionized water: 0.455 g of $Na_2SiO_3 \cdot 9\ H_2O$, 0.16 g of Na_2CO_3, 0.006 g of Na_2SO_4, 0.02 g of KOH, 0.118 g of $MgCl_2 \cdot 6H_2O$, 0.0081 g of $CaCl_2 \cdot 2H_2O$, 13.61 g of KH_2PO_4, 1.6 g of NaOH, 1.6 g of $NaNO_3$, 1.6 g of acetic acid, and 1 mL of trace element solution. The trace element solution contained per liter of deionized water 0.021 g of $LiCl_2$, 0.08 g of $CuSO_4 \cdot 5H_2O$, 0.106 g of $ZnSO_4 \cdot 7H_2O$, 0.6 g of H_3BO_3, 0.123 g of $Al_2(SO_4)_3 \cdot 18\ H_2O$, 0.11 g of $NiCl_2 \cdot 6H_2O$, 0.109 g of $CoSO_4 \cdot 7H_2O$, 0.06 g of $TiCl_4$, 0.03 g of KBr, 0.03 g of KI, 0.629 g of $MnCl_2 \cdot 4H_2O$, 0.036 g of $SnCl_2 \cdot 2H_2O$, 0.3 g of $FeSO_4 \cdot 7H_2O$. The pH of SGW medium was adjusted to 7.5 or 8.2 with 3N KOH, autoclaved at 121°C for 20 minutes, and transferred to an anaerobic glove box for degassing prior to use.

Pseudomonas sp. strain KC (ATCC deposit no. 55595, DSM deposit no. 7136) was grown in medium D or SGW medium under an atmosphere of nitrogen in pressure-tested 500 mL Wheaton bottles (Wheaton cat. no. 219819). The bottles were sealed with 30 mm Teflon™-faced butyl rubber septa held in place by screw-cap lids containing 12 mm openings for sampling (Wheaton no. 224174). All cultures were shaken at 150 to 200 rpm on a rotary shaker at 20 to 23°C. Culture manipulations were typically performed in the anaerobic glove box under an atmosphere of 97±2% N_2 and 3±2% H_2.

Analytical Methods

CT transformation assays were performed by injection of headspace gas onto a gas chromatograph equipped with an electron capture detector, as described by Tatara et al. (1993).

Preparation of Partially Purified Culture Supernatant

Partially purified culture supernatant was obtained by centrifuging actively transforming culture, decanting the supernatant, filtering the supernatant through a 0.2 µm filter, loading the resulting filtrate into a model 8400 Amicon ultrafiltration stirred cell, and forcing it through a PM 10 (10,000 molecular weight cutoff) ultrafiltration membrane (Amicon no. 13142). The 10,000 MW (molecular weight) filtrate was either used directly in CT transformation assays or further purified by filtration through a preconditioned YC 05 (500 MW cutoff) ultrafiltration membrane (Amicon no. 13042).

Experiments with Diverse Cell Types

Pseudomonas fluorescens (ATCC deposit no. 13525), *Escherichia coli* K-12 (ATCC deposit no. 10798), and *Bacillus subtilis* (ATCC deposit no. 6051) were obtained from the culture collection of the Microbiology Department at Michigan State University. *Pseudomonas stutzeri* strain EP3-071388 11G, an aquifer isolate from Seal Beach, California., was provided by H. Ridgway (Orange County Water District, California). An aquifer consortium, designated SC-1, was obtained by enrichment of organisms in a groundwater sample from a CT-contaminated aquifer at Schoolcraft, Michigan. The enrichment procedure was a 1% inoculum of CT-contaminated Schoolcraft groundwater into Nutrient Broth (Difco Co.). A second consortium, designated HC-14, was obtained from R. Skeen of Battelle Pacific Northwest Laboratories (Richland, Washington). HC-14 was derived from a sample of aquifer material at Hanford, Washington, as described by Skeen et al. (1994).

P. fluorescens, *P. stutzeri*, and SC-1 were grown aerobically at 20 to 23°C in pH 8.2 medium D or in pH 7.0 medium D supplemented with 10 µM $FeSO_4$. *B. subtilis* and in some instances *P. fluorescens* were grown at 20 to 23°C aerobically in nutrient broth (Difco Co.). *E. coli* was grown aerobically at 35°C in medium D, but with glucose (3 g/L) instead of acetate as the electron donor. HC-14 was grown under denitrifying conditions in medium SGW (pH 7.5).

Cultures were transferred to an anaerobic glove box and dispensed into centrifuge tubes. Cells were collected by centrifuging at 10,000 rpm for 5 minutes, wasting the culture supernatant, and resuspending the *Pseudomonas* KC, *P. fluorescens*, *P. stutzeri*, and SC-1 pellets in 4 mL of anoxic medium D at pH 8.2, or resuspending the HC-14 pellet in anoxic SGW medium at pH 7.5. Cultures of *E. coli* or *B. subtilis* were resuspended in anoxic medium D containing glucose as the carbon source. A 0.5 mL sample of cell suspension (approximately 10^8 to 10^9 CFU/mL) was added to 4.5 mL of filtered supernatant and assayed for CT transformation in 28-mL aluminum seal tubes, as described by Tatara et al. (1993).

CT Transformation Using Lyophilized Culture Filtrate

The 10,000 MW filtrate from strain KC grown in SGW medium was frozen at −20°C then freeze dried with Labconco freeze dryer (Labconco Co., Kansas City,

Missouri). The resulting powder was stored aerobically at –20°C. Freeze-dried powder was rehydrated in deionized water at one or two times its original concentration. Rehydrated powder was deoxygenated, dispensed into 28-mL aluminum seal tubes, mixed with *P. fluorescens* (10^8 to 10^9 CFU/mL), spiked with CT, and assayed for CT transformation. To obtain the transformation capacity of the powder, samples were respiked with CT until transformation stopped.

Transport of Secreted Factor(s) Through Aquifer Material

A Kontes® glass column (30 cm length, 2 cm i.d.) fitted with Teflon™ luer lock stopcocks was sanitized by soaking in a solution of 0.06% hypochlorite, rinsed with sterile deionized water, and packed aseptically with Schoolcraft aquifer material under a laminar flow hood. Schoolcraft aquifer material (courtesy of Brown & Root Environmental, Holt, Michigan) was combined with SGW medium containing 400 mg/L nitrate to create a slurry. Slurries of Schoolcraft aquifer material were poured directly into a column containing SGW medium and periodically tapped during filling to facilitate uniform packing and prevent the entrainment of air pockets. After packing, the column received three exchanges of SGW medium. The column was then connected to a Harvard syringe pump, and flow of SGW medium containing 10 μCi of ^3H_2O per liter was initiated at 2.5 mL/min. Breakthrough was monitored by collecting 5-mL fractions of column effluent and measuring the radioactivity in each fraction. The 300-μL effluent fractions were injected into 10 mL of liquid scintillation fluid and counted on a liquid scintillation counter.

Mobility of the secreted factor(s) was evaluated by pumping 10,000 MW filtrate from strain KC grown in SGW medium through the column at a rate of 2.5 mL/min. Column effluent was collected in 5-mL fractions, deoxygenated by passage through the interlock of an anaerobic glove box, mixed with cells of *P. fluorescens*, spiked with CT, and assayed for CT transformation.

RESULTS

Transformation of CT with Diverse Cell Types

As illustrated in Table 1, the partially purified supernatant transformed CT to a limited extent by itself, but its activity was highly variable, as indicated by the large standard deviation for this sample set. In the presence of viable cells, the rate and reliability of transformation increased dramatically. As shown in Table 1, CT transformation occurs when a wide range of cell types are incubated in the presence of the secreted factor(s) produced by strain KC. CT transformation was obtained using cells from related pseudomonads (*P. fluorescens*, *P. stutzeri*); *E. coli*; *B. subtilis*; a consortium enriched from CT-contaminated groundwater at Schoolcraft, Michigan (SC-1); and a consortium enriched from CT-contaminated aquifer solids from Hanford, Washington (Skeen et al. 1994).

TABLE 1. Half-lives for carbon tetrachloride in mixtures containing diverse cell types and the secreted factor produced by *Pseudomonas* sp. strain KC.

Cell Type	Growth Conditions[a]	Secreted Factor(s) Added?[b]	CT Half Life (min)[c]	n[d]
Secreted factor(s) alone (No cells present)	medium D, aerobic or anoxic	Yes	67 ± 65	13
Pseudomonas sp. strain KC	medium D, anoxic	Yes	4.0 ± 1.1	6
Pseudomonas fluorescens	medium D, aerobic	No	NT[e]	3
		Yes	3.4 ± 0.1	3
	nutrient broth	No	NT	3
		Yes	7.2 ± 0.2	3
Pseudomonas stutzeri	medium D, aerobic	No	NT	3
		Yes	4.5 ± 0.0	3
Escherichia coli K-12	medium D, glucose, aerobic	No	NT	3
		Yes	12 ± 1	3
Bacillus subtilis	nutrient broth	No	NT	3
		Yes	19 ± 3	3
Schoolcraft Consortium (SC-1)	medium D, aerobic	No	NT	3
		Yes	2.8 ± 0.1	3
	medium D, pH 7, 10 µm FE, aerobic	No	NT	3
		Yes	2.5 ± 0.3	3
Hanford Consortium (HC 14)	SGW pH 7.5 anoxic	No	NT	3
		Yes[f]	2.7 ± 0.2	3

(a) All cells were grown at pH 8.2 unless indicated otherwise.
(b) Secreted factor(s) added as 500-MW filtrate unless indicated otherwise.
(c) ± one standard deviation.
(d) n = number of samples.
(e) NT = no transformation was observed, therefore a half-life was not calculated.
(f) Secreted factor added as 10,000-MW filtrate.

HC-14 and *E. coli* K-12 were the only cultures used in this study reported to transform CT, however no transformation was observed during the time-course of the CT transformation assays. The single gram positive organism evaluated in this study (*Bacillus subtilis*) exhibited slower CT transformation upon combination with the supernatant factor(s) than the gram negative organisms (Table 1). This may be related to differences in membrane structure, but additional confirmation is needed. Also of interest was the finding that cultures that are combined with the secreted factor(s) do not need to be grown under iron-limiting conditions. Cultures of *P. fluorescens* and *B. subtilis* grown in nutrient broth and Schoolcraft consortium SC-1 grown at pH 7 with 10 µM iron were all able to transform CT in the presence of the secreted factor(s) (Table 1). Thus, while production of the secreted factor(s) by strain KC is dependent upon iron-limiting conditions (Tatara et al. 1993), regeneration of the factor(s) is not.

Chloroform (CF) was not detected during CT transformation by any of the supernatant/cell combinations tested (detection limit = 2 µg/L). This is consistent with previous reports (Criddle et al. 1990b; Lewis & Crawford 1993).

Stability of the Secreted Factor(s)

To assess the stability of the secreted factor(s) and to provide a possible means of concentrating it, 10,000 MW filtrate from strain KC grown in SGW was frozen at −20°C and lyophilized to dryness. Lyophilized powder was stored at −20°C for 6 days prior to rehydration. The lyophilized powder was rehydrated at one and two times its original concentration, then it was combined with cells of *P. fluorescens*. As shown in Table 2, CT transformation was most extensive when the powder was rehydrated to twice its original concentration. However, the increased mass of CT transformed was not proportional to the increase in powder concentration.

Transport of the Secreted Factor(s)
Through Aquifer Material

Transport of secreted factor(s) through aquifer material was evaluated by pumping 10,000 MW filtrate from SGW-grown strain KC through columns packed with aquifer solids from a CT-contaminated site in Schoolcraft, Michigan. Figure 1 illustrates the breakthrough profile of ^3H_2O and secreted factor(s) for Schoolcraft aquifer material as quantified by the CPM/CPM° ratio and the ratio of influent to effluent first-order bioassay rates using *P. fluorescens*. The breakthrough profile for the tritiated water was similar to that of secreted factor(s) demonstrating that the secreted factor(s) was not retarded in the aquifer material tested.

TABLE 2. **Transformation of carbon tetrachloride by rehydrated freeze-dried powder prepared from the supernatant of *Pseudomonas* sp. strain KC and combined with cells of *Pseudomonas fluorescens*.**

Test Sample	CT Removed after a 24-Hour Incubation (µg)[a]
(1) *Pseudomonas fluorescens* alone	0.08 ± 0.04
(2) *Pseudomonas fluorescens* + 1X rehydrated culture filtrate[b]	0.87 ± 0.03[d]
(3) *Pseudomonas fluorescens* + 2X rehydrated culture filtrate[c]	0.99 ± 0.01[d]

(a) ± one standard deviation.
(b) 1X indicates a freeze-dried powder rehydrated to its original concentration.
(c) 2X indicates a freeze-dried powder rehydrated to twice its original concentration.
(d) Initial CT concentrations were 0.2 µg of CT, and cultures were repeatedly spiked with this concentration until transformation stopped.

DISCUSSION

The present work presents a generally favorable outlook for use of the secreted factor(s) in field applications. Use of the secreted factor in combination with biostimulated indigenous populations may provide the benefits of both bio-augmentation and biostimulation. Use of the secreted factor(s) provides kinetic and pathway control. We detected no chloroform in any of the supernatant/cell combinations studied. In all likelihood, CT is transformed by the secreted factor(s) faster than chloroform can be produced by a parallel pathway. Because the cells required to regenerate the secreted factor(s) can be native to the contaminated site, many of the ecological and transport issues raised by the introduction of nonnative organisms may be avoided. We observed no difficulties in moving it through aquifer material (Figure 1). Other data supporting possible application of the secreted factor(s) are its ability to be concentrated by freeze-drying (Table 2), and its favorable storage properties after freeze-drying. Stability in water will need further evaluation, as only limited stability (4 to 6 days) was obtained in medium D and SGW medium (data not shown).

Additional studies are also needed to further understand the mechanism of transformation of CT by *Pseudomonas* sp. strain KC. Of primary importance is elucidation of the structure of the secreted factor(s). In addition, field-scale studies are needed to evaluate the possible use of the secreted factor(s) for CT remediation.

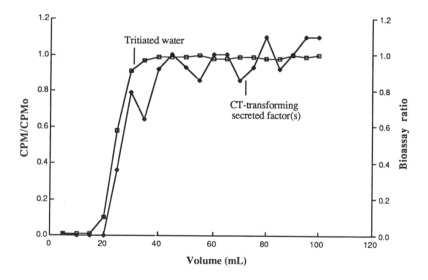

FIGURE 1. Breakthrough profile for tritiated water and the secreted factor(s) in a column packed with Schoolcraft, Michigan aquifer material. Water was pumped through the column at a flow rate of 2.5 mL/min. The bioassay ratio is the rate of CT transformation of the effluent fraction (combined with *P. fluorescens* cells) divided by the rate of CT transformation of the influent (combined with *P. fluorescens* cells).

ACKNOWLEDGMENTS

This work was supported by the National Science Foundation (NSF) Center for Microbial Ecology at Michigan State University under NSF grant BIR-9120006. Funding was also provided by the Office of Research and Development, U.S. Environmental Protection Agency under Grant R815750 to the Great Lakes and Mid-Atlantic Hazardous Substance Research Center. Partial funding of the research activities of the Center is also provided by the State of Michigan Department of Natural Resources. The content of this publication does not necessarily represent the views of either agency.

We thank Dr. Rod Skeen of the Battelle Pacific Northwest Laboratories for providing us with Hanford Consortium HC-14 and the recipe for SGW medium. We also thank Dr. Harry Ridgway of the Orange County Water District for providing us *Pseudomonas stutzeri* strain EP3-071388 11G. Finally, we thank Timothy J. Mayotte and Brown and Root Environmental (Holt, Michigan) for providing us with aquifer solids from the Schoolcraft site.

REFERENCES

Criddle, C. S., J. T. DeWitt, and P. L. McCarty. 1990a. "Reductive dehalogenation of carbon tetrachloride by *Escherichia coli* K-12." *Appl. Environ. Microbiol. 56*: 3247-3254.

Criddle, C. S., J. T. DeWitt, D. Grbić-Galić and P. L. McCarty. 1990b. "Transformation of carbon tetrachloride by *Pseudomonas* sp. strain KC under denitrification conditions." *Appl. Environ. Microbiol. 56*: 3240-3246.

Dybas, M. J., G. M. Tatara, and C. S. Criddle. 1994. "Localization and characterization of the carbon tetrachloride transforming activity of *Pseudomonas* sp. strain KC." Accepted for publication in *Appl. Environ. Microbiol.*

Egli, C., T. Tschan, R. Scholtz, A. M. Cook, and T. Leisinger. 1988. "Transformation of tetrachloromethane to dichloromethane and carbon dioxide by *Acetobacterium woodii.*" *Appl. Environ. Microbiol. 54*: 2819-2823.

Lewis, T. A. and R. L. Crawford. 1993. "Physiological factors affecting carbon tetrachloride dehalogenation by the denitrifying bacterium *Pseudomonas* sp. strain KC." *Appl. Environ. Microbiol. 59*: 1635-1641.

Skeen, R. J., K. M. Amos, and J. N. Peterson. 1994. "Influence of nitrate concentration on carbon tetrachloride transformation by denitrifying microbial consortia." *Water Resource. 28*: 2433-2438.

Semprini, L., G. D. Hopkins, P. L. McCarty, and P. V. Roberts. 1992. "In-situ transformation of carbon tetrachloride and other halogenated compounds resulting from biostimulation under anoxic conditions." *Environ. Sci. Technol. 26*: 2454-2461.

Sittig, M. (Ed.). 1985. *Handbook of Toxic and Hazardous Chemicals and Carcinogens*, 2nd ed. Noyes Publications, New York, NY.

Tatara, G. M., M. J. Dybas, and C. S. Criddle. 1993. "Effects of medium and trace metals on the kinetics of carbon tetrachloride transformation by *Pseudomonas* sp. strain KC." *Appl. Environ. Microbiol. 59*: 2126-2131.

Reductive Dechlorination of Pentachlorophenol by Enrichments from Municipal Digester Sludge

Victor S. Magar, Henning Mohn, Jaakko A. Puhakka,
H. David Stensel, and John F. Ferguson

ABSTRACT

Ten anaerobic digester sludges from municipal plants showed variable abilities for anaerobic dehalogenation of pentachlorophenol (PCP) and monochlorophenols (MCPs) in batch serum bottle tests. Acclimation for PCP dehalogenation was affected by sludge dilution and initial PCP concentration. During long-term bench-scale digester operation the PCP dehalogenation pathways changed, producing different end products including the loss of phenol production. Complete dechlorination of PCP was achieved in a continuously fed fluidized-bed reactor (FBR) and in serum bottles after combining PCP-to-3-MCP-dechlorinating and 3-MCP-dechlorinating enrichments.

INTRODUCTION

Since its introduction in the 1930s, pentachlorophenol (PCP) has been widely used as a biocide and wood preservative, and improper use and treatment processes have led to its release into the environment. Anaerobic cultures transform PCP and lower chlorophenols (CPs) via reductive dechlorination (for reviews, see Häggblom 1992; Mohn and Tiedje 1992). The extent of PCP dechlorination and the ability of anaerobic cultures to mineralize PCP varies, as cited in Nicholson et al. (1992), and for many cultures PCP dechlorination stops at MCPs. Enrichments of PCP-dechlorinating and mineralizing cultures may be useful for remediation of PCP-contaminated sites.

In this work, enrichment of methanogenic cultures from municipal digester sludges for PCP and MCP dechlorination was investigated using batch serum bottles, a bench-scale municipal digester, and an FBR. Ten municipal sludges were compared for their ability to dechlorinate PCP and MCPs. One objective of this research was to produce an enrichment to completely dechlorinate PCP. Dechlorinating enrichments were combined to promote complete

PCP dechlorination in batch serum bottles and in the FBR when it was not achieved after long-term PCP feeding.

METHODS

Sludge Survey

Dehalogenation of PCP was tested in 10 diluted municipal anaerobic digester sludge samples (Table 1), labeled as S1 through S10. Sludges were diluted about ten times with a reduced anaerobic mineral medium (RAMM) (Shelton and Tiedje 1984) to give final volatile solids (VS) concentrations of 1,900 mg/L. Batch degradation tests were conducted in 160-mL serum bottles containing 100 mL diluted sludge spiked with 2 mg/L PCP and supplemented with 2 mL municipal sludge (combined primary and secondary sludges) for an electron donor source. The bottles were sealed with butyl rubber stoppers and aluminum crimp caps. PCP removal and metabolite production were monitored in soluble samples. Soluble samples were obtained from centrifuged (12,000 g for 1 h) samples.

MCP dehalogenation and phenol degradation were surveyed in five sludges (S1, S2, S6, S8, S10); three of these sludges were capable of PCP dehalogenation. The sludges also were diluted 10 times in RAMM. One set of bottles was spiked with all three MCPs (10 mg/L each), and a second set was spiked with 10 mg/L phenol.

Digester Enrichment and PCP
Acclimation in Undiluted Sludge

Acclimation for PCP dehalogenation was examined using Renton, Washington undiluted sludge (S8), after PCP-dehalogenation was not observed during 3 months incubation of 10 times diluted sludge. Duplicate bottles were spiked at 14 μmol/L (3.8 mg/L) and 28 μmol/L (7.6 mg/L) PCP, establishing initial soluble concentrations of 1 μmol/L and 2 μmol/L, respectively. The bottles were fed 1 mL raw sludge on days 1 and 21. Soluble samples were monitored for PCP removal and metabolite accumulation.

A 4-L bench-scale anaerobic digester mixed with a magnetic stirrer was operated for 2 years to examine long-term PCP dechlorination. It was fed raw sludge with 5 mg/L PCP, establishing a 25-day solids retention time (SRT). The digester operated at pH 7.4, produced approximately 1.1 L gas/g VS, and destroyed about 50% of the feed VS (Ballapragada et al. 1994). Following 6 months of operation, soluble and total effluent samples were monitored for CPs for 9 months.

Fluidized-Bed Reactor Enrichments

PCP was dechlorinated to 3-MCP via 2,3,5-TCP and 3,5-DCP in a 460-mL FBR originally seeded with municipal sludge. The FBR was operated for

TABLE 1. PCP removal and dechlorination by municipal digester sludges.[a]

No.	Treatment Plant	% Removal	Metabolite Formation
S1	Southeast Plant, San Francisco, CA	100	235-TCP; 345-TCP
S2	Jackson, MI	100	235-TCP
S3	South Bay-Side, Redwood City, CA	100	345-TCP
S4	Union Sanitary District, CA	100	345-TCP
S5	West Point, Seattle, WA	97	345-TCP
S6	Chicago, IL	83	2345-TeCP; 235-TCP; 345-TCP
S7	91st Avenue, Tolleson, AZ	< 1	235-TCP; 2345-TeCP
S8	Renton, WA	< 1	235-TCP; 2345-TeCP
S9	LA County Sanitary Districts	< 1	345-TCP
S10	Terminal Island, CA	< 1	235-TCP
S11	Autoclaved Control	< 1	2345-TeCP

(a) Bottles were fed 2 mg PCP/L and were incubated for 23 days.

2 years without evidence of 3-MCP dechlorination. Complete dechlorination of PCP to phenol was developed in the FBR by seeding it with 30 mL of a 3-MCP dechlorinating enrichment developed from the sludge survey bottle test. The FBR employed 70 g Celite™ as the carrier and was operated at a 1-day hydraulic retention time (HRT) with a recycle ratio of 400:1, providing 50% expansion of the Celite™ media. The feed contained 200 mg chemical oxygen demand (COD)/L (glucose, butyrate, and methanol) as an electron donor and 1 mg/L yeast extract. The reactor was first operated in batch mode for 3 days and afterwards in continuous mode.

Chlorophenols (CPs) were determined by gas chromatography. An electron capture detector (ECD) was used for all the dichlorophenols (DCPs), trichlorophenols (TCPs), tetrachlorophenols (TeCPs), and PCP. MCPs and phenol were analyzed using a flame ionization detector (FID). Soluble-phase CPs were derivatized with acetic anhydride and extracted into hexane (Perkins et al. 1994). Total-phase CPs were derivatized using a modification of the procedure described by Lee et al. (1989), where samples were diluted into 50 mL deionized water, derivatized, and extracted into petroleum ether.

RESULTS

Sludge Survey

In the sludge survey experiment, the municipal anaerobic digester sludges were compared for their ability to dechlorinate PCP (Table 1). After 23 days of incubation, 100% transformation of PCP was found in four sludges (S1, S2, S3, and S4) and more than 80% in two sludges (S5 and S6). Little or no PCP dechlorination was observed in the other sludges. The most common metabolites were 2,3,5- and 3,4,5-TCP indicating predominantly *ortho*- and *para*-dechlorination, but no *meta*-dechlorination was observed.

The ability of the sludges to degrade phenol and MCPs did not appear to be related to PCP dehalogenation abilities. Table 2 shows that sludge S8, which did not dehalogenate PCP, degraded 2-MCP faster than sludges S1 and S6, which dehalogenated PCP. Degradation results for 3-MCP and 4-MCP were variable. Even though sludge S1 degraded 2-MCP and PCP, it did not degrade 3- and 4-MCP by day 136; sludge S10 did degrade 3-MCP by that time but did not degrade 4-MCP.

Digester Enrichment and PCP Acclimation in Undiluted Sludge

Renton sludge (S8) was used for enrichment of PCP dechlorination in undiluted cultures (Figure 1) and in the bench-scale digester. Immediate appearance of 2,3,4,5-TeCP and 3,4,5-TCP shows removal of *ortho*-chlorines without acclimation, but with no further dechlorination until day 60; 3-MCP

TABLE 2. Time for complete phenol or MCP removal.

No.	Treatment Plant	Phenol	2-MCP	3-MCP	4-MCP
S1	Southeast Plant, San Francisco, CA	54	18	>136[a]	>136[a]
S2	Jackson, MI	15	25	92	92
S6	Chicago, IL	32	25	N/A	74
S8	Renton, WA	15	15	67	39
S10	Terminal Island, CA	54	10	136	>153[b]
	Combined[c]	27	7	27	42

(a) Neither 3-MCP nor 4-MCP was removed within the 136-day study.
(b) 4-MCP was not removed within 153 days.
(c) All enrichments, except S6 (Chicago, IL), were combined and diluted 10 times on day 40. The reported times are after combining the enrichments.

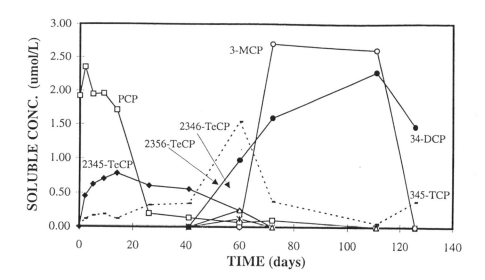

FIGURE 1. PCP dechlorination in unacclimated, undiluted sludge.

first appeared on day 72 and disappeared after day 111. Phenol (15 μmol/L) appeared on day 125, following the disappearance of 3-MCP. 2,3,4,6-TeCP and 2,3,5,6-TeCP appeared at low concentrations on day 60 and disappeared without further metabolites. PCP dehalogenation of the S8 sludge occurred more rapidly in undiluted sludge than in the diluted sludge used in the sludge survey, which was likely due to the increased biomass in the undiluted sludge. In both studies, dehalogenation occurred at the *ortho*-position. The undiluted cultures spiked with lower PCP (14 μmol/L) concentrations required longer acclimation; 345-TCP did not appear until day 60 and 3-MCP and 34-DCP did not appear during the 120-day study.

PCP was transformed to phenol and 3-MCP with traces of 2,3,5-TCP and 3,5-DCP after 6 months of continuous operation of the sludge digester. Transformation occurred via 2,3,4,5-TeCP to 2,3,5-TCP, 3,5-DCP, and 3-MCP; there was minor transformation via 2,3,5,6-TeCP. After 9 months, phenol no longer appeared and 3,4-DCP, 3-MCP, and 4-MCP accumulated, indicating that MCP-dechlorination was lost. After 1 year, PCP was dechlorinated to 2,5-DCP and 3-MCP, and in transient tests PCP dechlorinated via 2,3,5,6-TeCP, 2,3,5-TCP and 3,5-DCP. The 2,5-DCP was not further dechlorinated and the 3-MCP was produced from 3,5-DCP dechlorination only. These results show significant changes in pathways and extent of dechlorination over the experimental period.

Fluidized-Bed Reactor Enrichments

Establishment and maintenance of complete PCP dechlorination to phenol and subsequent phenol degradation was studied in a continuously fed FBR.

Prior to reseeding the FBR with a 3-MCP dechlorinating culture, the FBR culture dechlorinated PCP to 3-MCP via 2,3,5,6-TeCP, 2,3,5-TCP, and 3,5-DCP. Complete dechlorination was first observed on day 9 after reseeding (Figure 2), where 15% of the influent PCP was unaccounted for in the effluent. On day 21, approximately 25% of the influent PCP was unaccounted for, and this increased to 75% by day 73. The transient appearance of 3-MCP and phenol suggests that PCP was mineralized to CO_2, CH_4, and Cl; radiolabled carbon experiments were not conducted for confirmation. Enrichment of a PCP-mineralizing culture was occurring in the FBR over the 10-week period. These results were confirmed in serum bottle tests in which PCP-dechlorinating and 3-MCP-mineralizing cultures were combined to promote PCP mineralization. After 1 month, 1.5 mg PCP/L was completely removed within 24 hours, without the appearance of metabolites.

DISCUSSION

These studies show that variations in chlorinated phenol dehalogenation can be expected for different anaerobic sludge sources. The causes of the variations are not known. The results also show that some sludges could degrade MCPs but not PCP. Until bacteria or consortia that degrade chlorinated phenols are identified, the ability to select seed sources will be limited to trial and error.

Boyd and Shelton (1984) observed that 2-MCP was degraded by the S2 sludge without lag, while 3- and 4-MCP required approximately 4 weeks for degradation in diluted cultures. The longer lag times with the S2 sludge in this

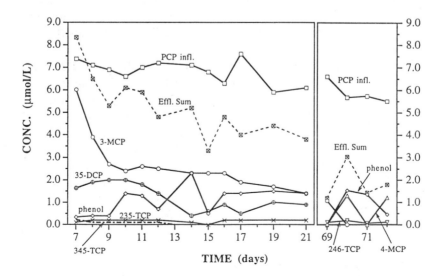

FIGURE 2. PCP mineralization in a fluidized-bed reactor (FBR) seeded with PCP-dechlorinating and 3-MCP-mineralizing cultures.

study may be due to long-term sludge storage before our tests or to changes in the sludge during the 10 years between our studies.

Acclimation was strongly affected by sludge dilution and initial PCP concentration. Undiluted cultures spiked at high PCP concentrations dechlorinated both *ortho*-chlorines without a lag and much more rapidly than diluted cultures or undiluted cultures spiked at low initial PCP concentrations. Reasons may be induction of dechlorination or higher dechlorinating biomass due to increased growth resulting from higher PCP concentrations and high initial biomass concentrations. During long-term operation of a digester, PCP pathways changed and produced different end products, including the loss of phenol production. These changes may be due to population variations over time and demonstrate lability of PCP dehalogenation.

PCP-mineralization was achieved in a continuously fed FBR and in serum bottles after combining PCP-dehalogenating and 3-MCP-mineralizing enrichments. Although the difficulty of MCP dechlorination has been frequently reported by others (Nicholson et al. 1992), Mikesell and Boyd (1986) combined 2-, 3-, and 4-MCP mineralizing cultures, which resulted in the mineralization of PCP. The separate enrichment for specific dechlorination activities, followed by combining enrichments, seems to be an expedient method for developing PCP-mineralizing cultures.

REFERENCES

Ballapragada, B.S., V.S. Magar, J.A. Puhakka, H.D. Stensel, and J.F. Ferguson. 1994. "Fate and Biotransformation of Tetrachloroethylene and Pentachlorophenol in Anaerobic Digesters." *Proceedings WEFTEC, Chicago, 1994.*

Boyd, S.A. and D.R. Shelton. 1984. "Anaerobic Biodegradation of Chlorophenols in Fresh and Acclimated Sludge." *Appl. Environ. Microbiol.* 47: 272-277.

Häggblom, M.M., 1992. "Microbial Breakdown of Halogenated Aromatic Pesticides and Related compounds." *FEMS Microbiol. Rev.* 103: 29-72.

Lee, J-B., et al. 1989. "Chemical Derivatization Analysis of Phenols. Part VI. Determination of Chlorinated Phenolics in Pulp and Paper Effluents." *J. Assoc. of Official Anal. Chem.* 72: 979-984.

Mikesell, M.D. and S.A. Boyd. 1986. "Complete Reductive Dechlorination and Mineralization of Pentachlorophenol by Anaerobic Microorganisms." *Appl. Environ. Microbiol.* 52: 861-865.

Mohn, W.W., and J.M Tiedje. 1992. "Microbial Reductive Dehalogenation." *Microbiol. Rev.* 56: 482-507.

Nicholson, S.L. Woods, J.D. Istok, and D.C. Peek. 1992. "Reductive Dechlorination of Chlorophenols by a Pentachlorophenol-Acclimated Methanogenic Consortium." *Appl. Environ. Microbiol.* 58: 2280-2286.

Perkins, P.S., S.J. Komisar, J.A. Puhakka, and J.F. Ferguson. 1994. "Effects of Electron Donors and Metabolic Inhibitors on Reductive Dechlorination of 2,4,6-TCP." *Wat. Res.* 28: 2101-2108.

Shelton, D.R. and Tiedje, J.M. 1984. "General Method for Determining Anaerobic Biodegradation Potential." *Appl. Environ. Microbiol.* 47: 850-857.

Effect of Nitrate Availability on Chloroform Production During CT Destruction

Juli L. Sherwood, James N. Petersen,
Rodney S. Skeen, and Brian S. Hooker

ABSTRACT

The biological degradation of carbon tetrachloride (CT) produces two potential end products: chloroform (CF) and carbon dioxide. Previous research indicated that, under denitrifying conditions, the percentage of CT that would be degraded to CF may be a function of the availability of electron acceptors. Hence, a set of experiments was completed using a denitrifying consortium capable of transforming carbon tetrachloride (CT) which was cultured from aquifer sediments from the U.S. Department of Energy's (DOE's) Hanford Site. These experiments allowed determination of the yield of CF from CT under two different conditions. In the first set of conditions, nitrate was added to the reactor on a daily basis; in the second, daily feeding of nitrate was suspended after 1 week of reactor operation so that the nitrate was depleted. The data presented here indicate that no direct link can be established between CF production and the availability of nitrate. In addition, CF production can be represented as a constant yield from CT depletion. Finally, when described using first-order reaction kinetics, the rate at which CT is degraded is not dependent on nitrate availability.

INTRODUCTION

In situ bioremediation of carbon tetrachloride (CT) by denitrifying microorganisms is currently being tested at the DOE's Hanford Site. Previous research has demonstrated that microbes from the Hanford aquifer are capable of transforming CT using nitrate with a variety of electron donors (Petersen et al. 1993). Both carbon dioxide and chloroform (CF) have been observed as end products of CT biotransformation by organisms isolated from this site (Hansen et al. 1994). The production of CF from CT has also been observed for other denitrifying consortia (Semprini et al. 1992; Bouwer and Wright 1988; Bouwer and McCarty 1983). The formation of CF poses a significant obstacle to exploiting this metabolism for remediation because CF is a known carcinogen, and evidence suggests that it is not further transformed by denitrifiers (Bouwer and McCarty 1983).

In a previous field test of CT degradation by denitrifying organisms, it was observed that altering the nutrient addition strategy could alter the amount of CF produced from CT (Semprini et al. 1990). Specifically, 55 to 67% of trans-formed CT appeared as CF when nitrate was continuously injected while acetate was fed in pulses. However, only 30 to 40% of the CT was observed to be trans-formed to CF after nitrate addition was terminated. The purpose of this paper is to examine the effects of altering the availability of nitrate on CF production by the Hanford consortium.

MATERIALS AND METHODS

Fed-batch experiments were performed using an experimental apparatus and procedures that have been described elsewhere (Skeen et al. 1994; Petersen et al. 1993). All experiments were designed such that acetate was in excess at all times, at a design level of 200 mg/L. Nitrate, however, was handled in two distinct fashions. In one set of duplicate experiments (set A), nitrate was meas-ured and returned to its design level of 100 mg/L on a daily basis for the dura-tion of the test. In the other set of duplicate experiments (set B), nitrate was handled as in set A for the first week (Phase I), while for the final 2 weeks nitrate was measured but not replenished (Phase II).

The inoculum for all experiments was grown following the protocol outlined previously (Petersen et al. 1993). After inoculation, the fed-batch reactors were placed in a constant temperature water bath maintained at 17°C, and were oper-ated for 7 days prior to the addition of CT to allow the bacterial consortium to acclimate. The reactors were then charged with enough CT to achieve approxi-mately 1.0 to 3.0 µg/mL CT in the aqueous phase.

Anion, CT, and CF concentrations were measured at least once per day using methods similar to those reported previously (Petersen et al. 1993). Biomass was measured daily using the BCA Protein Assay (Pierce). Calibration of this assay for this consortium showed that 62% of the dry weight of the biomass was measured as protein.

To determine the fraction of the CT that was degraded to CF, numerical tech-niques similar to those outlined by Hooker et al. (1994) were employed. A yield relationship was assumed to relate the rate of CF production to CT destruction:

$$\frac{d[CF]}{dt} = -y_{CF/CT}\frac{d[CT]}{dt}$$

The commercial program Simusolv was used to determine the value of the yield that best represented the experimental data. This analysis technique properly accounts for the effects of withdrawing samples and of adding nutrients upon CF and CT concentrations. In addition, Henry's law is used to account for the volatilization of the components into the gas phase in the reactor. Further, the model accounts for the kinetics of microbial growth via both assimilatory and dissimilatory pathways and for the kinetics of CT destruction. Previous work has demonstrated that this approach can be used to determine the best kinetic forms to describe microbial contaminant destruction, and to determine the values of the various parameters that should be used in these kinetic expressions (Hooker et al. 1994). Hooker et al. found that the rate of CT transformation could best be described as first order with respect to biomass and CT concentration, and inhibited by high nitrate concentrations.

RESULTS

Figure 1 shows both the experimental and predicted aqueous CT and CF concentrations as functions of time for the four experiments. On this figure, the

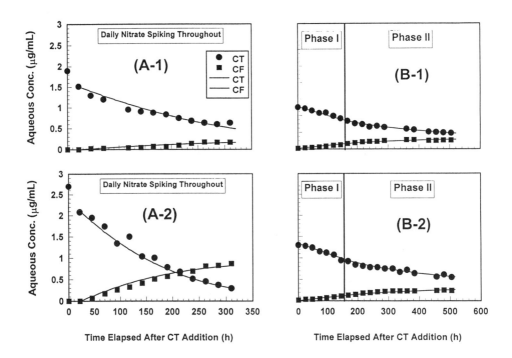

FIGURE 1. Aqueous carbon tetrachloride and chloroform concentrations as functions of time. In experimental set A (A1 and A2), nitrate was added to the reactor throughout the duration of the experiment. In experimental set B (B1 and B2), nitrate was added to the reactor during Phase I, but nitrate addition was terminated and the nitrate was depleted during Phase II. Solid lines are obtained using the kinetic expressions of Hooker et al. (1994) and assuming a constant CF/CT yield throughout the experiment.

time intervals in which the different nitrate spiking strategies were employed are indicated. These data show that CT destruction and CF production are obtained regardless of whether nitrate is spiked to the reactor on a regular interval. In addition, in all cases, a significant amount of CF was produced.

Several different analyses of the data were carried out. First, a constant-yield coefficient was determined to represent all the data from a particular experiment. Comparison of the model predictions and experimental data shown on Figure 1 demonstrates that the experimental data are well represented using this approach. However, to ensure that the yield was, in fact, constant, the experiments in set B were analyzed using a different yield coefficient for the two experimental phases. In addition, the experiments of set A were also analyzed assuming that the yield coefficient could be different after 1 week of operation to test if the yield coefficient changed with time. The yield coefficients determined in this fashion are shown on Table 1. Statistical analysis of these results indicates that at the 95% confidence level, these yield coefficients should be considered equal regardless of the treatment employed in the experiments. Hence, although experiment-to-experiment variations in the yield coefficient exist, these variations cannot be attributed to the availability of nitrate in the reactor.

In addition to determining the yield coefficient, the reaction rate for CT transformation also was determined for each of the experiments, using the CT transformation kinetic expressions developed previously by Hooker et al. (1994). These data are also shown on Table 1. Again, while there is some experiment-to-experiment variation in the fit reaction rate coefficient, statistical analysis shows that this variation cannot be attributed to the presence or absence of nitrate in the reactor. This result indicates that, on a per unit biomass basis, the rate of CT transformation is independent of the presence of nitrate. This is in contrast to the nitrate inhibition reported previously by Hooker et al. (1994) for this consortium. In this previous work, high nitrate greater than the inhibitory level

TABLE 1. Predicted CF yield coefficients and CT destruction reaction rate coefficients for the various experimental conditions.

	CF Yield Coefficient $y_{CF/CT}$		Reaction Rate Constant μ_{CT}	
Data Set Used in Parameter Determination	Value	Standard deviation	Value	Standard deviation
Full set average	0.500	0.173	0.045	0.025
Phase I of all experiments (first week)	0.458	0.192	0.044	0.025
Phase II, nitrate depleted	0.630	0.014	0.033	0.013
Phase II, nitrate added	0.775	0.148	0.060	0.040

of 200 mg/L was used. Here, such high levels were not employed, and so inhibitory effects are not expected. Additionally, the kinetic constant reported by Hooker et al. was somewhat higher than that reported here. However, this variation could be due to culture aging.

CONCLUSIONS

The yield of CF from CT transformation under denitrifying conditions is not a function of the availability of nitrate in solution. If chloroform production were directly linked to nitrate depletion, one would expect the percentage of chloroform produced during Phase II of set B, when no nitrate was available in the reactor, to be significantly different than the percentages obtained when nitrate was added to the reactor on a daily basis. Because this is not the case, one must conclude that no direct link between the CF production and nitrate depletion can be established. Surprisingly, the rate of CT degradation is also independent of the presence of nitrate. Therefore, it is apparent that the pulse feeding of nitrate at these levels will have no significant effect on the degradation of CT or on the production of CF. Hence, other routes must be explored to determine methods to enhance CT degradation and to either minimize CF production or to destroy produced CF.

REFERENCES

Bouwer, E. J., and P. L. McCarty. 1983. "Transformation of Halogenated Organic Compounds Under Denitrification Conditions." *Appl. Environ. Microbiol.* 45:1295-1299.

Bouwer, E. J., and J. P. Wright. 1988. "Transformations of Trace Halogenated Aliphatics in Anoxic Biofilm Columns." *J. Contam. Hydrol.* 2:155-169.

Hansen, E. J., D. L. Johnstone, J. K. Fredrickson, and T. M. Brouns. 1994. "Transformation of Tetrachloromethane under Denitrifying Conditions by a Subsurface Bacterial Consortium and Its Isolates." In R. E. Hinchee, A. Leeson, L. Semprini, and S. K. Ong (Eds.), *Bioremediation of Chlorinated and Polycyclic Aromatic Hydrocarbon Compounds*, pp. 293-297. Lewis Publishing, Chelsea, MI.

Hooker, B. S., R. S. Skeen, and J. N. Petersen. 1994. "Biological Destruction of CCl_4: II. Kinetic Modeling." *Biotechnol. Bioeng.* 44:211-218.

Petersen, J. N., R. S. Skeen, K. M. Amos, and B. S. Hooker. 1993. "Biological Destruction of CCl_4: I. Experimental Design and Data." *Biotechnol. and Bioengi.* 43:521-528.

Semprini, L., G. D. Hopkins, D. B. Janssen, M. Lang, P. V. Roberts, and P. L. McCarty. 1990. *In-Situ Biotransformation of Carbon Tetrachloride Under Anoxic Conditions.* U.S. Environmental Protection Agency Technical Report, EPA/2-90/060, R.S. Kerr Environmental Research Laboratory, Ada, OK.

Semprini, L., G. D. Hopkins, P. L. McCarty, and P. V. Roberts. 1992. "In-Situ Transformation of Carbon Tetrachloride and Other Halogenated Compounds Resulting from Biostimulation Under Anoxic Conditions." *Environ. Sci. Technol.* 26:2454-2461.

Skeen, R. S., M. J. Truex, J. N. Petersen, and J. S. Hill. 1994. "A Batch Reactor for Monitoring Process Dynamics During Biodegradation of Volatile Organics." *Environmental Progress* 13:174-177.

Development of Tetrachloroethene Transforming Anaerobic Cultures from Municipal Digester Sludge

Bhaskar S. Ballapragada, Jaakko A. Puhakka,
H. David Stensel, and John F. Ferguson

ABSTRACT

Three different systems were used to enrich for tetrachloroethene (PCE) and *cis*-dichloroethene (*c*-DCE) dechlorinators from municipal sludge digesters. In semicontinuously fed digesters with PCE, the enrichment accumulated vinyl chloride (VC). Batch enrichments that were repeatedly fed and spiked with lactate and PCE or *c*-DCE developed complete dechlorination to ethene. Fluidized-bed reactors (FBRs) fed with lactate and PCE or *c*-DCE dechlorinated to ethene with small residual VC in the effluent. Dechlorination always developed sequentially. Trichloroethene (TCE) was the predominant product for a short period, whereas *c*-DCE and especially VC were predominant for a longer time before being transformed to ethene. The rate of dechlorination correlated with chloroethene loadings. Feeding *c*-DCE instead of PCE did not significantly affect the time required for ethene generation. All dechlorinations occurred with methanogenesis.

INTRODUCTION

Chloroethenes have been widely used as solvents and chemical feedstocks and are common groundwater contaminants. Chloroethenes are toxic with many of them listed as U.S. Environmental Protection Agency priority pollutants. Extensive research is therefore being conducted to understand their microbial transformations and to develop effective treatment technology.

TCE, *c*-DCE, *trans*-dichloroethene (*t*-DCE), and VC can be degraded cometabolically by a variety of aerobic microorganisms (e.g. Tsien et al. 1989). However, PCE is highly oxidized and does not degrade aerobically. All chloroethenes degrade under anaerobic conditions (Barrio-Lage et al. 1986, Freedman and Gossett 1989, DeBruin et al. 1992, Tandoi et al. 1994). Anaerobic transformation of chloroethenes occurs via sequential replacement of chlorines with hydrogen. Thus, transformation of PCE usually occurs via TCE, *c*-DCE, and VC to ethene and ethane with one exception in which VC was converted to carbon dioxide (Vogel and McCarty 1985). Conversion of VC to ethene is desired because VC is more toxic than PCE.

Complete dechlorination of chloroethenes has been shown for batch-fed serum-bottle cultures and for a packed-bed reactor (Freedman and Gossett 1989, DeBruin et al. 1992). A variety of system designs are possible for anaerobic dehalogenation. The aim of this work was to evaluate different enrichment methods for dechlorination of chloroethenes.

EXPERIMENTAL METHODS

Three systems were used to enrich for PCE dechlorinators including (1) laboratory-scale anaerobic sludge digesters fed with PCE, (2) 160-mL batch bottles spiked repeatedly with lactate and PCE or c-DCE, and (3) FBRs fed with lactate and PCE or c-DCE.

Three laboratory-scale sludge digesters were semicontinuously fed at 35°C. All digesters received primary and secondary sludge from Metro Wastewater Treatment Plant at Renton, Washington. Digesters were operated at 40-day, 25-day, and 25-day solids retention times (SRTs) and received 5 mg/L, 5 mg/L, and 10 mg/L, respectively, of PCE in the feed along with four other chlorinated compounds (carbon tetrachloride, 1,2,4 trichlorobenzene, pentachorophenol, and polychlorinated biphenyls) (Ballapragada et al. 1994). Gas, liquid, and solid phases of the digester were monitored to determine chloroethene transformations and the degree of enrichment in the digesters.

A set of 160-mL serum bottles were inoculated with acclimated digester sludge and repeatedly spiked with 20 mg lactate and PCE or c-DCE at doses of 1 or 10 μmole/bottle. The bottles were usually respiked after significant transformation of the parent compound. Bottles were purged with N_2/CO_2 mixture (70% N_2 vs. 30% CO_2) before each respiking. The serum bottles consisted of 100 mL of liquid and 60 mL of gas and were sealed with Teflon™-lined mininert valves. Chloroethenes were measured from liquid samples, and the gas-phase concentrations were calculated using Henry's law constants. Ethene was measured in the gas phase only.

Two FBRs, with 300-mL empty-bed volume and 100-mL gas volume, were operated for more than 10 months at 0.5-day hydraulic retention time (HRT). Lactate was semicontinuously fed (15 seconds every 10 minutes) with a reduced anaerobic medium at 200 mg/L. One reactor received PCE and the other c-DCE at concentrations of 3.5 μmole/L. The concentrations were increased stepwise to a final concentration of 80 μmole/L. Gas was collected in a gas bag and regularly monitored. Chloroethenes were analyzed in the liquid phase, and ethene was analyzed in the gas phase. Batch tests were conducted in FBRs by spiking a known amount of PCE or c-DCE while feeding lactate at normal rates.

Chloroethenes were analyzed using purge and trap (Tekmar ALS-LSC) followed by a gas chromatograph (PE 8700), equipped with a capillary column (wide-bore Rtx 502.2 Restec) and a Hall detector. The temperature program was 5 minutes at 35°C, followed by a temperature ramp of 8°C/min up to 155°C. Ethene was analyzed on an HP 5840 gas chromatograph equipped with a 6-ft (1.8-m) HayeSep Q Supelco packed column and a flame ionization detector. The oven was isothermal at 80°C.

RESULTS

In bottles inoculated with sludge from anaerobic digesters and refed and respiked with PCE and c-DCE, the chloroethenes were sequentially dechlorinated (Figure 1). Development of cultures capable of transforming significant amounts of VC to ethene took more than 2 months, and "complete" transformation of PCE to ethene occurred after 6 months. In bottles fed with high concentrations of chloroethenes, transformation products appeared much more slowly, but the rate increased with repeated spikings. Methanogenesis was inhibited in the higher chloroethene-fed bottles. Chloroethene rates correlated with the amount of PCE during respike (Table 1).

The FBRs fed with PCE and c-DCE showed sequential transformations similar to batch enrichments and resulting in the formation of ethene (Figure 2). Increases of influent concentrations resulted initially in accumulation of VC. Gradually, the concentration of VC decreased. During continuous operation,

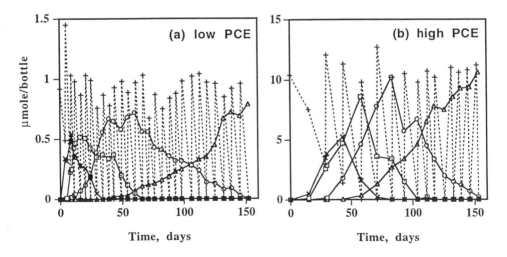

FIGURE 1. 160 mL serum bottles repeatedly spiked at two different concentrations of tetrachloroethene. 20 mg of lactate was fed during each respike. Bottles were purged with N_2/CO_2 before each respike. PCE (+), TCE (×), c-DCE (□), VC (O), and Ethene (△).

TABLE 1. Chloroethene loadings and dehalogenation rates in the anaerobic systems.

Parameter	Digester			Batch Enrichment Chloroethene		Fluidized-Bed Reactor Chloroethene	
	(1)	(2)	(3)	Low[a]	High[b]	Low	High
PCE loading rate, (μmole/L-day)	0.75	1.2	2.4	2.0	20.0	7.0	160.0
PCE degradation rate, (μmole/L-day)	7.1	8.6	9.1	3.0	120.0	45.0	190.0
c-DCE degradation rate, (μmole/L-day)	2.8	4.0	5.3	25.0	175.0	35.0	270.0
VC degradation rate, (L/day)	ND	ND	ND	1.2	3.5	ND	4.6
c-DCE loading (μmole/L-day)	NA	NA	NA	2.0	20.0	7.0	160.0
c-DCE degradation rate, (μmole/L-day)	NA	NA	NA	30.0	190.0	42.0	290.0
VC degradation rate (L/day)	NA	NA	NA	1.5	3.6	ND	5.2

ND = rates not determined; NA = not applicable.
(a) PCE and c-DCE fed at low conc. (10 to 20 μmole/L).
(b) PCE and c-DCE fed at high conc. (100 to 200 μmole/L).

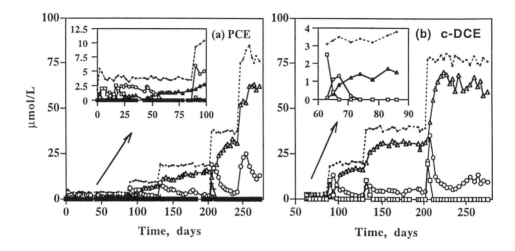

FIGURE 2. Performance of fluidized-bed reactor fed with (a) tetra-chloroethene (PCE) and (b) *cis*-dichloroethene (*c*-DCE) along with lactate (200 mg/L) at a hydraulic retention time of 0.5 day. PCE inf (···), *c*-DCE inf (- - -), PCE eff (+), TCE eff (×), *c*-DCE eff (□), VC eff (O), and ethene eff (Δ).

FBR effluent contained small amounts of VC. Typically, PCE or *c*-DCE was converted 80% to ethene and 20% to VC.

Table 1 shows the loading rates of PCE and *c*-DCE and dehalogenation rates of PCE, *c*-DCE, and VC in all three systems. In all digesters, PCE was sequentially dehalogenated to VC via TCE and *c*-DCE. Less than 5% of VC was converted to ethene. During the 2 years of continuous operation, none of the digesters enriched for VC dechlorination. Batch enrichments showed much higher transformation rates in higher chloroethene-fed bottles than in lower chloroethene-fed bottles. No difference in *c*-DCE or VC dehalogenation rates was observed in *c*-DCE-fed bottles compared to the PCE-fed bottles. In the FBRs, PCE and *c*-DCE conversions increased with the influent loading rates. In all three systems, dechlorination occurred in conjunction with methanogenesis.

DISCUSSION

Continuous feeding of the anaerobic digester resulted in VC accumulation, whereas VC was converted to ethene in batch bottle and FBR enrichments. Complete conversion of VC to ethene was not achieved during continuous FBR operations, but was shown in transient tests. These observations are consistent with the previously reported low first-order rate of transformation of VC (Tandoi et al. 1994).

In batch bottle enrichments, the biomass is retained. FBR retains biomass as biofilm and results in more efficient enrichment. Both systems have long SRTs. The sludge-fed digesters had lower SRTs and lower volumetric loading of PCE, and are expected to have a more complex microbial community than the lactate-fed enrichments. High solids content and gas production in digesters decrease soluble concentrations of chloroethene and, therefore, their bioavailability.

Enrichment with PCE vs. *c*-DCE resulted in similar VC dechlorination, indicating that different bacteria may be present. One would carry dechlorination of PCE to *c*-DCE and the other would carry dechlorination of *c*-DCE to ethene. If the dechlorinators obtain energy from dechlorination, increased chloroethene loadings should result in increased populations of dechlorinators, as indicated by increased dechlorination rates in our enrichments fed with higher concentrations of PCE or *c*-DCE. High spike concentrations usually inhibited methanogenesis, which may have benefited dechlorination through decreased competition for the electron donor.

Dechlorination occurred in conjunction with methanogenesis in nearly all tests. The kinetics of dehalogenation are reflective of conditions during in situ and ex situ treatment where dechlorinators are likely to compete with methanogens.

Rapid development of dechlorinating cultures is important for bioremediation. It was best accomplished by a batch enrichment technique minimizing production of toxic intermediate and retaining a maximum amount of enriched biomass. The FBR retains the enriched biomass and can potentially be used for on-site remediation using high chloroethene loading rates. However, complete dechlorination of VC is difficult to achieve during continuous operation. The FBR may be more useful when operated in a semicontinuous sequential batch mode to completely dechlorinate PCE.

CONCLUSIONS

Anaerobic digester dechlorinated PCE, TCE, and *c*-DCE but very little VC. Batch enrichment completely dechlorinated PCE or *c*-DCE after 150 days of repeated spikings and feedings. The FBRs partially converted PCE and *c*-DCE, with VC (20%) and ethene (80%) as final products. Increased chloroethene loadings result in higher rates of dechlorination. Dechlorination rates for *c*-DCE and VC were similar with PCE and *c*-DCE feedings. VC is the slowest dechlorination in all enrichments. Repeatedly spiking batch cultures with high concentrations of chloroethenes and feeding the FBRs at higher loadings result in increased transformation rates. FBRs, provide rapid enrichment, but do not completely dechlorinate PCE in the continuous mode of operation.

ACKNOWLEDGMENTS

We thank the Water Environment Research Foundation (Grant 91-TFT-3) and the King County Department of Metropolitan Services (Seattle Metro) for funding this study.

REFERENCES

Ballapragada, B. S., V. S. Magar, J. A. Puhakka, H. D. Stensel, and J. F. Ferguson. 1994. "Fate and Biotransformation of Tetrachloroethene and Pentachlorophenol in Anaerobic Digesters." *Proceedings WEFTEC, Chicago.*

Barrio-Lage, G., F. Z. Parsons, R. S. Nassar, and P. A. Lorenzo. 1986. "Sequential Dehalogenation of Chlorinated Ethenes." *Environ. Sci. Tech.* 20 (1): 96-99.

DeBruin, W. P., M. J. Kotterman, M. A. Posthusmus, G. Schraa, and J. B. Zehnder. 1992. "Complete Biological Reductive Transformation of Tetrachloroethene to Ethane." *Appl. Environ. Microbiol.* 58 (6): 1996-2000.

Freedman, D. L. and J. M. Gossett. 1989. "Biological Reductive Dechlorination of Tetrachloroethylene and Ethylene under Methanogenic Conditions." *Appl. Environ. Microbiol.* 55 (9): 2144-2151.

Tandoi, V., T. D. DiStefano, P. A. Bowser, J. Gossett, and S. H. Zinder. 1994. "Reductive Dehalogenation of Chlorinated Ethenes and Ethanes by a High-Rate Anaerobic Enrichment Culture." *Environ. Sci. Tech.* 28 (5): 973-979.

Tsien, H. C., G. A. Brusseau, R. S. Hanson, and L. P. Wackett. 1989. "Biodegradation of Trichloroethlene by *Methylosinus trichosporium* OB3b." *Appl. Environ. Microbiol.* 55(12): 3155-3161.

Vogel, T. M. and P. L. McCarty. 1985. "Biotransformation of Tetrachloroethylene to Trichloroethylene, Dichloroethylene, Vinyl Chloride, and Carbon Dioxide under Methanogenic Conditions." *Appl. Environ. Microbiol.* 49 (5): 1080-1083.

Molecular- and Cultural-Based Monitoring of Bioremediation at a TCE-Contaminated Site

Fred J. Brockman, William Payne,
Darla Workman, Adam Soong, Scott Manley,
Wuhua Sun, and Andrew Ogram

ABSTRACT

Sediment samples from a trichloroethylene (TCE)-contaminated zone 30 to 47 m below ground surface were analyzed after various nutrient addition strategies, including injection of air, 1% methane (in air), 4% methane (in air) pulsed with air, and multiple nutrients (4% methane pulsed with air and methane, and continuous nitrous oxide gas and triethyl phosphate gas; abbreviated N/P/pulsed 4% methane). Methanotroph most probable number (MPN) values in samples taken after the N/P/pulsed 4% methane campaign increased approximately two orders of magnitude compared to before-methane samples. The frequency of TCE biodegradative potential in methanotroph enrichments from samples taken after the N/P/pulsed 4% methane campaign was orders of magnitude greater than in all previous campaigns, indicating that in situ microbial activity was very strongly affected by adding nitrogen and phosphorus. The frequency of detection of the soluble methane monooxygenase (*mmoYBZC*) and methanol dehydrogenase (*moxF*) sequences was highest in sediment samples taken after the 1% methane campaign (78% and 56% of the samples, respectively). Comparing the culture-dependent method (i.e., enrichment MPN) with the culture-independent gene probe method showed that the culture method greatly underestimated the methanotrophic biomass in sediments throughout the demonstration.

INTRODUCTION

The use of horizontal wells to efficiently deliver nutrients to stimulate the growth and activity of indigenous microflora able to degrade TCE has been the focus of an Integrated Demonstration funded by the U.S. Department of Energy (DOE) at the Savannah River Site near Aiken, South Carolina. The bioremediation technology exploited the natural occurrence of methanotrophic microorganisms at the site (Phelps et al. 1988). Under aerobic conditions, these microorganisms oxidize both methane and TCE using the methane monooxygenase enzyme.

The majority of the TCE was located at 30 to 47 m below ground surface (bgs) in a stratum termed the tan clay zone (Eddy et al. 1991). The water table at the site was approximately 36.5 m bgs. The lower horizontal well was located in the aquifer 53 m bgs, and the upper horizontal well was located in the unsaturated zone 24 m bgs. Injection of gaseous nutrients through the lower well and a vacuum exerted on the upper well moved nutrients through the contaminated region to selectively promote the growth and activity of the methanotrophic microorganisms.

The first campaign of the demonstration was a control experiment, consisting of a 21-week-long injection of air only. The second campaign was a 15-week-long injection of 1% methane (by volume) in air. The third campaign, 8 weeks in length, consisted of 4% methane (by volume) in air. The fourth campaign, 17 weeks in length, consisted of injection of air alternated with injection of 4% methane in air. The final campaign, for 12 weeks, was identical to the fourth campaign, except that nitrous oxide gas and triethyl phosphate gas were injected continuously at a carbon:nitrogen:phosphorus stoichiometry of 100:10:1. Boreholes were drilled at the end of the first, second, fourth, and fifth campaigns to acquire sediment samples; 13 groundwater monitoring wells were sampled biweekly throughout the demonstration. A full description of the components and operation of the system is presented by Lombard et al. (1994).

Commonly, <1% of the total number of bacteria in aquifers and <0.1% of the bacteria in unsaturated zones can be cultured (i.e., grown) in the laboratory. Thus, the requirement for culturing in the standard MPN enrichment method usually underestimates the true number in the sample. A culture-independent method for estimating the density of a particular type of microorganism is the nucleic acid probe. The soluble methane monooxygenase genes and the methanol dehydrogenase gene were selected for use as nucleic acid probes to track the response of the target microbial type to injections of methane and nutrients, and to provide a direct measure of the density of methanotrophic bacteria that was independent of the requirement for growth of the bacteria in the laboratory.

PROCEDURES

Sampling

Figure 1 shows the location of boreholes from which sediment samples were analyzed. (Samples from 12 additional before-methane boreholes were analyzed by other researchers.) At 10-ft (3.05-m) intervals, 2-ft lengths of core

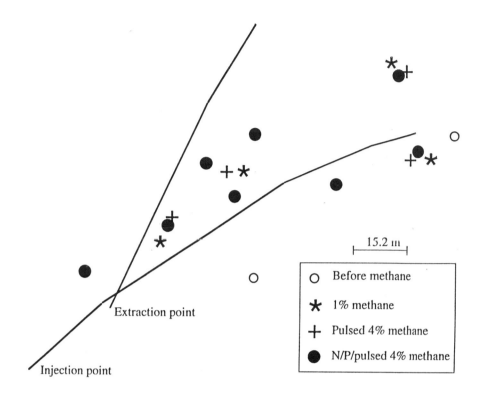

FIGURE 1. **Plan view of sampled boreholes and horizontal injection and extraction wells at the Area M bioremediation site, Savannah River Site.**

were homogenized, and approximately 200 g of this core was bagged for cultural analyses and shipped by overnight Federal Express on blue ice (approximately 4°C). A second 200-g aliquot for nucleic acid analyses was immediately frozen at –70°C, shipped by overnight Federal Express on blue ice, and stored in a –70°C freezer.

MPN Enumerations

The number of culturable methanotrophs (termed a MPN index) was estimated in enrichments set up in a 3-vial/dilution MPN format with 10 mL medium and 10 cc of headspace. Upon receipt of sediment samples, a 10-g aliquot of homogenized sediment was added to 95 mL 0.1% pyrophosphate (pH 7.0) and shaken at 180 rpm for 30 min on a reciprocating shaker before carrying out serial dilutions to headspace vials containing medium. The medium was Shelton's mineral salts (Shelton & Tiedje 1984) supplemented with 1 mL of a vitamin mixture (Wolin et al. 1963) per liter. After inoculation, methane was added to 25% of the headspace, and the vials were sealed with silicon septa and aluminum crimp closures. The presence of turbidity or a biofilm or pellicle after 10 weeks of incubation was scored as positive.

Biodegradative Potential

The biodegradative potential under methanotrophic enrichment conditions was assessed in triplicate vials with 10 mL medium and 10 cc of headspace. The medium and preparation of inoculum were as described above, with the exception that an additional medium was employed by omitting copper. A very low (<1 μM) to zero copper concentration results in expression of the soluble form of methane monooxygenase (sMMO), whereas higher copper concentrations result in expression of the membrane or particulate form of the methane monooxygenase (pMMO) (Tsien et al. 1989; DiSpirito et al. 1992). After inoculation, methane was added to 25% of the headspace (methane to TCE stoichiometry of 500 to 1) and vials were closed with 20-mm-thick Teflon™-lined rubber septa. The septa were lifted slightly to allow entry of the needle of a gastight syringe, which was used to deliver 10 μL of TCE (final concentration of 0.8 to 1.2 μg/mL) in methanol. The vials were immediately sealed. TCE was delivered in methanol and in water to different sets of vials for enrichments from the N/P/pulsed 4% methane campaign, to determine if methanol was interfering with TCE degradation by inhibiting methane utilization (Graham et al. 1993). Headspace in the vials was analyzed after 14 to 22 weeks using a Hewlett-Packard 5880A series gas chromatograph equipped with an electron capture detector and automatic headspace sampler. Removal, relative to sterile controls, of 75% of the contaminant in at least two of the triplicate vials was considered a positive result.

DNA Extraction and Hybridization

DNA was extracted from frozen sediment samples using a thaw-freeze-thaw procedure in the presence of sodium dodecyl sulfate, followed by dialysis and precipitation, as described previously (Brockman et al. 1994). The *mmoYBZC* genes (Tsien & Hanson 1992; Cardy et al. 1991a,b) and *moxF* gene (Machlin & Hanson 1988) were subcloned into in vitro transcription vectors (Stratagene Cloning Systems, La Jolla, California), and messenger RNA probes were generated using [alpha-^{32}P]UTP. The hybridization wash conditions yielded stringencies of Tm –21°C and Tm –24°C for the *mmoYBZC* and *moxF* probes, respectively, where stringency is expressed as the difference between the calculated melting temperature (Tm) of the hybrid during the final wash and the actual (lower) temperature used during the final wash.

RESULTS

Methanotroph MPN Index in
Response to Treatment Campaigns

The methanotroph MPN index was ≤30/g sediment (gs) in 86% of the sediments before the addition of methane (Figure 2). After the 1% methane campaign, 19% (n = 3) of the samples had measurable MPN indexes (700, 1,500,

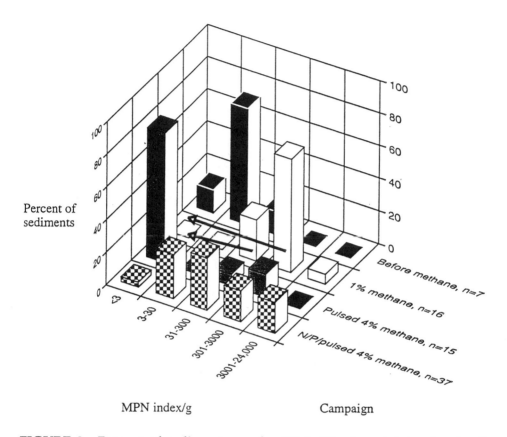

FIGURE 2. Percent of sediment samples (30 to 37 m) containing levels of methanotrophs by operating campaign. See text in Results section for clarification of values for the 1% methane campaign.

and 7,300/gs). However, the remaining 16 samples were overdiluted, resulting in 25% of the samples yielding a methanotroph MPN index of <300/gs and 81% of the samples yielding an MPN index of <3,000/gs. The arrows in Figure 2 indicate that the two highest vertical bars for the 1% methane campaign are maximum values; the actual MPN index values in samples could have been orders of magnitude lower. The percentage of samples with MPN indexes >3,000/gs was three times higher (19% versus 6%) for the N/P/pulsed 4% methane campaign compared to the 1% methane campaign. MPNs increased several orders of magnitude after the addition of nitrous oxide and triethyl phosphate compared to before the addition (pulsed 4% methane campaign). The increase in the methanotroph MPN index was greater in groundwater samples (typically four, and as much as seven, orders of magnitude; M. Enzien, S. Pfiffner, and T. Phelps, personal communication) than in sediment samples in campaigns with methane.

Frequency of Detection of Gene Sequences in Response to Treatment Campaigns

The detection level for the gene probes was determined by hybridizing the probe with known amounts of the target gene. The detection level of both gene probes was approximately 1 pg of target DNA/g sediment. A rating of + was assigned to sediments where 1 to 10 pg of probe hybridized to extracted DNA (on a per g of sediment basis). Ratings of ++ and +++ corresponded to >10 to 100 pg and >100 pg, respectively, of probe hybridizing to extracted DNA. Of the probe-positive cases, 95% were rated as (+), 3% as (++), and 1% as (+++). Frequency of detection is expressed as the percentage of all samples with detectable hybridization between the gene probe and DNA extracted from the sediment sample. The *mmoYBZC* and *moxF* gene probes most frequently detected homologous DNA in sediment samples from the 1% methane injection campaign and least frequently in sediment samples from the 4% methane injection campaign (Figure 3). The frequency of detection with the

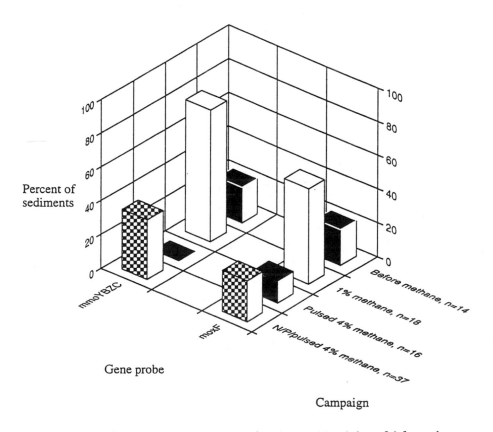

FIGURE 3. Percent of sediment samples (30 to 37 m) in which various sequences homologous to the gene probes were detected in the operating campaigns.

*mmo*YBZC and *mox*F gene probes after methane addition to the site increased a maximum of four times and two and one-half times, respectively, over the frequencies for the before-methane samples. The frequency of detection for both gene probes in samples from the N/P/pulsed 4% methane campaign was more similar to the before-methane samples than to the samples from the 1% methane campaign.

TCE Biodegradative Potential in Response to Treatment Campaigns

Greater than 75% removal of TCE in more than one of the triplicate vials under methanotrophic enrichment conditions was not observed prior to the N/P/pulsed 4% methane campaign (Table 1). Positive results may have been obtained in the 1% methane campaign samples in less dilute enrichments, i.e., with the equivalent of 100 mg or 1,000 mg sediment/vial. After the addition of nitrogen and phosphorus to the site, TCE degradation (copper omitted, TCE delivered in methanol) occurred in 68% of the sediment samples with an equivalent of 1 mg sediment/vial. Thus, the frequency of TCE biodegradative potential increased by several orders of magnitude in response to the addition of nitrogen and phosphorus.

The frequency of TCE degradation for the N/P/pulsed 4% methane campaign was similar regardless of whether TCE was added in water or in methanol. This was true for enrichments lacking exogenous copper (permissive of sMMO expression) and containing exogenous copper (permissive of pMMO expression). These results show that the use of methanol to deliver

TABLE 1. Percent of sediment samples (30 to 37 m) exhibiting TCE biodegradative potential in methanotrophic enrichments in the operating campaigns.

Campaign	Number of samples	Biodegradative potential%	Sediment equivalent[a]
Before methane	7	0	1,000 mg
1% methane	16	0	1 mg to 10 mg [b]
Pulsed 4% methane	15	0	100 mg
N/P/pulsed 4% methane	37	81	100 mg
N/P/pulsed 4% methane	37	68	1 mg

(a) Equivalent amount of sediment in vial, calculated from the dilution performed on the 10 g subsample of sediment.
(b) 10 mg for 30.5 and 33.5 m samples; 1 mg for 36.6, 39.6, and 42.7 m samples.

TCE to the enrichments did not significantly inhibit TCE degradation over the length of the incubation and with the criteria used for a positive result.

DISCUSSION

The Savannah River bioremediation demonstration represents the first use of horizontal well technology to deliver nutrients for bioremediation and the first time that carbon, nitrogen, and phosphorus nutrient sources have all been injected as gases. It is also one of the first times that nucleic acid-based methods have been used to document the performance of an engineered in situ bioremediation system in the field.

Our results are consistent with the overall bioremediation performance at the site, as determined by laboratory data from other participating laboratories and contaminant inventories from thousands of groundwater, sediment, and soil gas samples: performance during the 1% methane campaign was very good, performance during the pulsed 4% methane campaign was poor (resulting in the decision to add gaseous nitrogen and phosphorus to the site), and performance during the N/P/pulsed 4% methane campaign was very good.

The absence of significant amounts of TCE degradation in methanotrophic enrichments from the 1% methane campaign was somewhat suprising. Although the detection limit was high (300 or 3,000/gs depending on sample depth), 3 of the 16 samples had MPN indexes of 700, 1,500, and 7,300/gs, and *mmoYBZC* genes were detected in 14 of 18 samples, yet TCE degradation was not observed with an inoculum equivalent to 1 to 10 mg of sediment. Contributing factors for poor TCE degradation during the 1% methane campaign could be that cells were in a physiological state unconducive to TCE degradation under enrichment conditions, and the previously mentioned factors that could result in underestimation by enrichment methods. Thus, in sediment samples from the 1% methane campaign, the culture-based methods were a poor predictor of the success of bioremediation (discussed below).

In contrast, biodegradative potential was very high after the 12-week injection of gaseous carbon, nitrogen, and phosphorus sources, even though the frequency of detection of *mmoYBZC* and *moxF* sequences was lower than in the 1% methane campaign and the percentage of samples with methanotroph MPN indexes >3,000/g sediment was only three times higher than in the 1% methane campaign. The results suggest that adding nitrogen and phosphorus to the site improved the physiological status of methanotrophic bacteria, resulting in improved biodegradative potential in the subsequent enrichments, but may have had little effect on stimulating the growth of methanotrophic populations.

Poor bioremediation performance was indicated by all three methods in samples taken after the pulsed 4% methane campaign. These results are consistent with results from other laboratories conducting microbiological research at the site. The poor performance during the pulsed 4% methane campaign may have been due to very low inorganic nitrogen levels (T. Palumbo, personal communication), decreased bioavailable phosphate, and nutrient limitation-driven changes in the microbial community structure and physiological status.

It should be remembered that both MPN and gene probe methods of determining methanotroph populations are estimations. The 95% confidence limits for the MPN index value span one to one-and-one-half orders of magnitude. The MPN index may also be underestimated due to factors such as microcolonies that are not disrupted into component cells prior to dilutions, bacterial or viral predation during early stages of the enrichment, lack of growth under the laboratory culture conditions, and insufficient growth to visually detect turbidity. There are also factors that cloud the extrapolation of gene probe results to determining numbers of target microorganisms in environmental samples. Gene probes that target DNA may be detecting cells that are not metabolically active (dormant, dying, or dead), and do not preclude the detection of nontarget genes with short regions of high homology to the probe. On the other hand, nucleic acid extraction methods may not successfully lyse all the cells in a sediment sample, and a single gene probe is unlikely to detect all genes encoding for the target enzymatic function. (Gene copy number per genome is another consideration, but the genes used in this study appear to be single-copy genes.)

For both gene probes, the detection level corresponded to approximately 10^5 copies of the target genes/g sediment. Comparison of the culture-dependent method (MPN) with the culture-independent gene probe method suggested that the MPN underestimated, by orders of magnitude, the methanotrophic biomass throughout the demonstration. The degree of underestimation was largest in the before-methane samples and smallest in the samples from the N/P/pulsed 4% methane campaign. Although the gene probes suggest populations were much higher than could be determined by cultural methods, the gene probe results were only partially consistent with bioremediation performance at the site. Bioremediation performance was greatest in the 1% methane and N/P/pulsed 4% methane campaigns. The frequency of detection for both gene probes was highest in the 1% methane campaign, however, the frequency of detection in the N/P/pulsed 4% methane campaign was very similar to that in the before-methane samples. TCE degradation in methanotrophic enrichments was also only partially consistent with bioremediation performance at the site; degradation was not detected in the laboratory in the 1% methane campaign. The results highlight the difficulty of using only one approach for measuring bioremediation performance and indicate that a comprehensive and defensible interpretation of performance required multiple lines of evidence from different sources.

CONCLUSIONS

1. Taken together, methanotroph MPN enumerations and biodegradative potential in laboratory enrichments, and the frequency of detection of

mmoYBZC and *moxF* sequences. indicate that methanotrophs in sediment samples were enriched by methane injection and stimulated by the addition of nitrogen and phosphorus to the site.

2. Gene probes specific for the microbial group targeted for stimulation are valuable molecular diagnostic tools for the initial characterization of sites and for monitoring bioremediation performance. This was particularly evident in sediment samples from the 1% methane campaign, where TCE degradation in methanotrophic enrichments was a very poor predictor of bioremediation performance at the site.

ACKNOWLEDGMENTS

This research was funded by the Office of Technology Development, within the DOE's Office of Environmental Management, under the Non-arid Soils Volatile Organic Compounds Integrated Demonstration. Pacific Northwest Laboratory is operated by Battelle Memorial Institute for the U.S. DOE under Contract DE-AC06-76RLO 1830.

REFERENCES

Brockman, F. J., J. P. Bowman, J. T. Fleming, I. Gregory, A. Ogram, G. S. Sayler, W. Sun, and D. Zhang. 1994. "Nucleic Acid Technology Report for the In Situ Bioremediation Demonstration (Methane Biostimulation) of the Savannah River Integrated Demonstration Project DOE/OTD." Draft Technical Report, Pacific Northwest Laboratory, Richland, WA.

Cardy, D. L., V. Laidler, G. P. Salmond, and J. C. Murrell. 1991a. "Molecular analysis of the methane monooxygenase (MMO) gene cluster of *Methylosinus trichosporium* OB3b." *Molecular Microbiology* 5:335-342.

Cardy, D. L., V. Laidler, G. P. Salmond, and J. C. Murrell. 1991b. "The methane monooxygenase gene cluster of *Methylosinus trichosporium*: cloning and sequencing of the *mmoC* gene." *Archives of Microbiology* 156:477-483.

DiSpirito, A. A., J. Gulledge, A. K. Shiemke, J. C. Murrell, M. E. Lidstrom, and C. L. Krema. 1992. "Trichloroethylene oxidation by the membrane-associated methane monooxygenase in type I, type II and type X methanotrophs." *Biodegradation* 2:151-164.

Eddy, C. A., B. B. Looney, J. M. Dougherty, T. C. Hazen, and D. S. Kaback. 1991. *Characterization of the geology, geochemistry, hydrology and microbiology of the in situ air stripping demonstration site at the Savannah River Site (U).* Technical Report No. WSRC-RD-91-21. Westinghouse Savannah River Company, Aiken, SC.

Graham, D. W., J. A. Chaudhry, R. S. Hanson, and R. G. Arnold. 1993. "Factors affecting competition between type I and type II methanotrophs in two-organism, continuous-flow reactors." *Microbial Ecology* 25:1-17.

Lombard, K. H., J. W. Borthen, and T. C. Hazen. 1994. "The design and management of system components for in situ methanotrophic bioremediation of chlorinated hydrocarbons at the Savannah River Site," pp. 81-96. In R. E. Hinchee (Ed.), *Air Sparging for Site Remediation,* Lewis Publishers, Boca Raton, FL.

Machlin, S.M., and R.S. Hanson. 1988. "Nucleotide sequence and transcriptional start site of the *Methylobacterium organophilum* XX methanol dehydrogenase structural gene." *Journal of Bacteriology* 170:4739-4747.

Phelps, T. J., D. Ringelberg, D. Hendrick, J. Davis, C. B. Fliermans, and D. C. White. 1988. "Microbial biomass and activities associated with subsurface environments contaminated with chlorinated hydrocarbons." *Geomicrobiology Journal* 6:157-170.

Shelton, D. R. and J. M. Tiedje. 1984. "General method for determining anaerobic biodegradative potential." *Applied and Environmental Microbiology* 47:850-857.

Tsien, H.-C., G. A. Brusseau, R. S. Hanson, and L. P. Wackett. 1989. "Biodegradation of trichloroethylene by *Methylosinus trichosporium* OB3b." *Applied and Environmental Microbiology* 55:3155-3161.

Tsien, H.-C. and R. S. Hanson. 1992. "Soluble methane monooxygenase component B gene probe for identification of methanotrophs that rapidly degrade trichloroethylene." *Applied and Environmental Microbiology* 58:953-960.

Wolin, E.A., J.J. Wolin, and R.S. Wolfe. 1963. "Formation of methane by bacterial extracts." *Journal of Biological Chemistry* 238:2882-2886.

Design of an In Situ Carbon Tetrachloride Bioremediation System

Brent M. Peyton, Michael J. Truex,
Rodney S. Skeen, and Brian S. Hooker

ABSTRACT

A suite of simulation models were developed as a design tool in support of an in situ bioremediation demonstration at the Hanford site in Washington state. The design tool, calibrated with field- and bench-scale data, was used to answer four field-scale system design questions: (1) What are the important reaction processes and kinetics? (2) How will biomass distribute in the aquifer in response to injected substrate? (3) What well configuration best ensures proper nutrient transport and process control? (4) What operating and monitoring strategy should be used to confirm effective remediation? This paper does not describe the design tool itself, but describes how the design tool was used to optimize field site design parameters such as well spacing, hydraulic control, contaminant destruction, and nutrient injection strategies.

INTRODUCTION

In situ bioremediation of the chlorinated solvent, carbon tetrachloride (CCl_4), and nitrate is being performed at the U.S. Department of Energy's (DOE's) Hanford Site as part of the DOE Office of Technology Development's VOC Arid Integrated Demonstration. The CCl_4 is in groundwater in highly stratified sediments 76 m (250 ft) deep. In addition to its relative inaccessibility, CCl_4 is degraded by a cometabolic process that requires careful system control. The characteristics of both the contaminant and the site posed challenges in the design of the in situ process. The design tool was developed to help overcome these challenges. Of main concern were the four basic design questions, given in the abstract, which guided research and bench-scale test efforts.

The design tool was calibrated with field- and bench-scale data. Bench-scale batch tests with a native denitrifying consortium were used to determine reaction kinetics, including electron donor and acceptor consumption rates, CCl_4 degradation rate, inhibition constants, and by-product formation kinetics. After bacterial transport was investigated, nutrient injection strategies were tested in continuous-flow soil columns instrumented to continuously measure pH, pressure, and tracer concentrations. Field data included geological, hydrological, and microbiological characterization. In addition, the simulation design tool was developed to account for the interrelated effects of each process in various scenarios and to identify means to overcome rate-limiting steps in CCl_4 degradation. The design tool will also be an integral part of data evaluation during the field demonstration.

CCl_4 REACTION PROCESSES AND KINETICS

The demonstration at the Hanford Site will implement in situ bioremediation of CCl_4 and nitrate by supplying nutrients to indigenous microorganisms to stimulate metabolic activity and CCl_4 degradation. Biodegradation of CCl_4 using acetate as the carbon and energy source and nitrate as the terminal electron acceptor has been observed by Criddle et al. (1990); Lewis and Crawford (1993); Bae and Rittmann (1990); and Bouwer and Wright (1988). The denitrifying bacteria transform CCl_4 to carbon dioxide and chloride ions, while denitrification yields biomass, water, carbon dioxide, and nitrogen gas. Under some conditions, chloroform ($CHCl_3$) was produced as an undesirable by-product. Research is currently under way to determine its production and degradation kinetics.

Biodegradation of CCl_4 under denitrifying conditions is of particular interest at Hanford because both CCl_4 and nitrate are present in the unconfined aquifer. This, coupled with preliminary results showing CCl_4 destruction by subsurface microbes obtained from the Hanford Site (Brouns et al. 1990; Koegler et al. 1989), shows promise for successful in situ CCl_4 bioremediation.

A series of fed-batch experiments with a Hanford denitrifying consortium preceded the design of the in situ demonstration to formulate CCl_4 stoichiometry and validate kinetic expressions for CCl_4 degradation (Petersen et al. 1994; Hooker et al. 1994). The CCl_4 destruction expressions are first order in CCl_4 concentration and biomass and the rate is inhibited by the presence of more than 20 mg/L nitrate. Figure 1 shows the experimental data and model predictions of biomass, acetate, nitrate, and CCl_4 concentrations for a typical fed-batch treatability test. The resulting CCl_4 degradation kinetic expressions were used in the design tool.

BIOMASS DISTRIBUTION

In situ bioremediation relies on developing and maintaining microbial activity in contaminated regions. The main limitation to achieving this goal is the inability to evenly disperse rapidly reacting nutrients by recirculation wells. Radial groundwater flow patterns, in which fluid velocity diminishes rapidly with distance, cause injected nutrients to react near the well bore, thus limiting microbial activity at remote points in the flow field. This phenomenon occurs both in the laboratory (Cunningham and Wanner 1993) and in the field (Semprini et al. 1991). To extend the biologically active region, based on the work of others (Semprini et al. 1991; Shouche et al. 1993; Roberts et al. 1989), we focused on developing

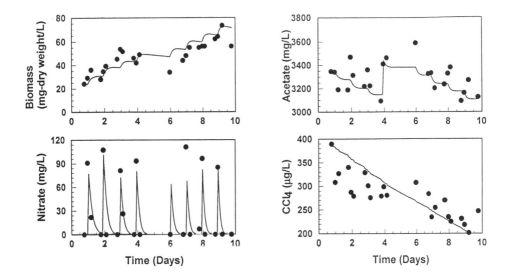

FIGURE 1. Experimental data and model predictions of biomass, acetate, nitrate, and CCl$_4$ concentrations for a typical batch treatability test.

nutrient feeding strategies that minimize near-well growth in order to optimize substrate transport.

The design tool predicts and soil column data indicate that the use of time-skewed acetate and nitrate pulses can reduce near-well biofouling. The design tool incorporates biofilm processes such as growth, attachment, and detachment. A sensitivity analysis determined that the detachment rate coefficient was a dominant parameter in predicting biomass distribution profiles (Peyton et al. 1994). To determine attachment and detachment rates, laboratory soil columns were fed acetate and nitrate continuously or in pulses. The 10.1-m (4-in)-diameter soil columns were packed with coarse sand, or in later tests, actual site sediments. The sediments were inoculated with approximately 1 × 10^7 colony-forming units of an indigenous Hanford denitrifying consortium per gram of soil. Comparisons of experimental results and design tool predictions indicate good agreement between effluent acetate and nitrate profiles. Biomass profiles, measured as protein, correlate well with predicted results (Figure 2). The primary independent variable that was manipulated to give agreement between the actual results and design tool predictions was the first-order biomass detachment rate coefficient, 0.003 min^{-1}, which was within 25% of that observed by Peyton et al. (1994).

WELL CONFIGURATION, NUTRIENT TRANSPORT, AND PROCESS CONTROL

A well configuration was selected which incorporated into the design three preexisting characterization wells. The configurations assessed with the design

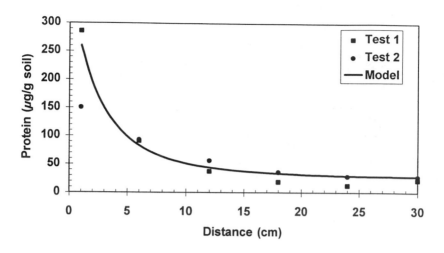

FIGURE 2. Biomass profiles in the soil columns, measured as protein, correlate well with simulation results.

tool included 2-well, 3-well, 4-well, and 5-well recirculation systems. Selection of the field design was based on design tool predictions of CCl_4 destruction which was a function of reaction kinetics and of hydraulic control. The amount of hydraulic control determines the amount of nitrate and CCl_4 that enters the reaction zone through the injection well. The desired process operation requires dispersion-induced mixing of acetate and nitrate pulses at a distance from the well bore to (1) keep nitrate concentration below 20 mg/L during acetate pulses and (2) limit biomass production, and thus biofouling, adjacent to the well bore. At 100% hydraulic control, no additional CCl_4 will enter the reaction zone, and therefore the reaction can be completely controlled by injection of the desired amounts of acetate and nitrate. At low hydraulic control, the amount of CCl_4 and nitrate entering the system from outside the reaction zone increases. The increased CCl_4 makes measuring CCl_4 destruction rates much more difficult and the increased nitrate permits biomass to grow adjacent to the well after an acetate pulse.

The design tool showed that steady-state CCl_4 destruction was essentially constant from about 70% to 95% hydraulic control. Destruction of CCl_4 was greatly enhanced with a hydraulic control near 100% and rapidly declined below 65%. Because 100% hydraulic control is not possible in the field, the design for the demonstration was selected to achieve a "simulated" hydraulic control of about 85%. Tracer tests indicate that the hydraulic control of the installed field system is approximately 90%.

Predictions from the design tool were combined with other characteristics of each well configuration to select the best design for the demonstration. Two primary issues were key to the selection: the ability to obtain measurable responses in monitoring parameters and the versatility of the design in testing strategies relevant to full-scale applications. These and other site-related criteria led to the selection of a 2-well recirculation configuration. This pattern provides adequate hydraulic control, so that reaction kinetics and biomass distribution can be controlled, and permits multiple well spacings between pumped wells with an acceptable number of monitoring points.

OPERATING AND MONITORING STRATEGY

Phase 1, abiotic recirculation, was performed from February 6, 1995 to March 29, 1995 and consisted of control operations with no addition of nutrients; the groundwater was continuously mixed and then sampled weekly. A hydraulic control of 90% was demonstrated with a bromide tracer. Data from this phase will be used to detect any abiotic removal of contaminant due to groundwater mixing and to calibrate the transport portion of the process model. Nutrient injection will be initiated in Phase 2, active bioremediation. An operation strategy based on design tool predictions and data from Phase 1 will be implemented and maintained for at least three months to collect sufficient information to assess the strategy. The strategy will be changed only if major problems occur, such as production of non-inert persistent by-products (chloroform and nitrite); undesirable changes in the hydraulic mixing pattern; insufficient or excessive biomass accumulation; or insufficient contaminant destruction. Phase 3 will implement a revised strategy to optimize performance.

The number of samples needed to demonstrate a statistical difference in CCl_4 concentration between the injection well and a monitoring well was calculated from the anticipated CCl_4 destruction rate of 15 g/day. Statistical analysis assumed a standard deviation of 25% of the mean for the field samples and an initial CCl_4 concentration of 2 mg/L. Phase 1 results to date indicate standard deviation for CCl_4 field samples is ~15%. At steady-state, the design goal results in a monitoring-well concentration of about 1.3 mg/L. Demonstration with 95% confidence that the steady state concentration is lower than the initial concentration requires 20 samples; however, to back calculate the actual CCl_4 destruction rate will require statistical proof that the steady-state concentration is lower by a specific amount. For an observed difference of 0.4 mg/L, 100 samples are needed to show that the difference is 0.3 mg/L with 90% confidence.

SUMMARY

A design tool, calibrated with field- and bench-scale data, was used to answer field-scale system design questions. Bench-scale batch tests with the native denitrifying consortium were used to determine reaction rate kinetics. Bacterial transport parameters were determined, and nutrient injection strategies were tested in continuous-flow soil columns. Key to determining the best design were the ability to obtain measurable responses in necessary monitoring parameters and the versatility of the design. The simulation design tool combined bench and field data to account for the interrelated effects of each process in various scenarios. A 2-well recirculation configuration was selected. Phase 1 of the operation strategy collected baseline data without nutrient injection; Phase 2 is the initial remediation phase; results from Phase 2 will be used to develop an optimized Phase 3 remediation strategy.

ACKNOWLEDGMENTS

This work was supported by the U.S. Department of Energy Office of Technology Development under the VOC Arid Integrated Demonstration. Pacific Northwest Laboratory is operated for the U.S. Department of Energy by Battelle Memorial Institute under contract DE-AC06-76RLO 1830.

REFERENCES

Bae, W. and B. E. Rittmann. 1990. In C. O'Melia (Ed.), *Environmental Engineering, Proceedings of the 1990 Specialty Conference*, American Society of Civil Engineers, New York, NY.

Bouwer, E. J., and J. P. Wright. 1988. "Transformations of Trace Halogenated Aliphatics in Anoxic Biofilm Columns," *J. Contam. Hydrol.* 2:155-169.

Brouns, T. M., S. S. Koegler, W. O. Heath, J. K. Fredrickson, H. D. Stensel, D. L. Johnstone, and T. L. Donaldson. 1990. *Development of a Biological Treatment System for Hanford Groundwater Remediation: FY 1989 Status Report*, PNL-7290, Pacific Northwest Laboratory, Richland, WA.

Criddle, C. S., J. T. DeWitt, D. Grbić-Galić, and P. L. McCarty. 1990. "Transformation of Carbon Tetrachloride by *Pseudomonas* sp. Strain KC under Denitrification Conditions," *Appl. Environ. Microbiol.* 56:3240-3246.

Cunningham, A. B. and O. Wanner. 1993. "Modeling Microbial Processes in Porous Media with Application to Biotransformation," *International Symposium on Hydrogeological, Chemical, and Biological Processes of Transformation and Transport of Contaminants in Aquatic Environments*, May 24-29, Rostov-on-Don, Commonwealth of Independent States (the former Soviet Union).

Hooker, B. S., R. S. Skeen, and J. N. Petersen. 1994. "Biological Destruction of CCl_4, Part II: Kinetic Modeling," *Biotechnol. Bioeng.* 44:211-218.

Koegler, S. S., T. M. Brouns, W. O. Heath, and R. J. Hicks. 1989. *Biodenitrification of Hanford Groundwater: FY 1988 Status Report*, PNL-6917, Pacific Northwest Laboratory, Richland, WA.

Lewis, T. A. and R. L. Crawford. 1993. "Physiological Factors Affecting Carbon Tetrachloride Dehalogenation by the Denitrifying Bacterium *Pseudomonas* sp. Strain KC," *Appl. Environ. Microbiol.* 59:1635-1641.

Petersen, J. N., R. S. Skeen, K. M. Amos, and B. S. Hooker. 1994. "Biological Destruction of CCl_4 I. Experimental Design and Data," *Biotechnol. Bioeng.* 43:521-528.

Peyton, B. M., R. S. Skeen, B. S. Hooker, R. W. Lundman, and A. B. Cunningham. "Evaluation of Bacterial Detachment Rates in Porous Media," Accepted in *Applied Biochem. Biotechnol.*

Roberts, P. V., L. Semprini, G. D. Hopkins, D. Grbić-Galić, P. L. McCarty, and M. Reinhard. 1989. *In Situ Aquifer Restoration of Chlorinated Aliphatics by Methanotrophic Bacteria*, EPA Report No. EPA 2-89/033 U.S. EPA, Ada, Oklahoma.

Semprini, L., G. D. Hopkins, D. B. Janssen, M. Lang, P. V. Roberts, and P. L. McCarty. 1991. *In Situ Biotransformation of Carbon Tetrachloride under Anoxic Conditions*, EPA Report No. EPA 2-90/060, U.S. EPA, Ada, Oklahoma.

Shouche, M., J. N. Petersen, and R. S. Skeen, 1993. "Use of a Mathematical Model for Prediction of Optimum Feeding Strategies for In Situ Bioremediation," *Applied Biochem. and Biotechnol.* 39: 763-779.

Metallic Iron-Enhanced Biotransformation of Carbon Tetrachloride and Chloroform Under Methanogenic Conditions

Lenly J. Weathers and Gene F. Parkin

ABSTRACT ━━━━━━━━━━━━━━━━━━━━━━━━━━━━━━━━━━━━━━

The transformation of carbon tetrachloride (CT) and chloroform (CF) was investigated in batch systems containing methanogenic cell suspension and iron filings. CT and CF transformation kinetics were fastest in iron-cell (IC) treatments compared to other treatments. A synergistic effect for CT and CF transformation occurred between iron and live cells, which significantly increased the transformation rates.

INTRODUCTION

The observed loss of halogenated organic compounds (HOC) in the presence of metal-containing well casing materials (Reynolds et al. 1990) has spurred an interest in the use of zero-valent metals to remediate HOC-contaminated groundwater. After confirming the potential for the process in laboratory batch reactors containing iron filings and various chlorinated methanes, ethanes, and ethenes (Gillham and O'Hannesin 1994), a field experiment at Canadian Forces Base Borden was conducted using a 1.6-m-thick, reactive barrier containing iron filings and sand constructed downgradient from a continuous perchloroethene (PCE) and trichloroethene (TCE) source. PCE and TCE were dechlorinated within the barrier, as shown by losses of PCE and TCE of 86% and 90%, respectively; a near-stoichiometric increase in dissolved chloride; and detection of dechlorination products (O'Hannesin 1993). Matheson and Tratnyek (1994) reported the nonstoichiometric, sequential reductive dechlorination of CT to CF and dichloromethane (DCM) by iron metal in laboratory batch reactors. Helland et al. (1995) observed reduced CT transformation rates by iron metal in the presence of molecular oxygen.

The research mentioned above focused on abiotic processes; however, methanogenic bacteria have demonstrated the ability to use metallic iron as an energy source for growth (Daniels et al. 1987; Lorowitz et al. 1992). When iron is immersed in water, the oxidation of Fe^0 to Fe^{2+} is coupled with the reduction of water-derived protons to form a hydrogen layer around the iron surface. Methanogenic bacteria grow on the hydrogen thus produced, thereby accelerating the corrosion process. This phenomenon, termed *biocorrosion,* is undesirable in most circumstances.

The microbial transformation of CT and CF has been observed in pure (Egli et al. 1987) and mixed, methanogenic cultures (Bouwer and McCarty 1983). In addition, hydrogen has proven to be an excellent electron donor for reductive dechlorination by anaerobic mixed cultures (DiStefano et al. 1992). It may be possible, then, to couple the biocorrosion of metals (such as elemental iron) with biodehalogenation to remediate HOC-contaminated groundwater. We report here the transformation of CT and CF in batch incubations containing metallic iron and methanogenic cell suspension.

MATERIALS AND METHODS

Chemicals

Saturated aqueous stock solutions of CT, DCM (certified American Chemical Society grade), and CF (HPLC grade) (Fisher Scientific Co., Pittsburgh, Pennsylvania) were used. Iron filings (40 mesh, degreased) (Malinckrodt) were used as a source of metallic iron.

Stock Culture Reactor

A mixed, methanogenic culture (8 L) was developed in a magnetically stirred, 9.5-L glass reactor. The reactor was maintained at 20°C on a 40-day hydraulic retention time such that a 200-mL cell suspension was removed daily and replaced with fresh anaerobic media, plus 1,600 μL glacial acetic acid. The volatile suspended solids (VSS) concentration of the stock reactor was 220 mg/L.

Experimental Design

Experiments were conducted in the dark on an orbital shaker table (200 rpm) at 20°C using 25 mL liquid volume in 38-mL serum bottles. Duplicates were used for all treatments. Iron cell (IC) and resting cell (RC) treatments were prepared as follows. For iron treatments, 1 g iron filings were weighed into bottles, which were then filled with deionized (DI) water, sealed with Teflon™-coated rubber septa (West Co., Phoenixville, Pennsylvania), and capped with aluminum crimp caps. Bottles were flushed with N_2/CO_2 gas (80:20, v/v) through the septa to purge the water. Bottles not receiving iron were purged in a similar manner. Cell suspension was transferred 48 h after the last feeding operation from the stock reactor to each bottle through the septum with a 100-mL glass, gastight syringe (Scientific Glass Engineering, Australia). Cell-free supernatant controls were prepared by centrifuging the cell suspension (10 min, 5,000 rpm), push-filtering the supernatant through a

0.45-μm pore size syringe tip filter, and then adding stock ferrous chloride and sodium sulfide solutions to replace iron sulfide precipitate that was removed by centrifugation and filtration. Iron-cell-free supernatant (IS) treatments were prepared in this manner using bottles containing 1 g iron filings. Anoxic water controls (pH 7.0) were prepared in an anaerobic glovebag (Coy Laboratory Products, Inc.) by adding 1 g $NaHCO_3$ to 200 mL autoclaved, DI water that had equilibrated with the glovebag atmosphere (N_2/H_2, 99:1, v/v), and then adjusting the pH with 1N HCl. The N_2/H_2 headspace of each water control bottle was purged by flushing with N_2/CO_2 gas (80:20, v/v) outside the glovebag.

Analytical Methods

CT, CF, DCM, and methane were measured by headspace analysis using 100-μL (CT and CF) or 500-μL samples (DCM and methane) taken with Pressure-Lok locking, gastight syringes (Precision Sampling Corp., Baton Rouge, Louisiana). CT and CF were analyzed using a Hewlett Packard (HP) model 5890 series II gas chromatograph (GC) equipped with a DB-5 capillary column (J&W Scientific, Folsom, California) and electron capture detector. DCM and methane were analyzed on a HP 5890 series II GC, equipped with a DB-WAX capillary column (J&W Scientific) and flame ionization detector. Detection limits for each compound were CT (0.03 μM), CF (0.08 μM), DCM (0.2 μM), and methane (1 μmol/bottle).

RESULTS AND DISCUSSION

Fig. 1 shows the time-course of a CT experiment (the mean of duplicates is given in all figures). CT was transformed fastest in the iron-cell (IC) treatment, followed by the iron-supernatant (IS) and resting cell (RC) treatments. Volatile losses in the DI control accounted for about 14% of the total CT loss, and transformation by cell-free supernatant in the absence of iron filings accounted for an additional 9% loss over the 68-h study.

Pseudo-first-order rate coefficients were calculated by applying linear regression to the integrated form of the pseudo-first-order rate equation, $\ln(M) = -k * t$, where M is the mass of CT or CF in the bottle (nmol), k is the pseudo-first-order rate coefficient (h^{-1}), and t is the time since the experiment began (h). Zero-order and second-order kinetic analyses produced smaller r^2 values than those in Table 1. Table 1 shows that a synergistic effect between the cells and metallic iron occurred as regards CT transformation: k for the IC treatment is 0.21 h^{-1}, which is 1.9 times the sum of the k-values for the resting cell and iron-supernatant treatments, 0.114 h^{-1}. If the two reactions were independent, the two rate coefficients should be additive.

CF, the hydrogenolysis product of CT, was produced and transformed most rapidly in the IC treatment (Fig. 2). At the end of the study at 68 h, although all treatments contained approximately the same mass of total chlorinated methanes, about 100 nmol, product distributions were different; the resting cell

Bioremediation of Chlorinated Solvents

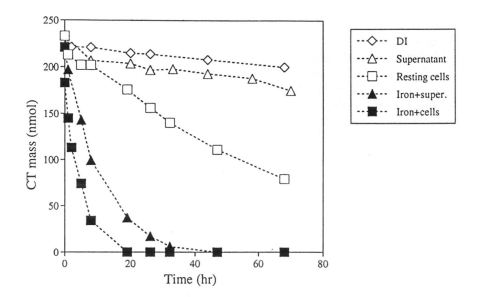

FIGURE 1. Transformation of CT in various treatments.

TABLE 1. Calculated pseudo-first-order transformation rate coefficients with CT or CF added.

Treatment	CT transformation data			CF transformation data		
	$k(h^{-1})$	half-life (h)	r^2	$k(h^{-1})$	half-life (h)	r^2
Resting cells (RC)	0.015	45	0.993	0.0033	207	0.837
Iron+cells (IC)	0.21	3	0.991	0.024	29	0.996
Supernatant (S)	0.0035	200	0.973	0.0015	472	0.647
Iron+supernatant (IS)	0.099	7	0.998	0.0036	190	0.924

treatment contained 83% CT and 17% CF, treatment IS contained 78% CF and 22% DCM, and treatment IC contained 32% CF and 68% DCM.

With CF was added, the pseudo-first-order CF transformation rate coefficient for the IC treatment was greater than for other treatments and a synergistic effect between iron and cells is apparent (Table 1); k for the IC treatment, 0.024 h^{-1}, is 3.4 times greater than the sum of the coefficients for the IS treatment and the RC treatment, 0.0070 h^{-1}.

DCM transformation was negligible in all treatments when added at an initial mass of 200 nmol, including a live cell treatment amended with 200 mg/L

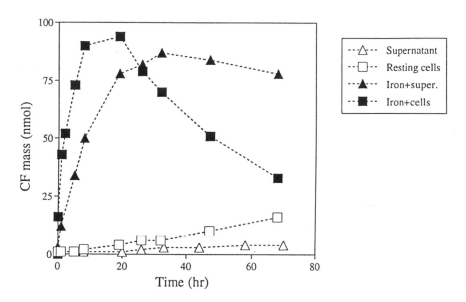

FIGURE 2. Production and transformation of CF in various treatments with CT added.

acetate (data not shown). Gillham and O'Hannesin (1994) and Matheson and Tratnyek (1994) also report the persistence of DCM in abiotic iron studies.

We propose that the enhanced CT and CF transformation rates in the iron-cell treatments may be caused by cometabolism by hydrogenotrophic methanogens. Methane production was negligible in CT- and CF-free controls containing cell suspension, but no iron filings, and in the IS treatment (data not shown), while methane was produced in the IC treatment. Also, although hydrogen was not analyzed, bubbles were seen rising from the surface of the iron in a bottle containing iron filings in DI water, and the production of hydrogen has been confirmed in similar experiments (Lorowitz et al. 1992, Gillham and O'Hannesin 1994, Matheson and Tratnyek 1994). Hydrogen produced by the corrosion of iron metal could be consumed by hydrogenotrophic methanogens, which then cometabolically channel reducing equivalents to CT or CF. Rates of DCM transformation are not enhanced because, although the methanogens may grow on hydrogen, they may not contain enzymes necessary to transform DCM. Significant DCM degradation under methanogenic conditions apparently requires a consortium enriched on DCM (Freedman and Gossett 1991).

Hydrogen is an excellent electron donor for methanogens, but the use of hydrogen as an in situ energy source for anaerobic microorganisms has been hampered by its low solubility, about 1.6 mg L^{-1} at 20C, based on a Henry's constant of 6.83×10^4 atm mol^{-1} (Metcalf and Eddy 1991). Using a porous reactive barrier containing iron metal may provide a way to continuously release hydrogen for consumption by cometabolizing anaerobes. Future work will address the role of hydrogen directly.

ACKNOWLEDGMENTS

This work was supported by grants from the Great Plains-Rocky Mountain Hazardous Substance Research Center (funded by U.S. Environmental Protection Agency), the National Institutes of Health, and the Iowa Center for Biocatalysis and Bioprocessing. Thanks to J.L. Schnoor for suggesting the use of metallic iron as a chemical reducing agent and to P.J.J. Alvarez for a thoughtful review of the manuscript.

REFERENCES

Bouwer, E.J. and P.L. McCarty. 1983. "Transformations of 1- and 2-carbon halogenated aliphatic organic compounds under methanogenic conditions." *Appl. Environ. Micro.* 45: 1286-1294.

Daniels, L., N. Belay, B.S. Rajagopal, and P.J. Weimer. 1987. "Bacterial methanogenesis and growth from CO_2 with elemental iron as the sole source of electrons." *Science.* 237: 509-511.

DiStefano, T.D., J.M. Gossett, and S.H. Zinder. 1992. "Hydrogen as an electron donor for dechlorination of tetrachloroethene by an anaerobic mixed culture." *Appl. Environ. Micro.* 58: 3622-3629.

Egli, C., R. Scholtz, A.M. Cook, and T. Leisinger. 1987. "Anaerobic dechlorination of tetrachloromethane and 1,2-dichloroethane to biodegradable products by pure cultures of *Desulfobacterium* sp. and *Methanobacterium* sp." *FEMS Microbiol. Lett.* 43:257-261.

Freedman, D.L. and J.M. Gossett. 1991. "Biodegradation of dichloromethane and its utilization as a growth substrate under methanogenic conditions." *Appl. Environ. Micro.* 57: 2847-2857.

Gillham, R.W. and S.F. O'Hannesin. 1994. "Enhanced degradation of halogenated aliphatics by zero-valent iron." *Ground Water.* 32: 958-967.

Helland, B.R., P.J.J. Alvarez and J.L. Schnoor. 1995. "Reductive dechlorination of carbon tetrachloride with elemental iron." *J. Haz. Mat.* In press.

Lorowitz, W.H., D.P. Nagle, Jr. and R.S. Tanner. 1992. "Anaerobic oxidation of elemental metals coupled to methanogenesis by *Methanobacterium thermoautotrophicum*." *Environ. Sci. Technol.* 26: 1606-1610.

Matheson, L.J., and P.G. Tratnyek. 1994. "Reductive dehalogenation of chlorinated methanes by iron metal." *Environ. Sci. Technol.* 28: 2045-2053.

Metcalf and Eddy, Inc. 1991. *Wastewater Engineering: Treatment, Disposal, Reuse.* 3rd ed., McGraw-Hill, New York, NY.

O'Hannesin, S.F. 1993. "A field demonstration of a permeable reaction wall for the in situ abiotic degradation of halogenated aliphatic organic compounds." M.Sc. Thesis, Univ. of Waterloo, Ontario, Canada.

Reynolds, G.W., J.T. Hoff, and R.W. Gillham. 1990. "Sampling bias caused by materials to monitor halocarbons in groundwater." *Environ. Sci. Technol.* 24: 135-142.

Accelerated Biotransformation of Carbon Tetrachloride and Chloroform by Sulfate-Reducing Enrichment Cultures

David L. Freedman, Matthew Lasecki,
Syed Hashsham, and Richard Scholze

ABSTRACT

The biotransformation of carbon tetrachloride (CT) and chloroform (CF) was examined with lactate- and acetate-grown sulfate-reducing enrichment cultures. Both cultures transformed CT, with approximately 50% reductively dechlorinated to CF and up to 10% to dichloromethane (DCM). Addition of cyanocobalamin (less than 0.1 mol % of the CT consumed) increased the rate of CT transformation more than 100-fold. The principal product from $[^{14}C]CT$ with cyanocobalamin added was carbon disulfide (CS_2); less than 3% was reduced to CF plus DCM. Autoclaved cultures that received cyanocobalamin were only one third as fast as their live counterparts, but produced similar amounts of CS_2. With CF, addition of cyanocobalamin to acetate- and lactate-grown cultures also increased the rate of transformation more than 100-fold. DCM was the principal transformation product until CF additions reached 270 mg/L, at which point almost no increase in DCM was observed. Thus, low levels of cyanocobalamin substantially accelerated the rate of CT and CF transformation and altered the distribution of products formed.

INTRODUCTION

CT and CF are among the most frequently encountered chlorinated aliphatic compounds in contaminated groundwater. The application of anaerobic biological processes for remediation has been limited, in part due to a lack of information on optimal methods for accelerating the rate of transformation, while ensuring that the products formed are nonhazardous. Sulfate-reducing conditions are prevalent in many anaerobic environments and can also be purposefully developed by the addition of sulfate and an appropriate electron donor (Beeman et al. 1994).

Biotransformation of CT under sulfate-reducing conditions has been demonstrated previously, but often at concentrations below 1 mg/L (Bouwer & Wright 1988; Cobb & Bouwer 1991). Substantially higher levels have been reported in contaminated aquifers (Riley et al. 1992; Petersen et al. 1994). Cobb and Bouwer (1991) observed complete removal of CT in mixed-culture continuous-flow reactors containing a sulfate-reducing zone, but it was not possible to discern what the transformation products were. Biotransformation of CT by pure cultures of sulfate-reducers has also been demonstrated (Egli et al. 1987, 1988, 1990), although reductive dechlorination was a major pathway leading to accumulation of CF and DCM. CF biodegradation under sulfate-reducing conditions has also been examined (Bouwer & Wright 1988; Cobb & Bouwer 1991), but at relatively low concentrations and without documentation of the products formed. Accumulation of DCM from CF has been observed consistently under methanogenic conditions (Egli et al. 1990; Mikesell & Boyd 1990).

Becker and Freedman (1994) recently demonstrated that the addition of cyanocobalamin (vitamin B_{12}) to a mixed methanogenic enrichment culture increased the rate of CF biotransformation approximately ten-fold. Addition of the transition metal coenzyme also increased the extent of CF oxidation to CO_2 and virtually eliminated accumulation of DCM. CF levels as high as 2.2 mM were readily transformed, at cyanocobalamin to CF molar ratios as low as 3%. Freedman et al. (1995) also observed much higher rates of CT biotransformation and less reductive dechlorination with the same methanogenic enrichment culture when supplemented with cyanocobalamin or other cobalamin homologs. The purpose of this study was to examine if low levels of cyanocobalamin can also be used to enhance transformation of CT and CF by lactate- and acetate-grown sulfate-reducing enrichment cultures.

MATERIALS AND METHODS

Chemicals

CT and CF ($\leq$99.9%; Aldrich) were added to cultures from either saturated basal medium (see below) solutions (approximately 6.9 mM CT, 65 mM CF) or as neat liquids (when the amount added exceeded 10 μL). Cyanocobalamin (99%; Sigma Chemical Co.) solutions (7.25 mM or 0.725 mM in deionized water) were stored in foil-wrapped bottles to protect them from light. The [^{14}C]CT (Dupont NEN Products) was diluted with distilled deionized water to 2.7 $\times$ 10^{-7} disintegrations per minute (dpm) per mL (2.6 mM CT). Scinti-Verse-E liquid scintillation cocktail was used (Fisher Scientific). CS_2 (98%) was obtained from EM Industries, carbon monoxide (CO; 99.5%) and chloromethane (CM; 99.5%) from Matheson, and methane (99.0%) from Linde. All other chemicals used were analytical grade.

Enrichment Cultures

Sulfate-reducing enrichment cultures were developed using sludge from an anaerobic digester at a municipal wastewater treatment plant (in Urbana,

Illinois). The inoculum was added to a basal medium (described in Widdel and Pfennig 1984) that includes a phosphate buffer, 4.0 g/L Na_2SO_4, and 0.12 g/L $Na_2S\cdot9H_2O$. One set of enrichments was developed using acetate (10 mM) as the electron donor, another set using lactate (10 mM). Several transfers were made prior to adding CT or CF. All experiments were conducted in 160-mL serum bottles containing 100 mL of the culture. Assuming complete consumption of the acetate or lactate, the maximum possible sulfide concentration prior to adding the chlorinated methanes was 10 or 15 mM, respectively.

The bottles were set up in an anaerobic glove box (Coy Laboratory Products, Inc.), then purged with 30% CO_2-70% N_2, sealed with slotted gray butyl rubber septa and aluminum crimp caps, and incubated at 35°C on a shaker table, in the dark, with the culture in contact with the septum. Loss of CT from water controls (serum bottles containing 100 mL of deionized water and CT) was minimal (less than 10% over a 30-day period), indicating the acceptability of using gray butyl rubber septa. Bottles that received supplemental cyanocobalamin were covered with aluminum foil to prevent photolysis.

Analytical Methods

CT, CF, DCM, CM, CH_4, and CS_2 were measured by gas chromatographic (GC; Perkin-Elmer Model 9000) analysis of a 0.5-mL headspace sample [flame-ionization detector, 1% SP-1000 on 60/80 Carbopack-B column (Supelco, Inc.)], as previously described (Freedman & Gossett 1989, 1991). In this study, we report the total mass of each compound present in a serum bottle. Based on Henry's constants [$(mol\cdot m^{-3}$ gas concentration)/($mol\cdot m^{-3}$ aqueous concentration)] at 35C of 33.1 for CH_4, 0.526 for CM, 0.128 for DCM, 0.235 for CF, 1.92 for CT, 46.9 for CO (from Freedman & Gossett 1989; Gossett 1987; Thibodeaux 1979), and 1.30 for CS_2 (determined in this study according to Freedman et al. 1993), 4.8% of the CH_4, 3.4% of the CO, 76% of the CM, 93% of the DCM, 88% of the CF, 46% of the CT, and 56% of the CS_2 were present in the 100 mL of liquid, with the balance in the 60-mL headspace. Confirmation of CS_2 as a product was obtained by GC/mass spectrometry (Hewlett Packard model 5890A; DB-5 capillary column).

Volatile ^{14}C-labeled compounds were analyzed by a GC-combustion technique described previously (Becker & Freedman 1994; Freedman & Gossett 1989). The SP-1000/Carbopack-B column was used to separate the chlorinated compounds and CS_2, whereas the nonchlorinated volatiles were separated on both Carbosieve S-II (3.2 mm × 3.2 m; Supelco) and GP 10% (3.2 mm × 3.2 m; Supelco) columns. Next, [14]CO_2 and ^{14}C-labeled nonvolatile compounds were measured as previously described (Becker & Freedman 1994; Freedman & Gossett 1989), once analysis of the ^{14}C-labeled volatile compounds was completed. Nonstrippable residue (NSR) is defined as the aqueous ^{14}C that remains after stripping at low pH (approximately 4). Soluble NSR represents the NSR that passes through a 0.45-μm filter; the retained ^{14}C represents nonsoluble NSR.

The total amount of sulfide in a 160-mL serum bottle was determined by measuring the concentration in a headspace sample (0.3 mL) and a liquid sample (0.5 mL), using a colorimetric method (American Public Health Association 1985). Lactate and acetate concentrations were determined by high-performance liquid chromatography with an ion-exchange column and UV/Vis detector, as previously described (Ehrlich et al. 1981).

RESULTS

Biotransformation of CT by Lactate and Acetate Enrichment Cultures

The establishment of sulfate reducing conditions in the enrichment cultures was determined by measuring organic acid consumption and sulfide production, prior to the addition of CT or CF. In the lactate-grown enrichments, approximately 0.8 mol of acetate was typically produced per mol of lactate consumed, with a nearly stoichiometric increase in sulfide. Methane output usually accounted for less than 4% of the lactate consumed. The lactate-grown enrichment cultures were dominated by gram-negative curved rods. In the acetate-grown enrichments, the increase in sulfide typically accounted for two-thirds of the acetate consumed, with the balance of substrate recovered as methane. In control bottles that received no lactate or acetate, a low amount of methane was produced but there was no increase in sulfides. Addition of CT and CF completely inhibited methanogenesis in both the lactate- and acetate-grown cultures. The CT and CF slowed but did not completely stop sulfate reduction.

Four experimental conditions were used to examine CT biotransformation with lactate- grown enrichment cultures: viable cultures alone; viable cultures supplemented with cyanocobalamin; autoclaved cultures alone; and autoclaved cultures supplemented with cyanocobalamin. Replicate bottles were used for each condition. All bottles received an initial CT dose of approximately 1 μmol/bottle. Viable cultures also received 0.5 mmol of lactate.

Over a 22-day period, two CT additions totaling 4.3 μmol were consumed by viable cultures that received no cyanocobalamin (Figure 1). The principal volatile product was CF (67% of the CT). No CS_2, CM, or CH_4, and only traces of DCM, were detected. Monitoring of these cultures was continued through day 70 (data not shown). The rate of CT transformation did not improve, despite addition of 1 mmol of lactate on day 28; and CF continued to be the only major volatile product.

Another set of viable cultures (4 bottles) received CT plus three additions of cyanocobalamin (0.40 μmol each), on days 0, 3, and 5. The results in Figure 2 show the dramatic effect of adding cyanocobalamin. Only 3 days were needed for consumption of the initial CT dose. The amount of CT added was gradually increased, so that by day 15, the aqueous concentration reached 350 mg/L (approximately one-third of the solubility limit for CT). Despite this

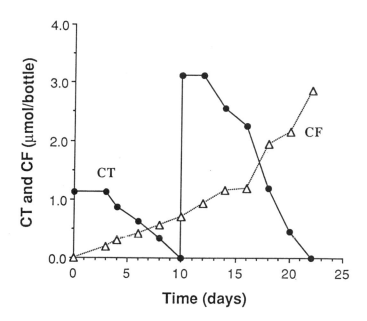

FIGURE 1. Transformation of CT in a lactate-grown enrichment culture.

high concentration, essentially complete removal was achieved in only 2 days. The principal volatile product formed was CS_2, with less than 2% accumulating as CF and DCM.

CT transformation also occurred in autoclaved cultures. Representative results for bottles receiving CT alone are presented in Figure 3. Similar to the live cultures without B_{12}, CF was a major product, but DCM was insignificant; unlike the live cultures, CS_2 formation was also significant. As shown in Figure 4, the addition of 1 μmol of cyanocobalamin on day 0 had a similarly dramatic effect as in the live cultures; the rate of CT transformation was significantly higher, and very little CT was reduced to CF. There was no indication of the CT transformation rate decreasing in these B_{12}-amended autoclaved cultures, even after 36 days of monitoring.

Then, [14C]CT was added to serum bottles covering three of the four experimental conditions, as indicated in Figures 2, 3, and 4. Once routine headspace analysis indicated that all (or nearly all) of the CT plus [14C]CT was consumed, the bottles were sacrificed to determine the product distribution. The loss of [14C] between the time of [14C]CT addition and analysis of products was proportional to the length of incubation. In bottles that received cyanocobalamin, the average [14C] loss was 16%, but jumped to 32% in bottles that received only CT. The results in Table 1 were normalized to the amount of [14C] identified at the time of analysis. On average, 90% (standard deviation = 7.3%) of the [14C] recovered at the time of analysis was accounted for as identifiable products. This recovery was defined as the total dpm in all compounds (CT + CF + DCM + CM + CS_2 + CH_4 + CO + CO_2 + soluble NSR + nonsoluble NSR) divided

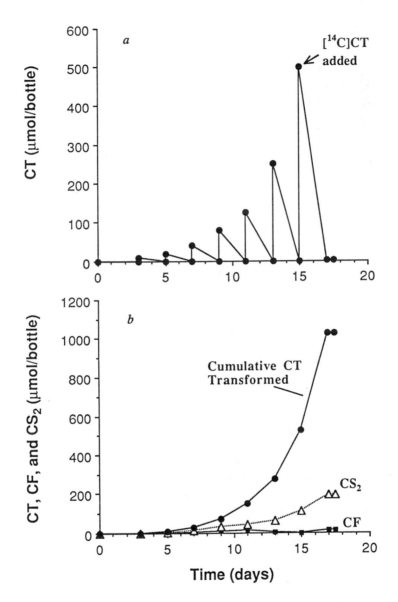

FIGURE 2. Transformation of CT in a lactate-grown enrichment culture supplemented with cyanocobalamin.

by the total dpm present in a serum bottle at the time of ^{14}C analysis. The denominator was determined by summing the dpm in the headspace (0.5-mL sample added to scintillation cocktail) plus liquid (100-μL sample added directly to scintillation cocktail).

Results for transformation of CT in live acetate-grown cultures, with and without supplemental cyanocobalamin added, were similar to those described

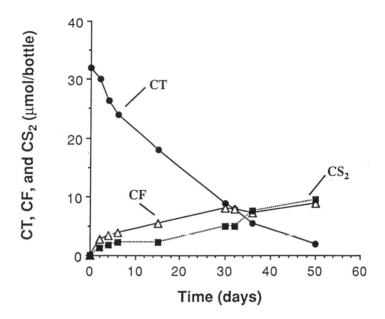

FIGURE 3. Transformation of CT in an autoclaved lactate-grown enrichment culture.

above for the lactate-grown enrichments. Duplicate bottles were used for each condition. Over a 71-day period, a total of 18 μmol of CT was transformed without supplemental cyanocobalamin, 54% of which accumulated as CF; an insignificant amount of DCM was observed, while CS_2 remained below the detectable limit (Figure 5). A much more rapid rate of CT transformation occurred in live cultures that received cyanocobalamin (0.2 μmol added on days 0, 2, 4, 6, and 12), with over 1,000 μmol consumed (Figure 6). Despite this high rate, less than 0.5% of the CT accumulated as CF plus DCM. CS_2 was the principal product for the first 40 days, but then it reached a plateau. Because [^{14}C]CT was not added to these bottles, it was not possible to determine what other products were formed beyond day 40.

Biotransformation of CF by Lactate and Acetate Enrichment Cultures

CF transformation was examined in live lactate- and acetate-grown enrichment cultures, with and without supplemental cyanocobalamin. Duplicate bottles were used for each condition. All bottles initially received 1 μmol of CF and 0.5 mmol of lactate or acetate.

As shown in Figure 7, the rate of CF biotransformation in lactate-grown cultures remained relatively constant during 100 days of incubation. Approximately 75% of the CF consumed was reductively dechlorinated to DCM, yet no CM or CH_4 formation was observed. Addition of cyanocobalamin

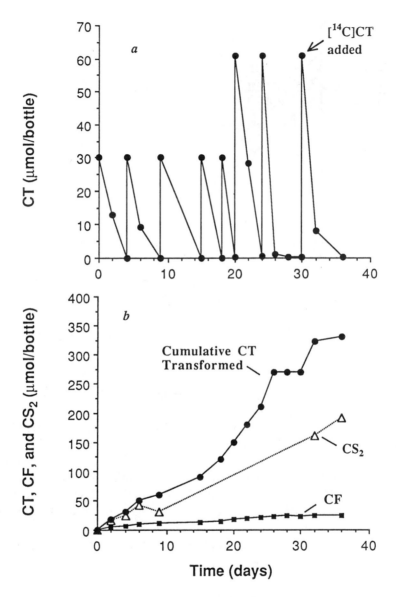

FIGURE 4. **Transformation of CT in an autoclaved lactate-grown enrichment culture supplemented with cyanocobalamin.**

(0.2 μmol on days 0, 16, 28, and 0.4 μmol on day 36) substantially improved the rate of CF transformation (Figure 8). By day 100, the cumulative amount of CF transformed reached 500 μmol. The last addition of CF, taking into account partitioning to the headspace, amounted to an aqueous-phase concentration of 270 mg/L. Prior to this addition, 40% of the CF consumed accumulated as DCM. With the last CF addition, only 9% was reductively dechlorinated, and

TABLE 1. Distribution of ^{14}C in bottles receiving [^{14}C]CT.

Experimental Condition	% of total dpm recovered as:[b]						Soluble	Nonsoluble
	CT	CF	DCM	CS$_2$	CO	CO$_2$	NSR	NSR
CT + cyanocobalamin, live cultures[c]	0.42 (±0.31)	2.7 (±0.60)	0.03 (±0.05)	73. (±2.8)	0.89 (±0.61)	9.1 (±0.61)	9.5 (±0.97)	4.3 (±2.1)
CT + cyanocobalamin, autoclaved cultures[d]	0.68 (±0.04)	4.6 (±0.48)	0.0	78. (±0.45)	3.6 (±0.08)	4.0 (±0.10)	5.7 (±0.34)	3.6 (±0.17)
CT alone, autoclaved cultures[d]	4.8 (±3.6)	38. (±4.1)	0.0	40. (±6.0)	2.0 (±1.6)	3.6 (±0.26)	6.6 (±0.02)	4.8 (±0.16)

[a]Numbers in parentheses are one standard deviation of replicate bottles.

[b]CH$_4$ and CM were analyzed for but not detected in any of the bottles.

[c]Four bottles analyzed.

[d]Two bottles analyzed.

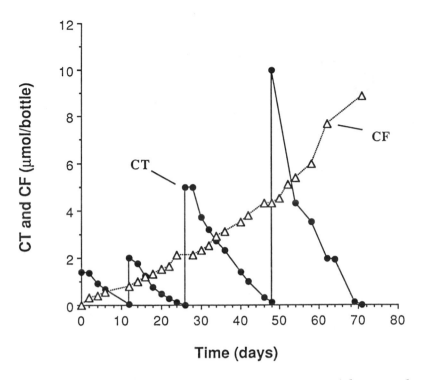

FIGURE 5. Transformation of CT in an acetate-grown enrichment culture.

the molar ratio of total cyanocobalamin present to CF added was less than 0.4%.

The same set of experiments described above with lactate-grown cultures was conducted with acetate-grown enrichments and essentially the same results were observed (figures not shown). Addition of cyanocobalamin substantially increased the rate of CF transformation, while at the same time reducing the extent that was reductively dechlorinated to DCM. Without addition of cyanocobalamin, only 2.3 μmol of CF was consumed over an 80-day period, and 80% was reduced to DCM. Over the same period, bottles that received supplemental cyanocobalamin (a total of 9 μmol) consumed 492 μmol of CF, 24% of which accumulated as DCM.

DISCUSSION

The results of this study demonstrated the potential of using low levels of cyanocobalamin to accelerate the rate of CT and CF transformation by lactate- and acetate-grown sulfate-reducing cultures. Addition of cyanocobalamin also reduced the accumulation of lesser-chlorinated methanes. Rapid transformation of CT and CF was obtained at aqueous-phase concentrations as high as 350 and 270 mg/L, respectively, without the need for an extensive acclimation

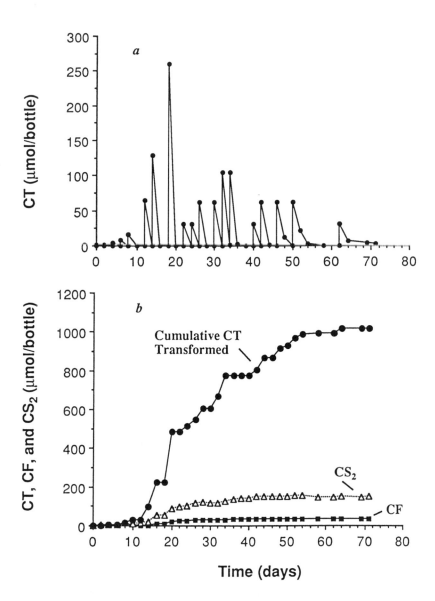

FIGURE 6. **Transformation of CT in an acetate-grown enrichment culture supplemented with cyanocobalamin.**

period. The positive effects we observed with addition of cyanocobalamin are consistent with numerous previous studies (e.g., Gantzer & Wackett 1991; Krone, et al. 1991; Holliger et al. 1992; Stromeyer et al. 1992; Assaf-Anid et al. 1994) that have documented the role of cobalamins in dehalogenation of chlorinated aliphatics.

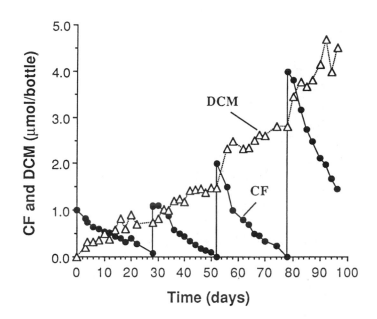

FIGURE 7. Transformation of CF in a lactate-grown enrichment culture.

Although addition of cyanocobalamin significantly increased the rate of CT transformation, the formation of CS_2 as a principal product is undesirable. CS_2 is not currently a regulated drinking water contaminant, yet it is a highly neurotoxic compound. Relatively little is known about the fate of CS_2 under anaerobic conditions. Several methanogens can use CS_2 as a sulfur source for growth (Rajagopal & Daniels 1986). Nevertheless, CS_2 tended to persist in our study, even with extended incubation. Aerobic oxidation of CS_2 and its use as a sulfur source by various *Thiobacillus* species has been demonstrated (Roths-child et al. 1969; Smith & Kelly 1988). This suggests that an anaerobic/aerobic system for CT remediation may be necessary to ensure no accumulation of hazardous products.

One other microbial study previously reported transformation of CT to CS_2, although only trace levels were detected (Criddle et al. 1990). The ^{14}C results from our study (Table 1) indicated that as much as three quarters of the CT was transformed to CS_2. Substantial CS_2 formation occurred only in cyanocobalamin-supplemented bottles, with roughly equal amounts in live and autoclaved cultures. This confirmed that transformation of CT to CS_2 is essentially an abiotic reaction. Kriegman-King and Reinhard (1992) previously demonstrated significant conversion of CT to CS_2 in aqueous solutions of dissolved hydrogen sulfide and biotite or vermiculite. Subsequent hydrolysis of CS_2 to CO_2 was postulated. Chiu and Reinhard (1995) recently observed rapid reductive dechlorination of CT, primarily to CF, in the presence of vitamin B_{12} and titanium citrate as the electron donor. They also observed slower reaction rates and less CF formation when using thio-reductants such as cysteine.

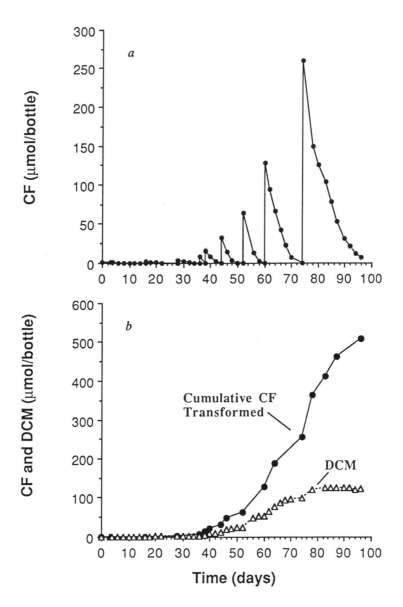

FIGURE 8. Transformation of CF in a lactate-grown enrichment culture supplemented with cyanocobalamin.

Comparing the ^{14}C results in Table 1 with the GC analyses shown in Figures 2, 3, and 4, there is a discrepancy in the extent of CT conversion to CS_2; the radiotracer results indicate that a higher percentage of CT was transformed to CS_2 than the GC headspace data. This prompted us to explore the possibility that a product in addition to CS_2 was being formed but happened to coelute with CS_2. However, GC/mass spectrometric analysis (using a different column) of several headspace samples containing CS_2 essentially ruled this out. A more probable explanation is that the GC measurements underestimated the amount of CS_2 present. Flame ionization was used for quantification, although this is not the method of choice for CS_2. Future analyses using a flame photometric detector are expected to result in better agreement between the two methods.

Accumulation of DCM is also unacceptable from a remediation standpoint. During the experiments with CF, addition of cyanocobalamin increased the rate of transformation, but it did not completely eliminate CF reduction to DCM. The reason for this is not yet known. Because [^{14}C]CF was not used in this study, we were not able to determine products other than DCM, CM, and CH_4. Headspace analysis by flame ionization GC indicated no detectable formation of CM or CH_4. The possibility of CF hydrolysis to CO and oxidation to CO_2 is being examined.

The findings from this study indicate that supplemental addition of cyanocobalamin to anaerobic environments significantly accelerates transformation of CT and CF, while minimizing the formation of lesser chlorinated methanes. Whether or not this approach will ever be viable for field-scale application depends on a number of critical issues, including the presence of other contaminants, the fate of CS_2, and the cost of using cobalamins. At the Hanford site in Washington, for example, high levels of nitrate are present along with CT (Petersen et al. 1994). Treatment under anoxic conditions would therefore be preferable, using a denitrifier such as *Pseudomonas* sp. strain KC (Criddle et al. 1990; Dybas et al. 1995). If sulfate-reducing conditions are developed, CS_2 formation will have to be dealt with, most likely using a subsequent aerobic treatment step.

Field use of cobalamins may be restricted primarily by their high cost (currently >$9 per gram), unless the amount used is kept low. Synthetic cobalamins are not a viable alternative, because they are actually more expensive. Adsorption of B_{12} to aquifer solids is a concern, although our preliminary results indicate B_{12} adsorbs very weakly to kaolinite, alumina, and sand (data not shown). The ratio of B_{12} to CT is also an important parameter. We observed enhancement of CT transformation at ratios below 0.5% (mol B_{12}/mol CT). However, the extent of incubation was not long enough to determine if continued additions of cobalamin are necessary. One way to help minimize the amount of B_{12} needed for in situ application would be to recirculate the treated groundwater, because relatively little consumption of B_{12} is anticipated. The concept of recirculating treated groundwater has been demonstrated by Beeman et al. (1994). They added sulfate and benzoate to recirculating water in order to stimulate in situ sulfate-reducing conditions and reductive dechlorination of tetrachloroethylene.

ACKNOWLEDGMENTS

This research was supported by a grant from the State of Illinois Water Resources Center.

REFERENCES

American Public Health Association. 1985. *Standard Methods for the Examination of Water and Wastewater.* APHA, AWWA, and WPCF.

Assaf-Anid, N., K. F. Hayes, and T. M. Vogel. 1994. "Reductive Dechlorination of Carbon Tetrachloride by Cobalamin(II) in the Presence of Dithiothreitol: Mechanistic Study, Effect of Redox Potential and pH." *Environmental Science & Technology* 28(2): 246-252.

Becker, J. G., and D. L. Freedman. 1994. "Use of Cyanocobalamin to Enhance Anaerobic Biodegradation of Chloroform." *Environmental Science & Technology* 28(11): 1942-1949.

Beeman, R. E., J. E. Howell, S. H. Shoemaker, E. A. Salazar, and J. R. Buttram. 1994. "A Field Evaluation of In Situ Microbial Reductive Dehalogenation by the Transformation of Chlorinated Ethenes." In *Bioremediation of Chlorinated and Polycyclic Aromatic Hydrocarbon Compounds*, R. E. Hinchee, A. Leeson, L. Semprini, and S. K. Ong (Eds.), Lewis Publishers: Boca Raton, FL. pp. 14-27.

Bouwer, E. J., and J. P. Wright. 1988. "Transformations of Trace Halogenated Aliphatics in Anoxic Biofilm Columns." *Journal of Contaminant Hydrology* 2: 155-169.

Chiu, P. C. and M. Reinhard. 1995. "Metallocoenzyme-Mediated Reductive Transformation of Carbon Tetrachloride in Titanium (III) Citrate Aqueous Solution." *Environmental Science & Technology* 29(3): 595-603.

Cobb, G. D., and E. J. Bouwer. 1991. "Effects of Electron Acceptors on Halogenated Organic Compound Biotransformations in a Biofilm Column." *Environmental Science & Technology* 25(6): 1068-1074.

Criddle, C. S., J. T. DeWitt, and P. L. McCarty. 1990. "Reductive Dehalogenation of Carbon Tetrachloride by *Escherichia coli* K-12." *Applied and Environmental Microbiology* 56(11): 3247-3254.

Criddle, C. S., J. T. DeWitt, D. Grbic-Galic, and P. L. McCarty. 1990. "Transformation of Carbon Tetrachloride by *Pseudomonas* sp. strain KC Under Denitrification Conditions." *Applied and Environmental Microbiology* 56(11): 3240-3246.

Dybas, M. J., G. M. Tatara, and C. S. Criddle. 1995. "Localization and Characterization of the Carbon Tetrachloride Transformation Activity of *Pseudomonas* sp. strain KC." *Applied and Environmental Microbiology* 61(2): 758-762.

Egli, C., R. Scholtz, A. M. Cook, and T. Leisinger. 1987. "Anaerobic Dechlorination of Tetrachloromethane and 1,2-Dichloroethane to Degradable Products by Pure Cultures of *Desulfobacterium* sp. and *Methanobacterium* sp." *FEMS Microbiology Letters* 43: 257-261.

Egli, C., S. Stromeyer, A. M. Cook, and T. Leisinger. 1990. "Transformation of Tetra- and Trichloromethane to CO_2 by Anaerobic Bacteria is a Non-Enzymatic Process." *FEMS Microbiology Letters* 68: 207-212.

Egli, C., T. Tschan, R. Scholtz, A. M. Cook, and T. Leisinger. 1988. "Transformation of Tetrachloromethane to Dichloromethane and Carbon Dioxide by *Acetobacterium woodii*." *Applied and Environmental Microbiology* 54: 2819-2824.

Ehrlich, G. G., D. F. Goerlitz, J. H. Bourell, G. V. Eisen, and E. M. Godsy. 1981. "Liquid Chromatographic Procedure for Fermentation Product Analysis in the Identification of Anaerobic Bacteria." *Applied and Environmental Microbiology* 42(5): 878-885.

Freedman, D. L., S. Chennupati, and B. Kim. 1993. "Variability of Henry's Constants with Temperature and the Effect on VOC Emissions at Wastewater Treatment Plants," *Proceedings of the 66th Annual Water Environment Federation Conference*, Anaheim, CA.

Freedman, D. L., and J. M. Gossett. 1989. "Biological Reductive Dechlorination of Tetrachloroethylene and Trichloroethylene to Ethylene under Methanogenic Conditions." *Applied and Environmental Microbiology* 55(9): 2144-2151.

Freedman, D. L., and J. M. Gossett. 1991. "Biodegradation of Dichloromethane and its Utilization as a Growth Substrate Under Methanogenic Conditions." *Applied and Environmental Microbiology* 57(10): 2847-2857.

Freedman, D. L., S. Hashsham, and R. Scholze. 1995. "Enhanced Biotransformation of Carbon Tetrachloride Under Methanogenic Conditions." Submitted to *Environmental Science & Technology*.

Gantzer, C. J. and L. P. Wackett. 1991. "Reductive Dechlorination Catalyzed by Bacterial Transition-Metal Coenzymes." *Environmental Science & Technology* 25(4): 715-722.

Gossett, J. M. 1987. "Measurement of Henry's Law Constants for C_1 and C_2 Chlorinated Hydrocarbons." *Environmental Science & Technology* 21(12): 202-208.

Holliger, C., G. Schraa, E. Stupperich, A. J. M. Stams, and A. J. B. Zehnder. 1992. "Evidence for the Involvement of Corrinoids and Factor F_{430} in the Reductive Dechlorination of 1,2-Dichloroethane by *Methanosarcina barkeri*." *Journal of Bacteriology* 174(13): 4427-4434.

Kriegman-King, M. R., and M. Reinhard. 1992. "Transformation of Carbon Tetrachloride in the Presence of Sulfide, Biotite, and Vermiculite." *Environmental Science & Technology* 26(11): 2198-2206.

Krone, U. E., R. K. Thauer, H. P. C. Hogenkamp, and K. Steinbach. 1991. "Reductive Formation of Carbon Monoxide from CCl_4 and FREONs 11, 12, and 13 Catalyzed by Corrinoids." *Biochemistry* 30: 2713-2719.

Mikesell, M. D., and S. A. Boyd. 1990. "Dechlorination of Chloroform by *Methanosarcina* Strains." *Applied and Environmental Microbiology* 56(4): 1198-1201.

Petersen, J. N., R. S. Skeen, K. M. Amos, and B. S. Hooker. 1994. "Biological Destruction of CCl_4: I. Experimental Design and Data." *Biotechnology and Bioengineering* 43: 521-528.

Rajagopal, B. S., and L. Daniels. 1986. "Investigation of Mercaptans, Organic Sulfides, and Inorganic Sulfur Compounds as Sulfur Sources for the Growth of Methanogenic Bacteria." *Current Microbiology* 14(3): 137-144.

Riley, R. G., J. M. Zachara, and F. J. Wobber (Eds.). 1992. *Chemical Contaminants on DOE Lands and Selection of Contaminant Mixtures for Subsurface Science Research*. U.S. Department of Energy, Office of Energy Research, DOE/ER-0547T. Washington, DC.

Rothschild, B., R. Butler, and J. R. Keller. 1969. "Metabolism of Carbon Disulfide by *Thiobacillus thiooxidans*." Abstract No. GP136, p. 64, 69th Annual Meeting of the American Society for Microbiology, Miami Beach, FL.

Smith, N. A., and D. P. Kelly. 1988. "Oxidation of Carbon Disulfide as the Sole Source of Energy for the Autotrophic Growth of *Thiobacillus thioparus* strain TK-m." *Journal of General Microbiology* 134: 3041-3048.

Stromeyer, S. A., K. Stumpf, A. M. Cook and T. Leisinger. 1992. "Anaerobic Degradation of Tetrachloromethane by *Acetobacterium woodii*: Separation of Dechlorinative Activities in Cell Extracts and Roles of Vitamin B_{12} and Other Factors." *Biodegradation* 3: 113-123.

Thibodeaux, L. J. 1979. *Chemodynamics, Environmental Movement of Chemicals in Air, Water, and Soil*. John Wiley & Sons, Inc., New York, NY.

Widdel, F. and N. Pfennig. 1984. "Dissimilatory Sulfate- and Sulfur-Reducing Bacteria." In J. G. Holt (Ed.), *Bergey's Manual of Systematic Bacteriology*, pp. 663-679, Williams & Wilkins, Baltimore, MD.

Removal of Tetrachloromethane by Granular Methanogenic Sludge

Miriam H. A. van Eekert, Maria C. Veiga,
Jim A. Field, Alfons J. M. Stams, and Gosse Schraa

ABSTRACT

Granular sludge grown on sucrose, methanol, or a volatile fatty acid mixture was found to readily degrade tetrachloromethane without any prior adaptation to this compound. Preliminary experiments showed that degradation rates were comparable for the three different sludges. Tetrachloromethane was degraded by granular sludge grown on methanol, forming chloroform as a transient intermediate which was rapidly degraded to dichloromethane and methyl chloride. Degradation by autoclaved controls indicated the involvement of biologically generated abiotic cofactors such as vitamin B_{12} and cofactor F_{430}. The dechlorination by the autoclaved sludge was inhibited by addition of 1-iodopropane, a selective inhibitor of vitamin B_{12}. Degradation by autoclaved granular sludge was also severely inhibited by addition of H_2O_2 to the medium, which oxidized all reducing equivalents.

INTRODUCTION

Acetogenic and methanogenic bacteria are able to degrade chlorinated compounds such as tetrachloromethane (CCl_4) by reductive dehalogenation (Egli et al. 1990; Gälli & McCarty 1989; Mikesell & Boyd 1990). The reaction is catalyzed by cofactors such as vitamin B_{12} and cofactor F_{430}, which are involved in metabolic pathways such as the acetyl-CoA pathway and the formation of methane. Vitamin B_{12} and cofactor F_{430} have been shown to degrade CCl_4 in abiotic experiments (Krone et al. 1989a, 1989b). The a-specific nature of the dechlorination reactions catalyzed by the biologically generated cofactors could provide broad spectrum applicability to halogenated compounds. The rates at which the xenobiotics are dechlorinated are dependent on cofactor concentrations (Krone 1989b) and therefore are related to the amount of biomass present. Granular anaerobic sludge has a high biomass density and as such might be a suitable source of dechlorinating activity in the bioremediation of contaminated sites.

The objective of this study is to determine the dechlorinating activity of unadapted anaerobic granular sludge and especially the involvement of the abiotic cofactors in the total dechlorinating activity of the sludge.

EXPERIMENTAL PROCEDURES AND MATERIALS

Three different granular methanogenic sludges were grown in upflow anaerobic sludge blanket (UASB) reactors on sucrose, methanol, and a volatile fatty acid (VFA) mixture [acetate : propionate : butyrate = 24:32:42 based on chemical oxygen demand (COD)]. The reactors were operated at a substrate loading rate of 10 kg COD/m^3/day with a hydraulic retention time of 12 hours. The biomass content of the reactors was around 20 g/L volatile suspended solids (VSS). Substrate removal efficiencies were at least 85%. In preliminary experiments, the unadapted granular sludges were screened for their dechlorinating activity using CCl_4 as a model compound.

Batch Experiments

Batch experiments were performed in 20 mL basal medium (pH 7.0 to 7.2) containing 120 mL serum flasks. After adding the sludge (2 g wet weight) the bottles were closed with a viton stopper and aluminum crimp cap. The sludges were washed two times with demineralized water and once with basal medium to remove residual substrate before they were used in the batch experiments. The atmosphere in the batches was N_2/CO_2 (4/1). The basal medium (concentrations given in mM) consisted of K_2HPO_4 (3.75), $NaH_2PO_4 \cdot 2H_2O$ (1.25), NH_4HCO_3 (5.60), $NaHCO_3$ (44.40), $CaCl_2 \cdot 2H_2O$ (0.75), and $MgCl_4 \cdot 4H_2O$ (0.50). Trace elements and vitamins (concentrations in the medium, in µM) were added sterile and anaerobically: EDTA (1.34), $FeCl_2 \cdot 4H_2O$ (10.06), $MnCl_2 \cdot 4H_2O$ (0.51), $CoCl_2 \cdot 4H_2O$ (0.80), $ZnCl_2$ (0.51), $CuCl_2$ (0.02), $AlCl_3 \cdot 6H_2O$ (0.04), H_3BO_3 (0.10), Na_2MoO_4 (0.15), $NiCl_2 \cdot 6H_2O$ (0.10), biotin (0.20), p-aminobenzoate (1.57), pantothenate (0.21), folic acid (0.04), lipoic acid (0.24), pyridoxine (0.59), nicotinamide (4.50), thiamine HCl (0.30), riboflavin (0.13), and cyanocobalamin (0.04). The appropriate cosubstrate (1.5 g COD/L) and tetrachloromethane were added. The batches were incubated statically at 30°C. The gas phase of the batches was analyzed periodically for chlorinated compounds, H_2, and CH_4. The loss of CCl_4 due to leaking through stoppers was checked in separate batches with medium (no addition of sludge).

If needed, the granular sludge was autoclaved in the basal medium (30 to 35 minutes, 120°C). The inhibitors, 1-iodopropane (1-IP) and H_2O_2, were added after autoclaving the sludge. The inhibitors (concentrations 100 mM and 3.5% in the water phase for 1-IP and H_2O_2, respectively) were allowed to react for 2 hours at 30°C before the chlorinated compound was added. The possible effect of 1-iodopropane and H_2O_2 on chemical transformation of CCl_4 was checked in separate controls without sludge.

Analysis

Chlorinated methanes were analyzed by injecting 200 µL gas from the headspace into a CP9000 gas chromatograph (GC) equipped with a flame ionization detector (FID) connected to a Sil 5CB column (25 m × 0.32 mm × 1.2 µm) and a splitter injector (ratio 1:50). Operating temperatures of the injector, column, and detector were 250, 50, and 300°C, respectively. Carrier gas was N_2 with an inlet pressure of 50 kPa. The retention times and peak areas were determined by a Spectra Physics SP 4290 integrator. The retention times were 5.3, 3.8, 2.5, and 1.7 minutes and the detection limits 30, 12, 38, and 30 nmol/batch for CCl_4, chloroform, dichloromethane, and methyl chloride, respectively.

Hydrogen and methane were determined by injecting 0.4 mL gas from the headspace in a 417 Packard GC equipped with a thermal conductivity detector (TCD) (100 mA) connected to a molecular sieve column (13X, 180 × ¼", 60 to 80 mesh). The temperature of the detector and the column were 100°C.

VFA concentrations were determined after centrifuging the sample at 13,000 rpm for 5 minutes, and diluting and conserving the supernatant with 3% formic acid solution. The (diluted) sample was injected into a 417 Packard GC using a FID connected to a glass column (10% Fluorad FC 431 coating, 100 to 120 mesh, 6 m × 2 mm). The temperatures of the injector, oven, and detector were 210, 130, and 230°C, respectively.

The COD for the sucrose, methanol, and VFA solutions was determined according to standard methods (American Public Health Association 1985). COD conversion factors (g/g) utilized were 1.07, 1.50, 1.07, 1.52, and 1.82 for sucrose, methanol, acetate, propionate, and butyrate, respectively.

The VSS of the sludges was determined by subtracting the ash-content from the dry weight after incubating the sludge overnight at 105°C. The ash content was determined after heating the dry sludge at 600°C for 90 minutes.

RESULTS

The degradation rates of CCl_4 by the unadapted granular sludges grown in the UASB reactors on different substrates are shown in Figure 1. Cosubstrate (glucose, VFA, and methanol for the granular sludge grown on sucrose, VFA and methanol, respectively) after addition was readily converted to methane. Because methanol-grown sludge appeared to be slightly faster in dechlorinating CCl_4, the following experiments were carried out mainly with sludge grown on methanol.

Methanol-grown granular sludge degrades CCl_4 to chloroform, which is readily converted to dichloromethane and methyl chloride (Figure 2). No cosubstrate was added in this case; this to determine to effect of the addition of cosubstrate. The absence of the cosubstrate is probably the reason for the incomplete degradation, because when cosubstrate was added complete degradation of CCl_4 by the sludge occurred. When added, all cosubstrate was rapidly converted to methane within 5 days. Hydrogen was not detected. Higher concentrations

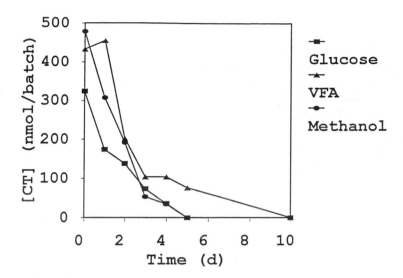

FIGURE 1. Degradation of CCl_4 (CT) by granular sludge grown on different substrates.

of CCl_4 (>0.2 mM) led to inhibition of methanogenesis, but degradation of CCl_4 continued.

To determine the importance of enzymatic cofactors in the dechlorinating process, the sludge was autoclaved and incubated with CCl_4. By autoclaving the sludge all biological activity in the sludge is eliminated. The abiotic cofactors, which are heat-stabile, remain active. Autoclaved controls also showed dechlorinating activity (Figure 3), with rates comparable to the rates of living sludge. Autoclaved sludge converted CCl_4 to chloroform, which was further degraded to unknown products. Dichloromethane and methyl chloride were not formed. CCl_4 was not degraded in batches without sludge.

The involvement of vitamin B_{12} was confirmed by deactivating this cofactor with 1-iodopropane, which resulted in a severe inhibition of CCl_4 dechlorination (data not shown). Dehalogenation was also severely inhibited after oxidizing the reducing equivalents with hydrogen peroxide.

DISCUSSION

The unadapted anaerobic sludges were able to degrade CCl_4 rapidly via reductive dehalogenation. Although no significant differences in dechlorination rates could be detected (so far) among the sludges, methanol-grown sludge was slightly more effective. Mass transfer limitations between the aqueous and the gas phase could have been responsible for the lack of differences in biodegradation rates between the different sludges, because the batches were incubated statically.

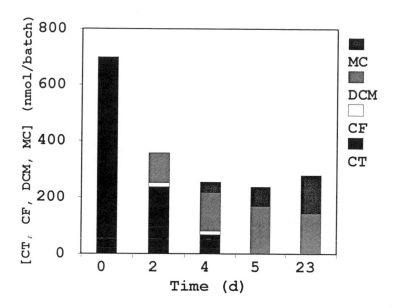

FIGURE 2.   Degradation of CCl_4 (CT) and formation of chloroform (CF), dichloromethane (DCM), and methyl chloride (MC).

Tetrachloromethane was degraded by the living sludges via chloroform and dichloromethane as transient intermediates to methyl chloride. Possibly some methyl chloride was converted to methane, but the amount formed would be too low to distinguish from the background methane production. Possible formation of CO_2 could not be assessed because a bicarbonate buffer was used. The autoclaved sludge also degraded CCl_4 and chloroform and some unknown products were formed. A possible degradation product formed by the autoclaved sludge could be CO_2. The formation of CO_2 from CCl_4 by abiotic oxidation reactions under anaerobic conditions has been described before (Egli et al. 1990).

Addition of a cosubstrate did not seem to have a stimulatory effect on the degradation rates of CCl_4. Degradation by the methanol sludge with cosubstrate (Figure 1) occurred as fast as the conversion by the consortium without added cosubstrate (Figure 2). We expected this, because reducing equivalents needed for dechlorination were most likely present in the sludge granules compared to the relatively low CCl_4-concentration used. Addition of extra cosubstrate might, however, have helped to degrade dichloromethane and methyl chloride more completely.

Autoclaved controls of methanol-grown sludge also showed dechlorinating activity, indicating the involvement of abiotic cofactors such as vitamin B_{12} and cofactor F_{430}. Spore-forming acetogenic bacteria in granular sludge, however, are known to survive pasteurization treatments such as 90°C for 30 minutes or 100°C for 10 minutes (Stams et al. 1992). Thus, the dechlorinating activity of the autoclaved sludge could also have been due to acetogenic bacteria arising

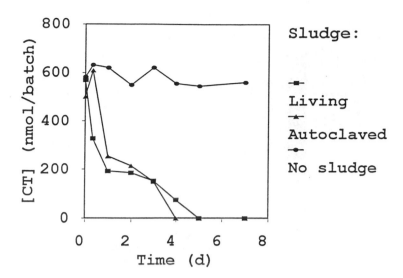

FIGURE 3. Degradation of CCl$_4$ (CT) by living and autoclaved granular sludge grown on methanol.

from spores that survived autoclaving. However, when spore-forming bacteria alone would have been involved in the degradation process, the dechlorinating activity would have been substantially lower compared to the sludge that had not been autoclaved, since it is very unlikely that all spore-forming bacteria survive the heat-treatment. Thus the abiotic co-factors must have at least been partly responsible for the degradation process. The possibility that spore-forming bacteria can even survive the autoclaving treatment is being investigated.

CONCLUSIONS

The degradation of tetrachloromethane by unadapted methanogenic consortia proceeds rapidly via reductive dechlorination to methyl chloride. Because autoclaved sludge also had dechlorinating activity, dehalogenation probably is caused in part by abiotic cofactors such as vitamin B$_{12}$ and cofactor F$_{430}$, that are present in the biomass of anaerobic sludge.

REFERENCES

American Public Health Association. 1985. *Standard methods for the examination of water and wastewater*, 16th ed. APHA, Washington, DC.

Egli, C., S. Stromeyer, A. M. Cook, and T. Leisinger. 1990. "Transformation of tetra- and trichloromethane to CO$_2$ by anaerobic bacteria is a non-enzymic process." *FEMS Microbiology Letters 68*: 207-212.

Gälli, R., and P. L. McCarty. 1989. "Biotransformation of 1,1,1-trichloroethane, trichloromethane, and tetrachloromethane by a *Clostridium* sp." *Applied and Environmental Microbiology 55*(4): 837-844.

Krone, U. E., R. K. Thauer, and H.P.C. Hogenkamp. 1989a. "Reductive dehalogenation of chlorinated C_1-hydrocarbons mediated by corrinoids." *Biochemistry 28*(11): 4908-4914.

Krone, U. E., K. Laufer, R. K. Thauer, and H.P.C. Hogenkamp. 1989b. "Coenzyme F_{430} as a possible catalyst for the reductive dehalogenation of chlorinated C_1-hydrocarbons in methanogenic bacteria." *Biochemistry 28*(26): 10061-10065.

Mikesell, M. D., and S. A. Boyd. 1990. "Dechlorination of chloroform by *Methanosarcina* strains." *Applied and Environmental Microbiology 56*(4): 1198-1201.

Stams, A.J.M., K.C.F. Grolle, C.T.M.J. Frijters, and J.B.M. van Lier. 1992. "Enrichment of a thermophilic propionate-oxidizing acetogenic bacterium in coculture with *Methanobacterium thermoautotrophicum* or *Methanobacterium thermoformicicum*." *Applied and Environmental Microbiology 58*: 346-352.

Surfactant-Enhanced Anaerobic Bioremediation of a Carbon Tetrachloride DNAPL

M. D. Lee, G. E. Gregory III, D. G. White,
J. F. Fountain, and S. H. Shoemaker

ABSTRACT

Two nonionic surfactants, Witconol 2722 and Tergitol 15-S-12, were used in a surfactant-flushing demonstration to remove carbon tetrachloride (CT) present as residual dense, nonaqueous-phase liquid (DNAPL) in a shallow water-bearing sand at a facility in Texas. A black coloration of the surfactant-flushed sand was noted during intermediate and final soil sampling. The color coincided with zones where almost all of the CT had been removed. This color change was attributed to the anaerobic biodegradation of the surfactants by sulfate-reducing bacteria. Subsequent groundwater analyses have shown an increase in the concentrations of dechlorination by-products of CT and a decrease in surfactant concentrations. Laboratory microcosm studies with site soil and groundwater showed that the surfactants increased the removal of CT and the conversion to daughter products including chloroform, dichloromethane, and carbon disulfide compared to poisoned controls. Reductive dechlorination accompanies the surfactant flushing process and enhances the removal of carbon tetrachloride.

INTRODUCTION

Soil flushing with surfactants offers the potential to increase the quantity of trapped DNAPL that can be removed over that removed by pump-and-treat remediation strategies (West & Harwell 1992). Surfactants solubilize DNAPLs by forming micelles into which the DNAPL partitions, thereby allowing the DNAPL to move with the groundwater. Surfactants can also decrease the capillary and surface tension forces that trap DNAPLs in soil pores, which then mobilize the DNAPLs.

Surfactants can also promote the anaerobic dechlorination of chlorinated compounds by making them more soluble and potentially more available for microbial attack. Van Hoof and Rogers (1992) found that at critical micellar concentrations (CMC) or lower, Tween 80, a nonionic surfactant, enhanced the anaerobic dechlorination of hexachlorobenzene. At concentrations well above the CMC, Tween 80 initially promoted dechlorination of hexachlorobenzene, but the activity ceased after 40 days. Other nonionic surfactants, including Tween 85 and Brij 30, did not enhance dechlorination. Abramowicz et al. (1993) determined that the surfactant Triton X-705 decreased the lag period and increased the extent of polychlorinated biphenyl dechlorination in Hudson River sediments over the live controls. Anaerobically biodegradable surfactants could be used as an organic substrate to support reductive dechlorination of chlorinated solvents.

The objective of this study was to conduct a pilot demonstration of surfactant-flushing of a chlorocarbon DNAPL. Field observations led to the conclusion that the surfactant was being biodegraded and that the surfactant biodegradation supported the degradation of the DNAPL. Laboratory studies were initiated to confirm the observations from the field.

FIELD PILOT

A surfactant-flushing pilot study was carried out by Fountain and Waddell-Sheets (1993) at a former chlorocarbons manufacturing facility where the shallow sand unit is contaminated with residual-phase CT. A 3.7-m-thick fractured clay unit overlies the 3.7-m-thick sand unit, which in turn overlies a 20-m-thick clay aquitard. The sand unit in the area of the pilot is composed of a very well-sorted, fine-grained sand with clay contents ranging between 5 to 15% clay in the less permeable upper 0.9 m and less than 5% clay in the lower, more permeable 2.6 m. The hydraulic conductivity of the upper sand is 1.3×10^{-3} cm/s and the lower sand is 3.9×10^{-3} cm/s. The pilot was conducted in a 6.1-m by 9.1-m plot. The equipment used in the pilot consisted of six delivery wells on the outer edges of the plot, two extraction wells in the interior of the plot, five monitoring wells, tanks for the surfactant solutions, and two air strippers. The extracted contaminants were removed from the groundwater by the air strippers and captured on activated carbon. The surfactant solutions were reinjected after the surfactant concentrations were adjusted to 1%.

Witconol 2722, a nonionic sorbitan monooleate surfactant, was used initially (Fountain & Waddell-Sheets 1993). The CMC for Witconol is approximately 0.001%. Very little of the 1% solution of the Witconol was detected in the extraction wells after the equivalent of nine pore volumes (approximately 76,000 L) of the surfactant solution had been pumped through the aquifer. The Witconol did not increase the CT concentrations in the extracted groundwater. The Witconol was apparently biodegraded and/or sorbed onto the aquifer solids.

Another surfactant, Tergitol 15-S-12, was substituted for the Witconol (Fountain & Waddell-Sheets 1993). Tergitol is a nonionic, straight-chain alcohol ethoxylate, which had a lower sorption affinity for the aquifer solids based on laboratory soil column experiments. The 1% solution of Tergitol reached the extraction well after 15 days and initially increased the CT concentrations in the extracted water from 410 to 680 mg/L. By contrast, up to 20,000 mg/L of CT was solubilized in a 1% Tergitol solution in laboratory experiments with free-phase CT. The small increase in the concentrations of CT in the extracted groundwater was attributed to the limited distribution of residual-phase CT in the flushed zone and consequent dilution of the extracted CT in the overall groundwater flow regime.

Soil samples were collected from three cores in the pilot plot during the surfactant flushing with Tergitol and at the end of the study (Fountain 1993). High CT concentrations and low total organic carbon (TOC) concentrations were found in the samples from the upper sand section of Core C3, which had a higher clay content and had been less extensively flushed with the surfactant. Elevated TOC concentrations indicate the presence of the surfactant. CT concentrations were below 10 mg/kg in the majority of the samples from the lower sand section of Core C3 which contained less clay. Two zones in core C3 of less than 0.6 m in thickness contained CT at residual-phase levels. Core A, collected at the end of the test in an area that had received an estimated three pore volumes of surfactant, showed complete removal of CT in the lower sand below 4.8 m. Core B, collected in an area flushed with both surfactants, contained 2,200 mg/kg CT above the surfactant-flushed zone and less than 40 mg/kg CT in the surfactant-flushed zone.

A black coloration of the surfactant-flushed sand indicating highly reduced conditions was noted and coincided with many of the zones where almost all of the CT had been removed. This color change was attributed to the anaerobic biodegradation of the surfactants by sulfate-reducing bacteria and the production of a ferrous sulfide precipitate. Immunoassay testing found greater than 10^6 sulfate-reducing bacteria per mL in groundwater taken from the surfactant-flushed zone.

Groundwater analyses of the wells in the pilot area over the 19 months since the pilot ended show average decreases of 74% and 98% in the concentrations of CT and the surfactant, respectively (Table 1). Chloroform (CF) and dichloromethane (DCM), which are dechlorination by-products of carbon tetrachloride, increased to an average of 0.19 millimolar (mM). In addition, an average of 0.11 mM of carbon disulfide (CDS) was detected at the last sampling event. Kriegman-King and Reinhard (1994) reported abiotic transformation of CT to CDS under anaerobic conditions in the presence of pyrite. Pyrite is an iron sulfide mineral produced under sulfate-reducing conditions (Atlas & Bartha 1993). The reductive dechlorination of CT to CF, DCM, and CDS may be catalyzed by the pyrite rather than being a direct biological reaction. However, the dechlorination reaction involves microbial processes because the activity of sulfate-reducers in necessary to produce the pyrite.

LABORATORY EXPERIMENTS

Methods

Duplicate 556-mL serum bottle microcosms were set up with CT, 10% soil, and 90% groundwater from the surfactant-treated zone of the site. The following treatments were evaluated: 0.1% Witconol 2722, 0.1% Tergitol 15-S-12,

TABLE 1. Average concentrations of CT, CF, DCM, CDS, and surfactant in pilot area wells following surfactant flushing.

Month Units	CT mM	CF mM	DCM mM	CDS mM	Surfactant %
0	0.86				1.0
9	0.42	0.057	0.001		0.20
19	0.22	0.15	0.036	0.11	0.016

poisoned control (0.1% mercuric chloride), and unamended with no additional surfactant. The concentrations of CT, CF, DCM, CDS, total organic carbon (TOC), and sulfide were followed over 118 days by subsampling the microcosms.

Results

Table 2 summarizes the results of the microcosm tests. The average removals of CT in the duplicates of the unamended, 0.1% Tergitol-amended, and 0.1% Witconol-amended treatments were greater (73 to 78 %) than the control treatment (52%). Volatilization of CT into the headspace in the bottles and losses during sampling were responsible for a portion of the CT losses observed in all treatments. The unamended treatment contained 140 mg/L of TOC at the beginning of the study. This TOC represents residual surfactant in the groundwater at the pilot area and supported CT degradation in the unamended treatment.

Much more of the CT was converted to daughter products including CF and CDS in the Witconol- and Tergitol-amended treatments than the control treatment at the end of the study. The unamended treatment showed less conversion to daughter products (34 %). Dichloromethane was detected at times during the study in all treatments, but was not found in any of the Day 118 samples.

Sulfate-reducing bacteria were found in all treatments based upon the detection of sulfide. Sulfide was detected in the control treatment (a maximum of 0.03 mM), which suggested that adding 0.1% mercuric chloride did not completely sterilize it. The sulfide concentrations were higher in the unamended treatment than the Tergitol-amended or Witconol-amended treatments.

CONCLUSIONS

Field and laboratory data showed that nonionic surfactants used for surfactant-flushing promoted the removal and anaerobic degradation of a carbon tetrachloride DNAPL at the facility. The combination of surfactant flushing

TABLE 2. Average % initial CT removed, average % initial CT converted to daughter products (CF, DCM, and CDS), and maximum sulfide produced during microcosm study.

Treatment	% CT Removed	% Daughter Products Formed	Maximum Sulfide (mM) Produced
Control	52	2	0.03
Unamended	76	34	0.23
0.1% Tergitol 15-S-12	78	44	0.18
0.1% Witconol 2722	73	40	0.05

and anaerobic biodegradation has considerable potential for remediation of DNAPL-contaminated sites. Additional studies are needed to understand this process. The selection of the surfactant is of particular importance because DNAPL removal efficiencies will vary with the surfactants. The criteria for selecting surfactants to use for enhancement of the anaerobic biodegradation of DNAPLs should include the following: anaerobically biodegradable, nontoxic, nonionic or anionic, readily solubilize the DNAPL constituents, moderate to low adsorption to the soil, moderate interfacial tension reduction, nonfoaming, and low cost.

REFERENCES

Abramowicz, D. A., M. J. Brennan, H. M. van Dort, and E. L. Gallagher. 1993. "Factors Influencing the Rate of Polychlorinated Biphenyl Dechlorination in the Hudson River Sediments." *Environmental Science and Technology* 27(6): 1125-1131.

Atlas, R. M. and R. Bartha. 1993. *Microbial Ecology. Fundamentals and Applications.* 3rd ed. The Benjamin/Cummings Publishing Company, Inc., Redwood City, CA.

Fountain, J. C. 1993. *Extraction of Organic Pollutants Using Enhanced Surfactant Flushing: Pilot Scale Field Trial at Corpus Christi, Texas. Project Summary.* New York State Center for Hazardous Waste Management, SUNY at Buffalo, NY.

Fountain, J. C. and C. Waddell-Sheets. 1993. "A Pilot Field Test of Surfactant Enhanced Aquifer Remediation: Corpus Christi, Texas." *Preprint Extended Abstracts.* pp. 943-946. Presented at the I&EC Special Symposium, American Chemical Society, Atlanta, GA, September 27-29, 1993.

Kriegman-King, M. R. and M. Reinhard. 1994. "Transformation of Carbon Tetrachloride by Pyrite in Aqueous Solution." *Environmental Science and Technology* 28:692-700.

Van Hoof, P. L. and J. E. Rogers. 1992. "Influence of Low Levels of Nonionic Surfactants on the Anaerobic Dechlorination of Hexachlorobenzene." In *Biosystems Technology Development Program. Bioremediation of Hazardous Wastes.* EPA/600/R-92/126. pp. 105-106. U.S. EPA, Washington, DC.

West, C. C. and J. H. Harwell. 1992. "Surfactants and Subsurface Remediation." *Environmental Science and Technology* 26:2324-2330.

PCE Treatment in Saturated Soil Columns: Methanogens in Sequence with Methanotrophs

Sam Fogel, Ronald Lewis,
Daniel Groher, and Margaret Findlay

ABSTRACT

Perchloroethylene (PCE) was reductively dechlorinated under anaerobic conditions in a continuous-flow, saturated soil column. The column was inoculated by a culture known to dechlorinate PCE to ethylene. Methanol was used as the source of carbon and energy. Column conditions were modified to optimize the reaction time and conversion rate of PCE to ethylene. After 200 days of continuous operation, PCE was converted to vinyl chloride (60%) and ethylene (40%) after 2 h of contact time. Sequential treatment by an aerobic, methanotrophic column was used for treatment of vinyl chloride produced during reductive dechlorination of PCE. Field application for the use of this technology for in situ treatment of PCE-contaminated saturated soils and aquifers is discussed.

INTRODUCTION

Biological treatment methods provide a potentially low-cost alternative to conventional methods for the remediation of saturated soils contaminated with chlorinated solvents. PCE is among the most common groundwater contaminants in the United States, due to its wide use in dry cleaning and as an industrial degreasing agent. Although PCE is not biodegraded by aerobic bacteria, it is biotransformed anaerobically by reductive dechlorination to trichloroethylene (TCE), 1,2-dichloroethylene (DCE), vinyl chloride (VC), ethene, and ethane (Bouwer and McCarty 1983, Fathepure and Boyd 1988, Danna et al. 1989, DiStefano et al. 1991, Hollinger et al. 1993, and Tandoi et al. 1994).

A number of studies have focused on the extent of biological dechlorination of PCE. For example, studies by Freedman and Gossett (1989), using a mixed culture, obtained both VC and ethylene as products of anaerobic PCE degradation. Complete dechlorination of PCE to ethylene in an enrichment culture was reported by DiStefano et al. (1991), whereas complete dechlorination of PCE to ethane was achieved in a continuous-flow, fixed-bed column by de Bruin et al. (1992).

Using a high-rate dechlorinating culture provided by Dr. J. Gossett (Cornell University) in a continuous-flow, saturated soil column, we measured the formation of PCE dechlorination products and modified column conditions in an attempt to optimize the conversion rate of PCE to ethylene. We employed a second aerobic-methanotrophic column for treatment of VC produced during reductive dechlorination of PCE.

EXPERIMENTAL PROCEDURES AND MATERIALS

The bench-scale soil column treatment system is displayed in Figure 1. Two 5-cm-diameter, 46-cm-long glass columns were packed with a mixture of medium and coarse sand (0.125 mm to 0.25 mm, 0.25 mm to 0.5 mm). Anaerobic conditions were created in the first column to simulate an anaerobic zone in an aquifer, and aerobic conditions were fostered in the second column to simulate a downgradient aerobic zone in the aquifer.

A PCE-degrading mixed culture derived from a sewage treatment plant digester was obtained from Dr. James Gossett. This culture could degrade a high concentration (100 mg/L) of PCE to ethylene within several days in a serum bottle (35°C). This study was designed to determine the capabilities of this culture at ambient temperatures (20°C) in a soil matrix. The culture was initially added to the anaerobic soil column by periodically injecting 10 mL of

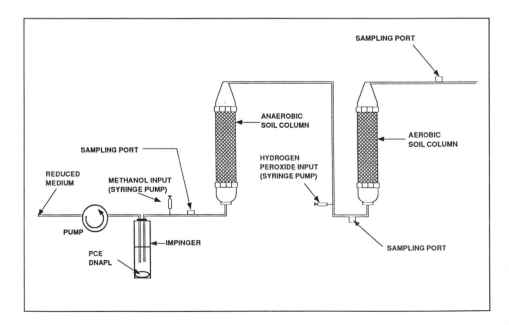

FIGURE 1. Schematic of sequential anaerobic/aerobic test apparatus.

concentrated culture to the column influent over several weeks during which the column was operated in a recirculation mode. After the culture was established in the anaerobic soil column, column operation was switched from recirculation to a continuous flowthrough mode. The second column was seeded with a methanotrophic culture by mixing into the soil during packing. To aerobically degrade the anaerobic products in the second column, oxygen was introduced by adding hydrogen peroxide to the anaerobic column effluent.

Analytical Procedures

Aqueous concentrations of PCE, TCE, DCE, and VC were determined by U.S. Environmental Proctection Agency (EPA) Method 8010/8020 (purge and trap with Hall and photoionization detection) with associated quality assurance procedures outlined in the EPA standard of work (SW) 846. Aqueous samples (5 μL to 500 μL) were collected by syringe, diluted to 5 mL with laboratory water and analyzed immediately, without storage or preservation. Gaseous methane and ethylene were analyzed by direct injection of a 100-μL sample of headspace equilibrated with liquid effluent, via a gastight syringe onto a gas chromatograph with flame ionization detection (GC/FID) (HP 5880, Hewlett Packard Co.). A packed glass chromatographic column, 183 cm by 4 mm i.d., with Carbosieve II support (Supelco, Inc.) was used for separation of methane and ethylene. Calibration was carried out using a 1% calibration gas (Scott Specialty Gases).

Concentrations were determined by external standard method. The concentrations of methane and ethylene were determined by filling a 1.5-mL vial with the water sample, sealing the vial with a screw top cap with septum, then removing 0.5 to 1.0 mL of sample and displacing the volume with air. The vial was shaken for several minutes to allow partitioning of the dissolved gases between the water and the headspace, after which the headspace was analyzed as described above. The aqueous concentrations were calculated using experimentally determined Henry's law constants of 23.3 and 7.6 for methane and ethylene, respectively.

RESULTS

Batch/Serum Bottle Tests

It was important to demonstrate that the dechlorinating culture obtained from Dr. Gossett would, in our laboratory, dechlorinate PCE to ethylene at room temperature. The dechlorinating characteristics of the test culture using methanol as the electron donor at 20°C is shown in Figure 2. PCE was converted to VC and ethylene with no detectable TCE or DCE. By 50 h, ethylene was the dominant transformation product. Vinyl chloride accounted for approximately 10%.

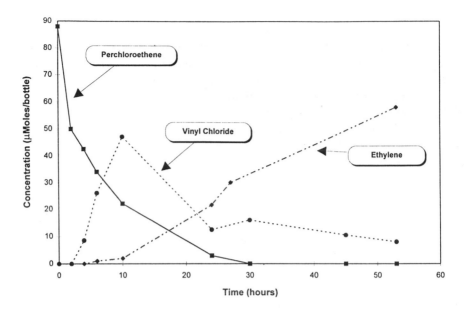

FIGURE 2. PCE dechlorination in a serum bottle over 52 h.

Electron Donors

Four electron donors (methanol, acetate, glucose, and hydrogen) were tested for their effect on PCE degradation and product formation on the test culture. When fed methanol, PCE was degraded to VC and then to ethylene, whereas feeding acetate resulted in the production of only TCE and DCE and some VC. Feeding hydrogen and acetate initially produced a mixture of VC and ethylene; after 10 days, ethylene production ceased and TCE and DCE began to accumulate. Glucose was able to sustain ethylene production for awhile, but at a slower degradation rate than methanol. After 40 days, glucose-fed cultures no longer produced ethylene, but began to accumulate TCE and DCE.

Anaerobic Soil Column: Startup Recirculating Conditions

The purpose of this period was to acclimate the culture to soil and ambient temperature conditions. The anaerobic column was inoculated with PCE degraders and operated under recirculating conditions with intermittent additions of PCE and methanol for 90 days at 20°C. DCE was the sole transformation product at the end of this period.

Anaerobic Soil Column: Continuous-Flow Conditions

At the end of the startup period, groundwater containing PCE and methanol was pumped continuously through the column. Figure 3 shows the molar balance of PCE and its dechlorinating products during the first 2 months

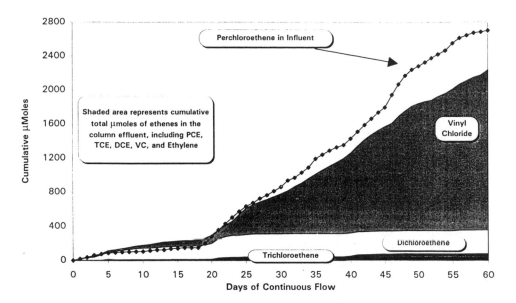

FIGURE 3. Cumulative μmoles of chlorinated ethenes detected in influent and effluent of anaerobic column.

of continuous-flow operation with a contact time of 8 h. The accumulative moles of influent PCE are compared to the accumulative moles of effluent ethenes. PCE, TCE, *cis*-DCE, and VC were detected in the column effluent for the first 17 days. By day 30, VC was the only chlorinated ethene detected in the column effluent and by day 50, PCE was completely accounted for by VC (90%) and ethylene (10%). After 200 days of operation, the dechlorinating efficiency of the column increased until 20 mg/L PCE could be converted to 60% VC and 40% ethylene with 2 h of contact time. During this period, PCE degradation to VC and ethylene was achieved at aqueous PCE concentrations of 5 to 150 mg/L.

Aerobic Soil Column

When the effluent from the anaerobic column was oxygenated and pumped through the aerobic column under methanotrophic conditions, the PCE breakdown products were degraded. Hydrogen peroxide injection was initiated at a concentration of 400 mg/L. At the start of the test, the anaerobic column effluent contained PCE, TCE, *c*DCE, and VC. The sequential anaerobic/aerobic soil columns were operated for 60 days, with contact times of 8 h each. After 30 days, only VC was still detected in the anaerobic soil column effluent and 40 to 50% of the VC was removed by the aerobic column. Evidence of methanotrophic degradation was indicated by the periodic detection of *cis*-DCE epoxide in samples in which higher concentrations of *cis*-DCE were present. The *cis*-DCE epoxide is a transient intermediate product of *cis*-DCE oxidation by methanotrophs (Leahy et al. 1985). The methane produced by the methanogens in the anaerobic column appeared to be sufficient to maintain the methanotrophic bacteria in the aerobic column. The results from the aerobic column operation (8 h aqueous retention time) are given in Figure 4.

Column Culture Characteristics

The "Gossett Culture" was readily acclimated to soil and maintained its dechlorinating ability when grown under continuous-flow conditions for a period of more than 200 days. We have also obtained evidence that methanotrophic bacteria were present in the anaerobic column and that the PCE-degrading culture, after months of acclimation in the column, was not readily killed by exposure to air. Evidence for both aerotolerance of methanogens and the coexistence of methanotrophs with methanogens is based on incubation of effluent from the anaerobic column in serum bottles for 8 days in the presence of oxygen. Methane concentrations in the headspace initially fell which is consistent with the presence of methanotrophs. When methanol and PCE were added to this serum bottle, methane was produced and PCE was dechlorinated to VC over a period of 20 days. Similar observations were reported by Enzien et al. (1994).

DISCUSSION

PCE was degraded to a nonhazardous end product (ethylene) with acceptable reaction rates for in situ treatment. PCE was dechlorinated to 10%

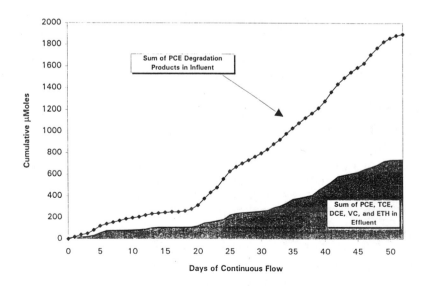

FIGURE 4. Cumulative μmoles of chlorinated ethenes detected in influent and effluent of aerobic column.

ethylene and 90% VC with a contact time of 8 h, 60 days after inoculation. Assuming first order kinetics for the conversion of VC to ethylene, 22 days of contact time would be required for 99.9% conversion to ethylene. After one year of continuous operation, 40% of the VC could be converted to ethylene in 2 h contact time, giving a calculated contact time required for 99.9% conversion to ethylene of 27 h. The increase in the dechlorinating efficiency of the column is likely attributable to a progressive establishment of the microbial community within the soil.

Potential Field Application

The "sequential" nature of the anaerobic and aerobic treatments can be applied either in "time" or in "space." *Time-sequence* treatment was simulated in a laboratory-saturated soil bioreactor (Fogel 1988) (Danna et al. 1989), in which the same soil was treated first anaerobically under recirculating conditions to convert PCE completely to DCE, then, with the introduction of oxygen and continued recirculation of water, treated aerobically by methanotrophs to mineralize the DCE. This approach would be appropriate for batch treatment of contaminated soils.

Space-sequence treatment, referred to as "Two Zone Plume Interception," could be used to stop the forward migration of a PCE plume by providing an anaerobic zone downgradient of the leading edge of the plume, and an aerobic zone further downgradient. Each treatment zone would be established by a line of closely spaced wells placed perpendicular to the expected path of the plume, the first injecting the carbon source and minerals for methanogens, and the second injecting the oxygen required for the methanotrophs.

ACKNOWLEDGMENT

This research was funded under grant # CR- 816820-01-0 from U.S. EPA's Superfund Innovative Technology Evaluation (SITE) Emerging Technology Program to ABB-Environmental Services, Wakefield, Massachusetts. Alex Vira carried out the day-to-day operations and provided important input to the experimental design and data analysis for this project. Deborah Hogan also provided daily laboratory assistance and maintained all methanogenic and methanotrophic cultures.

REFERENCES

Bouwer, E.J., and P.L. McCarty. 1983. "Transformations of 1- and 2-carbon halogenated aliphatic compounds under methanogenic conditions." *Appl. Environ. Microbiol.* 45:1286.

Danna, M., S. Fogel, and M. Findlay. 1989. "Sequential anaerobic/aerobic biodegradation of chlorinated ethenes in an aquifer simulator." *International Symposium on Processes Governing the Movement and Fate of Contaminants in Subsurface Environments*, Stanford University, Stanford, CA. July 23.

de Bruin, W.P., M. Kotterman, M.A. Posthumus, G. Schraa, and A. Zehnder. 1992. "Complete biological reductive transformation of tetrachloroethene to ethane." *Appl. Environ. Microbiol.* 58:1996.

DiStefano, T.D., J.M. Gossett, and S.H. Zinder. 1991. "Reductive dechlorination of high concentrations of tetrachloroethylene to ethene by an anaerobic enrichment culture in the absence of methanogenesis." *Appl. Environ. Microbiol.* 57:2287.

DiStefano, T.D., J.M. Gossett, and S.H. Zinder. 1992. "Hydrogen as an electron donor for dechlorination of tetrachloroethene by an anaerobic mixed culture." *Appl. Environ. Microbiol.* 58:3622.

Enzien, M.V., F. Picardal, T.C. Hazen, R.G. Arnold, and C.B. Fliermans. 1994. "Reductive Dechlorination of Trichloroethylene and Tetrachloroethylene under Aerobic Conditions in a Sediment Column." *Appl. Environ. Microbiol.* 60:2200.

Fathepure, B.Z., and S.A. Boyd. 1988. "Reductive dechlorination of perchloroethylene and the role of methanogens." *FEMS* 49:149.

Findlay, M., A.C. Leonard, S. Fogel and W.H. Mitchell. 1995. "A Laboratory Biodegradation Test to Define In Situ Potential for PCE Transformation." In R.E. Hinchee, C.M. Vogel, and F.J. Brockman (Eds.), *Microbial Processes for Bioremediation.* Battelle Press, Columbus, OH. (In Press)

Fogel, S. (1988). *Feasibility of Biodegradation of Tetrachloroethylene in Contaminated Aquifers.* Report to National Science Foundation, Award # ISI-8760424.

Fogel (Findlay), M., A.R. Taddeo, and S. Fogel. 1986. "Biodegradation of chlorinated ethenes by a methane-utilizing mixed culture." *Appl. Environ. Microbiol.* 50:720-724.

Freedman, D.L., and J.M. Gossett. 1989. "Biological reductive dechlorination of tetrachloroethylene and trichloroethylene to ethylene under methanogenic conditions." *Appl. Environ. Microbiol.* 55:2144.

Hollinger, C., G. Schraa, A.J. M. Stams, and A.J.B. Zehnder. 1993." A highly purified enrichment culture couples the reductive dechlorination of tetrachloroethene to growth." *Appl. Environ. Microbiol.* 59:2991.

Leahy, M.L., M. Dooley-Danna, M.S. Young, and M. Fogel. 1985. "Epoxide intermediate formation from the biodegradation of *cis*- and *trans*-1,2-dichloroethylene by a methanotrophic consortium." Presented at the 87th ASM Meeting, Atlanta, GA; Abstract No. K178.

Major D.W., E.W. Hodgins, and B.J. Butler. 1991. "Field and laboratory evidence of in situ biotransformation of tetrachloroethene to ethene and ethane at a chemical transfer facility in North Toronto." In R.E. Hinchee and R.F. Olfenbuttel (Eds.), *On-Site Bioreclamation Processes for Xenobiotic and Hydrocarbon Treatment.* Butterworth-Heinemann, Stoneham, MA, 1991, pp. 147-171.

Tandoi, V., T.D. DiStefano, P. A. Bowser, J.M. Gossett, and S.H. Zinder. 1994. "Reductive dehalogenation of chlorinated ethenes and halogenated ethanes by a high-rate anaerobic enrichment culture." *Environ. Sci. & Technol.* 28:973.

U.S. EPA. 1992. "Two-zone plume interception in situ treatment strategy." In *The Superfund Innovative Technology Program: Technology Profiles,* 5th ed. p. 208.

Vogel, T.M., B.Z. Fathepure, and H. Selig. 1989. "Sequential anaerobic/aerobic degradation of chlorinated organic compounds." *International Symposium on Processes Governing the Movement and Fate of Contaminants in the Subsurface Environment,* Stanford University, Stanford, CA, July 23.

Effect of Trichloroethylene Loading on Mixed Methanotrophic Community Stability

Stuart E. Strand, Gunter A. Walter, and H. David Stensel

ABSTRACT

There is concern that trichloroethylene (TCE) degradation by methanotrophic cultures cannot be maintained due to metabolite toxicity. The purpose of this research was to understand the effects of metabolite toxicity on methanotrophic cultures and to define operating conditions that might minimize these effects. A mixed methanotrophic culture was maintained with continuous supply of methane and nutrients as TCE loading was increased from 4 to 10 μg TCE/(mg protein-d). The maximum sustainable ratio of TCE degraded to methane consumed was 6 μg TCE/mg methane. Biomass, methane consumption, and TCE degradation fell precipitous ly when the loading exceeded 10 μg TCE/(mg protein-d). Reactor failure was preceded by a fall in the soluble methane monooxygenase (sMMO) activities. Phospholipid fatty acid analyses of the biomass before and after failure indicated a shift from a type II to a type I methanotrophic culture, a change that suggests a decreased ability to cometabolize TCE. These results suggest that continued exposure of a mixed methanotrophic culture to levels of TCE that damage cell viability can result in loss of cometabolic capabilities and a change in the population of the methanotrophic culture.

INTRODUCTION

Methanotrophs have a limited ability to degrade TCE, which is defined as the transformation capacity (T_c) (mg TCE degraded/mg biomass inactivated) (Alvarez-Cohen et al. 1991b). This limit results from intermediate toxicity, the damage caused by TCE oxidation, and the limited availability of intracellular reducing power, nicotinamide adenine dinucleotide (NADH) to initiate the oxidation of TCE.

The addition of formate may partially overcome reducing power limitations. The addition of formate improves the T_c approximately two-fold over endogenous conditions (Alvarez-Cohen and McCarty 1991b). However, as

TCE degradation increases at high loadings, cellular damage increases, preventing the reuse of the methanotrophic biomass (Alvarez-Cohen et al. 1991a).

Intermittent feeding of methane and TCE allowed activity to recover after TCE degradation by a mixed methanotrophic culture (Treat 1993). Long-term TCE degradation was possible, and degradation rates compared favorably to formate-fed cultures. The presence of methane induced increased levels of soluble methane monooxygenase (sMMO) and provided NADH for transformation of TCE.

The objectives of this research were to determine the maximum TCE loadings that can be sustained by a continuously growing, mixed culture of methanotrophs with nutrient addition and daily solids wasting. Population shifts and changes in enzyme activity that may occur with long-term exposure to TCE were investigated. The methane feeding rate necessary to sustain the maximum TCE loading rate was also determined.

EXPERIMENTAL PROCEDURES AND MATERIALS

The bioreactor consisted of a 2-L flask with a manometric gas feeding apparatus (Strand et al. 1990). The bioreactor was operated with continuous feeding of substrate and removal of biomass to provide a 3.7-day cell age. The methane and oxygen percentages in the headspace were 10% and 20%, respectively. These percentages were maintained by the removal of carbon dioxide and supply of methane and oxygen at the stoichiometric ratio of 1:1.6. TCE addition was continuous from a saturated aqueous solution, providing loadings from 0.045 to 0.175 mg TCE/(L-hr), as shown in Figure 1.

TCE concentration was determined by pentane extraction and gas chromatography using internal and external standards (Strand et al. 1990). The sMMO activity was determined by naphthalene oxidation (Brusseau et al. 1990) and by diazo dye absorbance at 528 nm. Potential and endogenous sMMO activities were assayed with and without formate addition, respectively. The sMMO activities were normalized to biomass.

Biomass was determined as protein using the Lowry assay (Lowry et al. 1951). Phospholipid fatty acids were extracted and assayed according to a procedure described by Microbial Identification, Inc. (Sasser 1990).

RESULTS

As the TCE loading rate increased, the biomass in the reactor did not decrease until the fifth loading period, allowing the specific TCE loading rate to remain relatively constant (Figure 1). Biomass in the bioreactor declined rapidly during the fifth loading period when the specific TCE loading exceeded

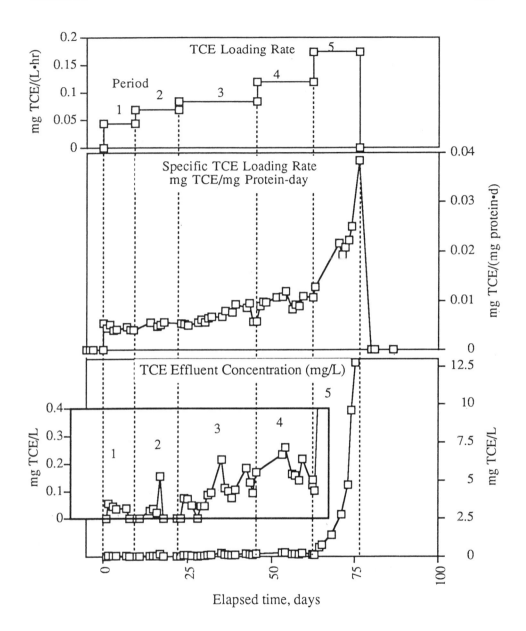

FIGURE 1. TCE-specific loadings and effluent TCE concentrations.

0.01 mg TCE/(mg protein-d). As biomass fell, the process failed and the effluent TCE concentration rose.

At least 95% of the influent TCE was degraded at specific TCE loading of
less than 0.013 mg TCE/(mg protein-d) (Figure 2). The bioreactor did not maintain TCE degradation at greater loadings. Methane consumption also declined
sharply during this period.

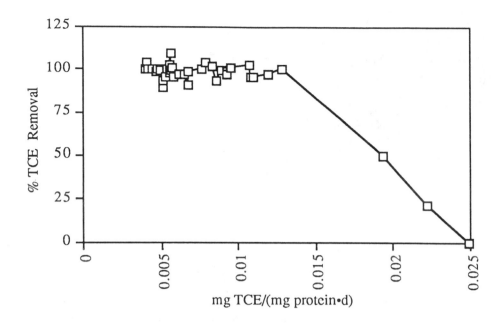

FIGURE 2. Effects of specific TCE loading on percent TCE removal.

During the fourth loading period, the culture attained the highest sustainable ratio of TCE degraded to methane consumed, 0.0064 mg TCE/mg methane (0.0007 standard deviation). Thus, at least 167 mg methane was required to maintain a culture per mg TCE degraded (Figure 3).

The biomass yields were calculated as the methane consumed divided by the biomass wasted. The yields averaged 0.18 mg protein/mg CH_4 (0.13 standard deviation) and did not decline with increased TCE loading during periods 1 through 4.

The potential sMMO activity declined by about 50% following the first addition of TCE in period 1 (Figure 3, P < 1% by Student's t-test). The endogenous sMMO activity declined by about 67% during loading period 1 (P < 1% by Student's t-test) and remained at that level during increased loadings in periods 2 through 4. When the potential sMMO activity declined to the level of endogenous sMMO activity, the culture was enzyme-limited rather than energy-limited. During loading period 5, potential sMMO activity declined below endogenous activity, corresponding to process failure.

Type II methanotrophs, which express sMMO, the enzyme necessary for rapid TCE degradation, contain predominantly 18-carbon fatty acids. However, Type I methanotrophs, which have lower levels of sMMO, contain more 16-carbon fatty acids in their cell membranes. Before TCE exposure, the bioreactor culture contained fatty acids with 70% 18-carbon fatty acids (percent of total chromatographic area, Figure 4). After process failure and regrowth, the culture contained more 16-carbon fatty acids and only 34% 18-carbon fatty acids.

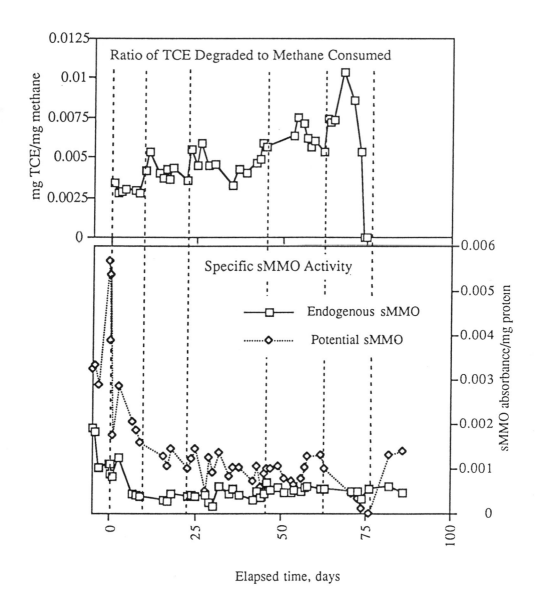

FIGURE 3. Ratio of TCE degraded to methane consumed and sMMO activity.

DISCUSSION

Aerobic cometabolism of TCE by methanotrophs in a continuous TCE-fed system is an unstable process. The competitive effects between TCE and methane prevent the efficient uptake of methane. Methane is necessary to recover from the damage of TCE oxidation and to provide reducing power

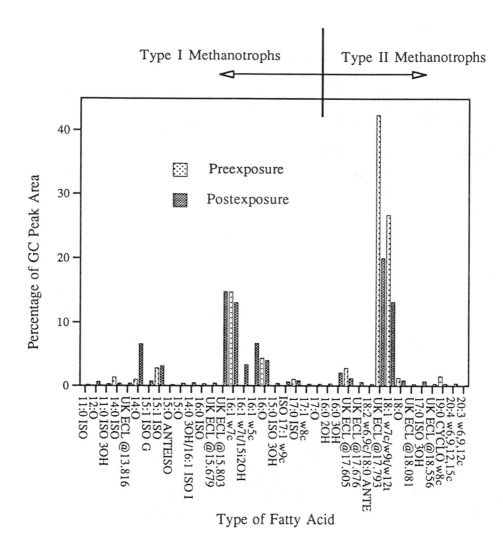

FIGURE 4. Phospholipid fatty acid composition of the methanotrophic mixed culture, before and after TCE exposure.

(NADH). The decline in endogenous sMMO activity after the initial addition of TCE in period 1 (Figure 3) is evidence of decreased intracellular availability of reducing power. Decreased availability of reducing power due to cometabolic degradation of TCE negatively affects both TCE and methane oxidation. Despite the reduced activity TCE degradation continued at a high rate during periods 1 through 4.

The biomass yield did not change with increasing TCE degradation. A lower yield would be expected if the toxic intermediates were causing the methanotrophs to be less efficient in their use of methane. Thus, it appears that decreased availability of reducing power during TCE degradation at loadings less than 0.1 mg TCE/(mg protein-d) were not detrimental to culture maintenance.

Degradation of TCE at high loadings in continuous flow systems leads to process failure and a population shift to Type I methanotrophs, which have a decreased ability to degrade TCE (Figure 4). Despite the lack of copper in the medium, Type I methanotrophs were selected by high TCE loadings and process failure. Following such process failure, recovery of efficient TCE degradation may be difficult, requiring reintroduction of a Type II methanotrophic inoculum.

REFERENCES

Alvarez-Cohen, L. and P. L. McCarty. 1991a. "A cometabolic biotransformation model for halo-genated aliphatic compounds exhibiting product toxicity." *Environmental Science & Technology* 25: 1381-1387.

Alvarez-Cohen, L. and P. L. McCarty. 1991b. "TCE transformation by a mixed methanotrophic culture—effects of toxicity, aeration and reductant supply." *Appl. Envir. Microbiol. 57*: 228-235.

Brusseau, G. A., H.-C. Tsien, R. S. Hanson and L. P. Wackett. 1990. *Optimization of trichloroeth-ylene oxidation by methanotrophs and the use of a colorimetric assay to detct soluble methane monooxygenase activity.* Gray Freshwater Biological Institute, University of Minnesota, Navarre, MN,

Lowry, O. H., N. J. Rosebourgh, A. L. Farr and R. J. Randall. 1951. *J. Biol. Chem. 193*: 265-275.

Sasser, M. 1990. *Identification of Bacteria by Gas Chromatography of Cellular Fatty Acids.* Microbial Identification, Inc.

Strand, S. E., M. D. Bjelland and H. D. Stensel. 1990. "Kinetics of chlorinated hydrocarbon degradation by suspended cultures of methane-oxidizing bacteria." *Research Journal of the Water Pollution Control Federation 62*: 124-129.

Treat, T. P. 1993. "Toxic Effects of Trichloroethylene on a Mixed Methanotrophic Culture." M. S. Thesis, University of Washington, Seattle, WA.

In Situ Bioremediation of Chlorinated Hydrocarbons Under Field Aerobic-Anaerobic Environments

Babu Z. Fathepure, Greg A. Youngers,
Dave L. Richter, and Charles E. Downs

ABSTRACT

A pilot study demonstrated destruction of chlorinated compounds in an open pond by stimulating the indigenous microflora under natural aerobic-anaerobic conditions. Ten biocells (of 500 L capacity) were installed in the pond and fed casamino acid and/or acetate as the source of carbon and energy. Results showed that more than 44 to 80% of initially present chemicals were degraded within 6 to 8 weeks. Active degradation was seen even in the presence of high concentrations of target compounds (total concentration > 440 mg/L). The degradation rate fell after 6 to 8 weeks. The exact reason was not known; however, it could be due the lack of appropriate nutrient(s) and/or growth-inhibiting substances present in the bottom sludge. Headspace analysis showed the formation of methane, ethane, ethene, chloromethane, vinyl chloride, and other chlorinated volatile organic compounds (VOCs). Maximum accumulation of VC and other chlorinated VOCs occurred only in the first 2 weeks of operation and thereafter the concentrations rapidly fell to near detection limit. Degradation occurred with both acetate and casamino acid substrates. Adding external microorganisms (bioaugmentation) had no overall effect on the reactor performance. Nearly complete removal of chlorinated compounds occurred in biocells provided with a physical barrier (sand or bentonite) between bottom sludge and upper water, indicating the toxic effect of high concentrations of chlorinated compounds and/or growth-inhibiting substances in the bottom sludge.

BACKGROUND AND INTRODUCTION

The pilot study area is located in the Gulf Coast region of the United States. The test site includes two surface impoundments that were used for the disposal of various industrial waste liquids in the late 1960s and early 1970s. These impoundments, the East and West Ponds, have a combined area of 5.2 acres (2.1 hectares) and originally contained approximately 21 million L of waste liquids comprised of mainly liquid chlorinated hydrocarbons (LCH). Figure 1 shows the location of the East and West Ponds of the test site.

The initial remedial activity included removing of LCH prior to removing water and sludge to minimize the mixing of free-phase LCH and sludge. A

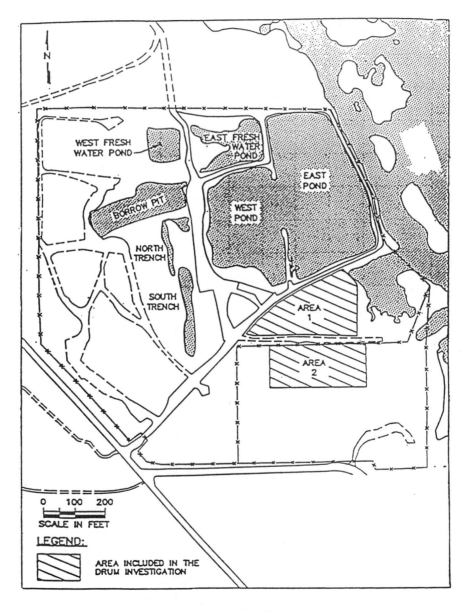

FIGURE 1. Location of East and West ponds.

total of 6 million L of LCH were removed from the pond and recycled (from June 1990 to February 1992). In addition to the LCH, more than 15 million L of contaminated sludge have been removed from the ponds. The removal activity was considered completed, because no recoverable quantities of LCH were identified in February of 1992.

According to the pond closure feasibility study, the major remaining remediation challenges included the treatment of pond water, residual sludge, and soil beneath the pond. The first phase of the pond closure effort is treating the contaminated water, because it is both a continuous source for the groundwater contamination and an odor problem. It is estimated that approximately 41 to 45 million L of contaminated water exists in the pond to a depth of 4 to 5 ft (1.2 to 1.5 m).

Analyses indicated that the concentration of chlorinated compounds in pond water and sludge ranges from 8 to 12 mg/L and 13,000 to 48,000 mg/kg, respectively. Table 1 summarizes the concentration and composition of chlorinated compounds present in pond water and sludge. Analyses show the presence of chlorinated ethanes, ethenes, and methanes at varied concentrations.

TABLE 1. Chlorinated compounds in pond water and sludge and their environmental fate.[a]

Chemical	Concentration (mg/L)		Degradation Environment[b]
	Water	Pond	
Chloromethane	BD[c]	BD	A, R
Vinyl chloride	0.11	200.5	A, M, R
Chloroethane	1.24	3,775.6	A, H, R
1,1-Dichloroethene	BD	BD	M, R
Methylene chloride	BD	BD	A
t-1,2-Dichloroethene	0.24	1,155.8	M, R
1,1-Dichloroethane	0.18	821.8	A
c-1,2-Dichloroethene	0.2	835.1	M, R
Chloroform	1.48	6872.5	M, R
1,2-Dichloroethane	2.79	8,166.9	A, R
1,1,1-Trichloroethane	RD[c]	RD	C, H, R
Carbon tetrachloride	0.27	4,155.1	C, R
Trichloroethene	BD	BD	M, R
1,1,2-Trichloroethane	1.56	9,350.6	C, H, R
Tetrachloroethene	BD	267.5	C, R
1,1,1,2-Tetrachloroethane	RD	RD	C, H, R
1,1,2,2-Tetrachloroethane	0.15	980.5	C, H, R

(a) Chemicals are listed in the order of their GC elution. The concentration represents an average value.

(b) From literature; A = Aerobic; C = Chemical (Abiotic); H = Hydrolysis; M = Methanotrophic; R = Reductive dechlorination.

Among the detected compounds, the 1,2-dichloroethane (1,2-DCA), 1,1,2-trichloroethane (1,1,2-TCA), chloroform (CF), and chloroethane (CA) constitute over 70 to 80% (by mass) of all measured compounds. In addition, small concentrations of tetra- and trichloroethene, dichloroethenes, methylene chloride, vinyl chloride, and other chlorinated aliphatic chemicals were detected.

Transformation of chlorinated compounds in the natural environment can occur both chemically and biologically (Bhatnagar and Fathepure 1991; Bouwer 1994; Mohn and Tiedje 1992; McCarty 1988; Vogel et al. 1987; Wackett et al. 1992). Chemical degradation (abiotic) usually results in only a partial transformation of a compound and may lead to the formation of a totally new compound. Hydrolysis and dehydrohalogenation (removal of a hydrogen and a halogen atom) are the two most common pathways of abiotic transformation occurring under both aerobic and anaerobic conditions.

Biological transformation (biotic) of chlorinated compounds occurs under aerobic or anaerobic conditions. The degradation products are different under aerobic and anaerobic environmental conditions. Biotic aerobic degradation proceeds via oxidation, and the degradation products are more water soluble and generally nontoxic. Biotic anaerobic degradation of chlorinated compounds occurs mainly via a sequential dechlorination pathway leading to the formation of less chlorinated compounds and ultimately may result in the removal of all chlorine atoms (Bhatnagar and Fathepure 1991; Bouwer 1994; McCarty 1988; Mohn and Tiedje 1992; Vogel et al. 1987; Wackett et al. 1992).

It is now a well-established fact that highly chlorinated compounds such as tetrachloroethene (perchloroethene = PCE), trichloroethene (TCE), 1,1,2-TCA, CF, and carbon tetrachloride (CT) can be readily transformed by anaerobic organisms via reductive mechanisms to less chlorinated products. The rate of dechlorination decreases markedly with the removal of subsequent chlorine atoms. As a result, under an anaerobic environment, compounds with one or two chlorines tend to accumulate. However, many aerobic organisms are capable of degrading compounds with fewer chlorine atoms on the molecule (McCarty 1988; Bouwer 1994; Vogel 1994). This oxidative process often results in complete mineralization to carbon dioxide. Therefore, the combination of an anaerobic process followed by an aerobic process has promise for bioremediating an environment contaminated with a mixture of chlorinated organics.

The overall goal of the present field study was to evaluate the feasibility of creating a mixed redox environment by adding appropriate nutrients to the open pond for the degradation of the mixture of chlorinated hydrocarbons. The specific objectives were to,

- Demonstrate degradation of chlorinated compounds by stimulating the growth of indigenous bacterial populations under field conditions
- Monitor and quantify the release VOCs such as vinyl chloride, ethene, ethane, and methane during the active biodegradation phase
- Evaluate the effect of the bioaugmentation on the rate and extent of biodegradation

- Evaluate the effect of substrates (casamino acid vs. acetate) on bio-degradability of the target compounds for economical purposes.

A pilot study involving several bioreactors (biocells) directly placed in the pond was conducted to accomplish the above tasks. Appropriate amounts of casamino acid and/or acetate were added to the biocells to induce the growth of indigenous populations and to create a mixed redox environment. The study also included the addition of mixtures of aerobic and anaerobic organisms to certain biocells to test the effect of bioaugmentation on the rate and extent of contaminant removal.

MATERIALS AND METHODS

Feasibility Study: Construction of Field Biocells

Each field biocell was constructed of a galvanized corrugated steel pipe 2 ft in diameter and 15 ft in length (0.61 × 4.57 m). Ten biocells were driven verti-cally into the sediment of the East Pond for the pilot study. A catwalk con-necting all the biocells was constructed for sampling and other maintenance purposes. A tarp was placed over all of the biocells to protect them from rain. To quantify the emission of VOCs during biodegradation, some of the biocells were equipped with an airtight lid, a sampling port for headspace gases, and a pressure gauge to monitor the total volume of gases produced. Figure 2 is a schematic diagram of a closed biocell. For some biocells, a physical barrier of sand or bentonite was placed in the bottom to separate water and sludge to evaluate the effect of sludge on biodegradation of chlorinated compounds dis-solved in water.

Treatment Plan

All the biocells, except the control (A1), were fed 0.2% casamino acid and/or 0.2% anhydrous sodium acetate as an electron donor (substrate). Sodium phosphate (mono and dibasic at equimolar ratio) was added to all the active cells as a source of phosphorus in the ratio of 100:1 (C:P). Biocells received no additional source of nitrogen other than casamino acid itself. Both substrate(s) and inorganic salts were added as concentrated stock solutions prepared in a minimum amount of pond water. Table 2 summarizes the experimental plan and various treatment schemes to evaluate the feasibility of in-pond biodegra-dation of chlorinated compounds. Biocell A1 served as an unamended control, while the pond itself acted as a background control. All the biocells, including A1, were maintained under neutral pH.

Biocells A4 and B4 were inoculated with mixed anaerobic and aerobic microorganisms obtained from a municipal wastewater treatment plant. Approximately 1 to 1.5% (v/v) thickened activated sludge and anaerobic sludge were added at equal proportions to A4 and B4 as an inoculum.

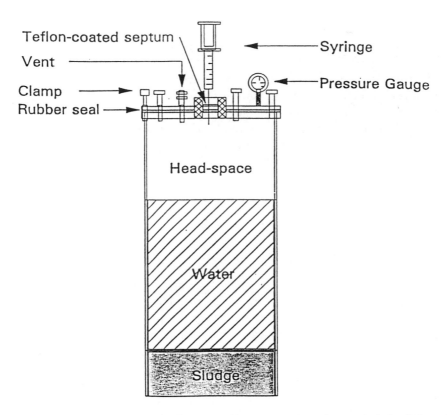

FIGURE 2. The schematic of a biocell equipped with an airtight lid and a headspace sampling port.

Biocells A5 and B5 contained a physical barrier of sand or bentonite separating the bottom sludge from the upper water. The results from these cells will indicate whether high concentrations of chlorinated compounds in sludge have an inhibitory effect on the rate and extent of biodegradation in liquid.

Biocells A1, A3, B3, and B5 were equipped with an airtight lid. These biocells were closed for approximately 24 h prior to the sampling of headspace. The headspace gases released during biodegradation were withdrawn via a custom-built injection port and a 1-L capacity gastight syringe (Hamilton Co., Reno, Nevada). The gases were collected in 1-L capacity Tedlar™ bags (SKC Inc., Eighty Four, Pennsylvania). The samples were analyzed for the accumulation of VOCs including vinyl chloride, chloroethane, chloromethane, ethene, ethane, and methane.

Sampling and Analyses

All the biocells, plus the pond, were sampled in the beginning and at the end of 2 or 4 weeks after the biocells had been properly installed in the pond

TABLE 2. Field biocells and treatment plan.

Biocell[a]	Amendment	Treatment
Pond	No substrate No pH Adjustment	Background control
A1	No substrate pH adjusted	Experimental control
A2 & B2	CA + A	Biodegradation
A3 & B3	CA + A	Biodegradation and VOCs measured
A4 & B4	CA + A	Bioaugmentation
A5	CA + A	Sand barrier
B5	CA + A	Bentonite barrier
A6[b]	A	Biodegradation

(a) Biocells A1, A3, B3, and B5 were equipped with an airtight lid and monitored for VOCs.
(b) All the biocells, except A1 and the pond, were amended with 0.2% casamino acid (CA) and/or acetate (A). Biocell A6 was amended with 0.2% acetate alone.

sediment and the pH of the water had been adjusted to neutral. Dissolved oxygen (DO), redox (Eh), pH, temperature, and conductivity of water were measured on a weekly basis using the Hydrolab instrument (HYDROLAB® Corporation, Austin, Texas).

Using a plastic bailer, 5 water-column samples of 0 to 3 ft (0.91 m) deep were withdrawn from each biocell or the pond and composited into separate 1-L capacity bottles. Similarly, five sludge samples (50 to 100 mL) from each biocell or pond were withdrawn using a custom-built sampling device, and the samples were composited in separate bottles. The composites were thoroughly agitated and transferred into duplicate 40-mL VOA bottles (Baxter Healthcare Corp, Grand Prairie, Texas) with no headspace. Both water and sludge samples were shipped overnight in ice to the laboratory for the analysis of chlorinated compounds and total organic carbon (TOC).

The headspace from the closed biocells and the atmospheric air over the pond were collected in separate Tedlar™ bags at the end of every 4 weeks and shipped overnight in ice to the laboratory for the analysis of VOCs.

Analytical Procedures

The water was analyzed for TOC by Clean Water Act Method 415.1, using the Astro 2000 carbon analyzer (Astro International, League City, Texas). The anions were assayed using ion chromatography (Dionex DX-100, Dionex Corporation, Sunnyvale, California), employing Standard Methods for the Examination of Water and Wastewater (Method No. 429). Chlorinated aliphatic compounds were assayed according to the U.S. Environmental Protection Agency (EPA) method 8240 using a purge-and-trap equipped with 4460A concentrator (O.I. Analytical, College Station, Texas) connected to a Varian gas chromatograph (Varian 3400, Sunnyvale, California) and Finnigan mass spectrometer (San Jose, California). Samples were separated using a DB-624 capillary column 30 m long × 0.53 mm ID (J&W Scientific, Folsom, California). The method detection limit was 0.002 mg/L.

Chemicals present in sludge were quantified as follows; Approximately 2 g of composited sludge was first dissolved in 10 mL of methanol by handshaking for 5 min, and the extract was serially diluted several times using methanol (100,000-fold). The sufficiently diluted sample was analyzed using the purge-and-trap method specified above (EPA 8240).

Headspace gases were analyzed for methane, ethane, and ethene using an HP 5890 gas chromatograph equipped with a flame ionization detector and DB-1 capillary column 30 m long × 0.53 mm ID. Similarly, chlorinated VOCs were analyzed using HP 5890 gas chromatograph equipped with ELCD detector and RESTEX-502.2 column. The temperature program for both the assays was 35°C for 5 min then to 180°C at 20°C/min. The injector and detector temperatures were 150° and 200°C, respectively. Helium was used as a carrier gas.

RESULTS AND DISCUSSION

Anaerobic-Aerobic Process

Both anaerobic and aerobic microorganisms are necessary to degrade the target compounds (Table 1), because the pond contains mixtures of highly chlorinated and lesser chlorinated aliphatic compounds. A gradient of oxic-anoxic (aerobic-anaerobic) conditions was established inside the cells through the addition of electron donors, such as casamino acid and/or acetate. In the presence of an adequate amount of added substrate(s), most of the dissolved oxygen should be rapidly consumed by aerobic bacteria in the top few inches or feet, preventing the diffusion of oxygen to lower layers of water. As a result, a gradient of oxic-anoxic conditions should be created in the water column so that dissolved oxygen concentrations decrease with depth. Periodic monitoring of the DO using the Hydrolab revealed that the DO concentration decreased to less than 1.0 mg/L below 0.8 to 1 ft from the surface.

Highly chlorinated compounds undergo transformation via reductive dechlorination, hydrolysis, and other biotic and abiotic mechanisms to less-chlorinated and completely dechlorinated products (Bouwer 1994; Mohn and Tiedje 1992; Vogel 1994; Wackett et al. 1992) under an anaerobic environment. Thus formed, less-chlorinated compounds and compounds with no chlorine would diffuse to the surface due to their low density and high aqueous solubility where they are subsequently degraded by aerobic/microaerophilic organisms in the upper layers yielding carbon dioxide and other innocuous end products (Armenante et al. 1992; Fathepure 1991; McFarland et al. 1992; Phelps et al. 1991). Figure 3 depicts the dynamic process that is believed to occur in all the biocells supplemented with substrate(s).

Biotransformation

Figures 4 and 5 show the fate of chlorinated compounds in the active and control biocells. Results from biocell B3 show that more than 70% of initially present chlorinated compounds were removed within the first 6 to 8 weeks of operation. Minimal degradation occurred in the control (biocell A1) and the pond during the first 6 to 8 weeks. A similar trend of decrease in the concentration of target compounds (> 44 to 80%) was seen in all the biocells amended with the substrates. An initial concentration increase was noted in both the

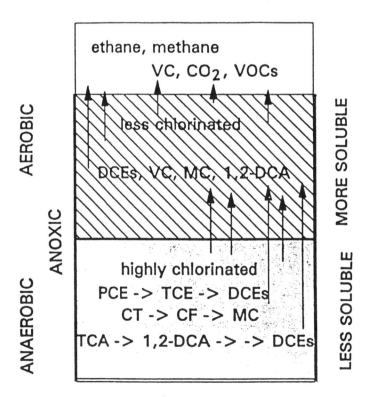

FIGURE 3. Dynamics of biodegradation under an aerobic-anaerobic environment.

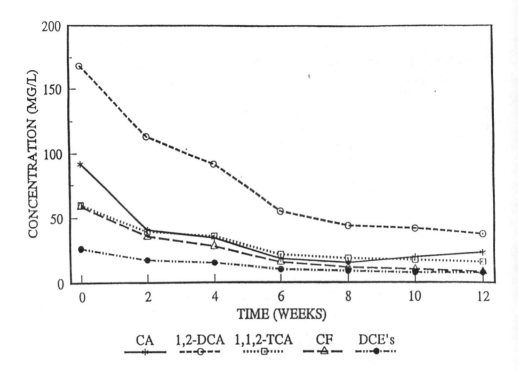

FIGURE 4. **Biodegradation of chlorinated compounds in an active biocell (B3). For clarity, only major contaminants are shown. DCEs = 1,1-DCE, *cis*-DCE, and *trans*-DCE.**

controls (pond and biocell A1). This could have been due to some physical disturbances of the bottom sludge, in which case a similar increase in concentration might have also occurred in all the biocells. If so, these transient increases in concentrations may have not been detected because of rapid degradation rates in the active biocells. The data indicate that rate of removal decreased after 6 to 8 weeks in all the active biocells. This could have been due to several reasons including lack of appropriate nutrients, low seasonal temperatures (12 to 15°C), and unknown growth-inhibiting chemicals from the sludge.

The initial concentrations of target compounds in all biocells were different due to an accidental agitation of the bottom sludge during initial startup of the project. The present study reveals that biotransformation of target compounds has occurred at all the initial concentrations ranging from 13 to 440 mg/L, suggesting that the indigenous microorganisms are capable of degrading pollutants over a wide range of concentrations (Table 3). This is important, because the concentrations of target compounds in the pond water often fluctuate drastically due to the presence of high levels of chemicals in the sludge.

Biocells also were monitored for concentration changes in the sludge. Small changes in concentrations due to any bio/abiotic removal could not be

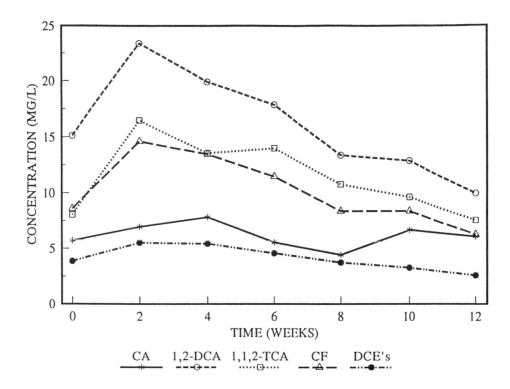

FIGURE 5. **Biodegradation of chlorinated compounds in the control biocell
(A1). For clarity, only major contaminants are shown. DCEs = 1,1-DCE,
cis-DCE, and *trans*-DCE.**

detected (data not shown) because of the natural heterogeneity and the presence of extremely high levels of chemicals in the sludge.

Monitoring headspace gases showed that relatively high levels of chlorinated VOCs accumulated only during initial phases of the project (second
week); the VOC concentration fell significantly after 2 weeks. The concentration profile of total chlorinated VOCs and VC is shown in Figures 6 and 7,
respectively. The headspace analysis indicated that total VOC concentration
varied from <100 to 1,000 mg/m^3/day in all the tested biocells except biocell B3
during week 2. Similarly, VC concentration varied from <0.01 to 30 mg/m^3/day,
except for week 2 analysis in biocell B3. The high concentrations of chlorinated
VOCs detected in biocell B3 may have been due to higher initial concentrations
of target compounds. In general, the concentrations of headspace VOCs
including VC, ethene, ethane, chloromethane, and chloroethane decreased
with time. This could have been due to increased aerobic and methanotrophic
oxidation of the above products to carbon dioxide, chloride, and other water-
soluble products (Bhatnagar and Fathepure 1991; Fogel et al. 1986; McCarty
1988; Phelps et al. 1991.).

TABLE 3. Percent removal of chlorinated compounds at varied initial
concentrations.

Biocell/ Pond	Concentration 0 Wk (mg/L)[a]	Concentration 6 Wk (mg/L)	Percent Removal
Pond	8.2	11.7	−42.3(b)
A1	45.9	60.4	−31.7(b)
A2	154.4	59.5	61.5
B2	96.4	35.0	63.7
A3	108.0	60.5	44.0
B3	442.9	134.5	69.6
A4	69.2	38.5	44.6
B4	125.8	77.8	38.2
A5	22.1	6.3	71.7
B5	12.9	2.1	83.8
A6	81.4	26.8	67.1

(a) Due to an initial accidental agitation, the startup concentration
    of chlorinated compounds in each biocell was different.
    Maximum care was taken to minimize any further agitation of
    the biocell content.
(b) The concentrations of chlorinated compounds in the una-
    mended biocell (A1) and the background control (pond)
    increased after the start of the project. This could be due to
    natural disturbances of the bottom sludge. Similarly, increased
    concentrations were expected in all biocells, but due to rapid
    degradation rates, no net increase in concentrations was seen.

In the present study, < 50 to 200 mg/m^3/day methane was measured in the
headspace gases throughout the experimental period (data not shown).
Methane was also detected in the unamended biocell, A1. This may have been
due to the metabolism of dissolved organic matter present in the water and/or
sludge. The formation of methane indicates the growth of strict anaerobic
microorganisms. Anaerobic conditions are necessary for reductive dechlorination of highly chlorinated compounds such as TCA, PCE, TCE, CT, and CF. The
detection of chloromethane, ethene, and ethane in headspace gases indicated
that extensive dechlorination of the target compounds occurred.

The methane formed in anaerobic zones can support the growth of methane-oxidizing, aerobic organisms in the surface waters. After 2 weeks, decreasing concentrations of methane were measured in the headspace, which
coincided well with decreasing concentrations of chlorinated VOCs. This
strongly suggests that methane may have been consumed by methane-oxidizing
aerobic organisms. Methanotrophic bacteria were shown to cometabolize

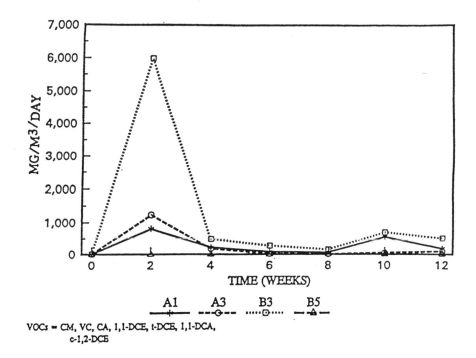

FIGURE 6. Headspace chlorinated VOCs in active and control biocells.

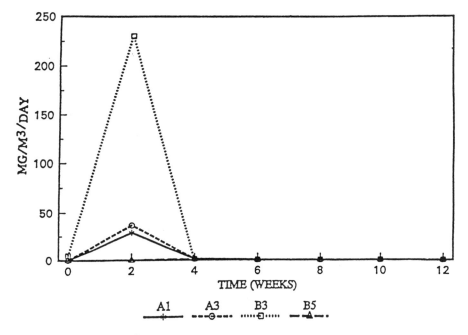

FIGURE 7. Headspace vinyl chloride concentrations in active and control biocells.

several less chlorinated compounds including TCE, *cis*-DCE, *trans*-DCE, and VC. Recently, McFarland et al. (1992) have shown that methane-oxidizing organisms can decompose various lightly chlorinated compounds to water-soluble products such as formic acid, glyoxylic acid, and acetic acid, which further degrade to carbon dioxide by mixed microorganisms. In addition, the oxidation of 1,1-dichloroethane, 1,2-DCA, methylene chloride, VC, ethene, and ethane by aerobic heterotrophs was expected.

Acetate vs. Casamino Acid

Biocells were amended with casamino acid (CA) and/or acetate to evaluate which substrate would be appropriate for maximum degradation of target compounds. Our previous bench-scale research had indicated that both CA and acetate stimulated good dechlorination activity. In addition, CA may be used as a source of nitrogen, and acetate is a commonly detected organic compound in anaerobic habitats. Results indicate (Figure 8) that degradation occurred without delay in the presence of both the tested substrates. These results are in agreement with the previously published information that acetate is a good substrate for dechlorination (Fathepure and Vogel 1991). Also, the fact that we could use acetate as a primary carbon source to degrade chlorinated pollutants is significant due to its low cost.

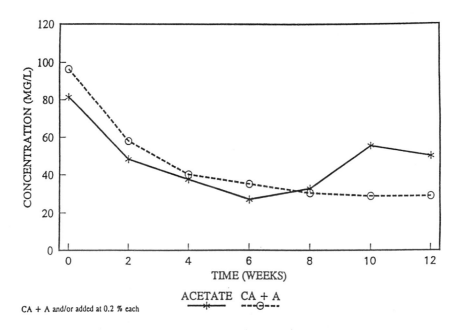

FIGURE 8. **Biodegradation of chlorinated compounds in the presence of added casamino acid and/or acetate.**

Effect of Sludge

Biodegradation in liquid was tested by adding either a sand or bentonite blanket to the bottom of the biocell to minimize the direct contact between the sludge and the aqueous phase, because the sludge contains extremely high levels of chlorinated compounds (4 to 5%). Figure 9 shows that 70 to 80% of the initially present chemicals were degraded within 6 weeks in the presence of sand or bentonite. The extent of degradation was maximum in the presence of added bentonite. The total residual concentrations were 6 and 2 mg/L in the presence of sand and bentonite respectively, at the end of 6 weeks. The reason sand was a poor barrier could be its greater permeability and lower cation exchange capacity. Also, bentonite could have facilitated the biological and/or chemical dechlorination reactions of some target compounds.

Bioaugmentation

The application of external microorganisms in remedial operations would be beneficial when contaminants resist biodegradation by indigenous microflora, where evidence of toxicity exists, or when populations of indigenous microorganisms are absent. The test pond meets all the above criteria. For example, acidic water (pH ~3), the presence of a variety of chlorinated toxic

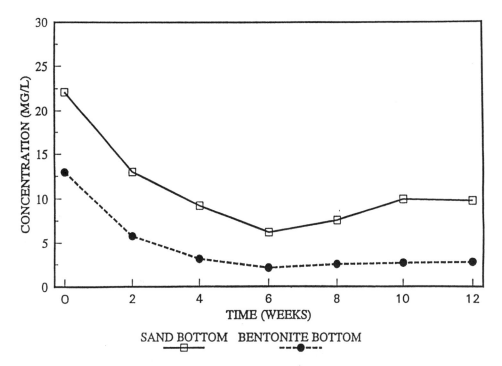

SAND BOTTOM   BENTONITE BOTTOM

FIGURE 9. Biodegradation of chlorinated compounds in the absence of sludge.

chemicals, high concentrations of solvents in the sludge, and low numbers of bacteria indicated a need for bioaugmentation.

Mixed cultures of aerobic and anaerobic organisms were added to biocells A4 and B4. Results showed that the rate and extent of degradation were not significantly higher than when not added (data not shown). However, no inhibition of degradation by native bacteria was evident by the seed culture. The competition for substrate, the lack of specific degradative microorganisms in the inoculum, and the hostile environments may have been the main cause that affected the survival and biodegradation potential of the inoculants.

CONCLUSIONS

The biodegradation of 1 and 2 carbon chlorinated compounds with varying degrees of chlorination was evaluated under fieled conditions. To achieve complete degradation of these compounds, a mixed-redox environment that supports the growth of aerobic heterotrophic and anaerobic organisms is essential. In the present pilot study, such an environment was created by adding appropriate amounts of electron donor(s), such as casamino acid and/or acetate, to open biocells. Measurement of DO and Eh at various depths in each biocell indicated the presence of aerobic, anoxic, and perfectly anaerobic (Eh = −250 to −300 mV) environments.

The field study clearly showed that 44 to 80% of the chlorinated target compounds present in water was degraded within 6 to 8 weeks in all the fed biocells. The accumulation of VC, chloromethane, ethene, and ethane suggests that extensive dechlorination occurred. An effective mass balance was not possible due to the complexity of the system including the presence of already dechlorinated products and high levels of chlorinated compounds in the bottom sludge. In addition, aerobic and methanotrophic oxidation of less-chlorinated compounds to CO_2 and water-soluble metabolic intermediates make the mass balance difficult. The following list shows specific observations made during this pilot work:

- From 44 to 80% of the initially present chlorinated compounds in pond water was removed within 6 to 8 weeks under aerobic-anaerobic environments.
- Little or no removal of target compounds was detected (at the end of 6 to 8 weeks) either in the pond or in the unamended biocell, indicating the need for appropriate growth substrate.
- The headspace analysis showed the formation of methane, ethane, ethene, and chloromethane, indicating extensive dechlorination.
- The concentration of VOCs in the headspace decreased significantly after 2 weeks, suggesting increased aerobic activity. After 2 weeks, VC concentrations did not rise above 1 mg/m^3/day.

- No accumulation of partially dechlorinated compounds occurred. This could be due to complete mineralization of the target compounds to carbon dioxide and other water-soluble products under aerobic-anaerobic conditions.
- Rapid degradation occurred immediately after the addition casamino acid and/or acetate as an electron donor. This indicates the presence of already acclimated, carbon-limited (starving) bacterial populations.
- Degradation proceeded without a lag in all the substrate-amended biocells containing concentrations of target compounds ranging from 12 to 440 mg/L, suggesting robust indigenous bacterial populations.
- The addition of bacteria (bioaugmentation) did not significantly enhance either the rate or extent of degradation. This could be due to the lack of specific degraders in the seed, the hostile new environment, and/or competition with native bacteria for the same substrate.
- More than 80% of the initially present chemicals were degraded in biocells containing sand or bentonite as a barrier. In the biocell with bentonite, the individual residual concentrations were < 0.6 mg/L.
- The fact that acetate can be used as a carbon source is economically significant due to its low cost.

ACKNOWLEDGMENTS

The following individuals are gratefully acknowledged for their help and support during this project: John Bredthauer, Jerry Spetseris, Tim Champagne (Browing-Ferris Industry), Bill Muldoon, Dan DiFeo, and the entire analytical staff of Environmental Services Division. Also, thanks are due to Michael Lee and Martin Odom for useful discussion.

REFERENCES

Armenante, P. M., D. Kafkewitz, G. Lewandowski, and C-M, Kung. 1992. "Integrated anaerobic-aerobic process for the biodegradation of chlorinated aromatic compounds." *Environmental Progress.* 11: 113–122.

Bhatnagar, L., and B. Z. Fathepure. 1991. "Mixed cultures in detoxification of hazardous waste." In J. G. Zeikus and E. Johnson (Eds.), *Mixed Cultures in Biotechnology.* pp. 293–340. McGraw-Hill, New York, NY.

Bouwer, E. J. 1994. " Remediation of chlorinated solvents using alternate electron acceptors." In R.D. Norris at al. (Eds.), *Handbook of Bioremediation*, pp. 149–175. CRC Press, Inc., Boca Raton, FL.

Fathepure, B. Z, and T. M. Vogel. 1991. "Complete degradation of polychlorinated hydrocarbons by a two-stage biofilm reactor." *Appl. Environ. Microbiol.* 57: 3418–3422.

Fogel, M.M., A.R. Taddeo, and S. Fogel. 1986. "Biodegradation of chlorinated ethenes by a methane-utilizing mixted culture." *Appl. Environ. Microbiol.* 50(4): 720–724.

McCarty, P. L. 1988. "Bioengineering issues related to in-situ remediation of contaminated soils and groundwater." In G.S. Omenn (Ed.), *Environmental Biotechnology*, pp. 143–162. Plenum Publishing Corp., New York, NY.

McFarland, M. J., C. M. Vogel, and J. C. Spain. 1992. "Methanotrophic Cometabolism of trichloroethylene (TCE) in a two stage bioreactor system." *Water Resources* 26: 259–265.

Mohn, W.W, and J.M. Tiedje. 1992. " Microbial reductive dehalogenation." *Microbiol. Rev. 56*: 482–507.

Phelps, T. J., K. Malachowski, R. M. Schram, and D.C. White. 1991. "Aerobic mineralization of vinyl chloride by a bacterium of the order Actinomycetales." *Appl. Environ. Microbiol.* 57(4): 1252–1254.

Vogel, T. M. 1994. "Natural bioremediation of chlorinated solvents." In R.D. Norris et al. (Eds.), *Handbook of Bioremediation*, pp. 201–225. CRC Press, Inc. Boca Raton, FL.

Vogel, T. M., C. S. Criddle, and P. L. McCarty. 1987. "Transformation of halogenated aliphatic compounds." *Environ. Sci. Technol. 21*: 722–736.

Wackett, L.P, M. S. P. Logan, F.A. Blocki, and C. Bao-li. 1992. "A mechanistic perspective on bacterial metabolism of chlorinated methanes." *Biodegradation 3*: 19–36.

Cometabolic Degradation of Chlorinated Solvents: Bacterial Inhibition, Inactivation, and Recovery

Roger L. Ely, Kenneth J. Williamson,
Michael R. Hyman, and Daniel J. Arp

ABSTRACT

This paper summarizes an approach for quantifying degradation kinetics and bacterial activity changes during cometabolic degradation of chlorinated solvents, results from trichloroethylene (TCE) degradation experiments, and a mathematical model addressing fluctuations in activity caused by enzyme inhibition, inactivation, and respondent enzyme synthesis. Using *Nitrosomonas europaea* as a slow-growing exemplar capable of effecting cometabolic transformations, quasi-steady-state ammonia oxidation was established in a small bioreactor. A chlorinated solvent was injected to perturb the system, and bacterial activity and solvent degradation were monitored. At TCE concentrations to about 3.5 mg/L, from slight to nearly complete ammonia monooxygenase (AMO) inactivation occurred without causing immediate cell death. Results suggested cellular injury was limited primarily to AMO, most metabolic systems remained functional, and bacterial recovery processes, independent of cell growth, were initiated while degrading TCE. The TCE strongly inhibited ammonia oxidation; parameter estimates showed TCE to have about four times greater affinity than ammonia for AMO, and the maximum TCE oxidation rate was about 1/100th that of ammonia. AMO inactivation was proportional to TCE oxidation, and bacterial recovery effects were statistically significant in most experiments. Bacterial capability to recover from injury suggests that sustainable cometabolism of chlorinated solvents may be achievable under properly controlled conditions.

INTRODUCTION

Cometabolic transformation of some substances, e.g., trichloroethylene (TCE), is known to lead to decreases in bacterial culture activity (e.g., Wackett and Householder 1989; Fox et al. 1990; Alvarez-Cohen and McCarty 1991;

Oldenhuis et al. 1991; Rasche et al. 1991). Oldenhuis et al. (1991), observed that TCE-associated inactivation of the methanotroph, *Methylosinus trichosporium* OB3b, (in the absence of methane) was proportional to TCE oxidation. They interpreted inactivation as a decrease in culture specific activity, i.e., changes in active enzyme level within live cells, and suggested that the remaining culture specific activity after a period of TCE degradation was equal to the initial specific activity minus an "inactivation constant" multiplied by the amount of TCE degraded. In experiments in which both methane and TCE were present, they found that "...a deviation from Michaelis-Menten kinetics occurred, which made it impossible to determine an inhibition constant."

In similar experiments without methane, Alvarez-Cohen and McCarty (1991) interpreted inactivation of a mixed methanotrophic culture as a decrease in "active biomass" that was proportional to the amount of TCE transformed. They proposed a model, based on Monod growth kinetics, which assumed that losses in culture activity were attributable to partial death of the bacterial population, i.e., loss of active biomass, while culture specific activity remained constant. The model subsequently was developed further (Criddle 1993), but the assumption of constant specific activity was not modified.

In experiments with *Pseudomonas putida* F1, a toluene-oxidizing bacterium, Wackett and Householder (1989) showed that active toluene dioxygenase was required for TCE-related inactivation to occur, and that *P. putida* F1 could recover from and/or cope with the harmful effects of TCE exposure. Following 1-, 2-, or 3-h exposure to TCE, cell doubling times, calculated from exponential growth curves, were 1.5, 1.6, and 1.7 h, respectively, within 3 h of the removal of TCE, compared to 1.5 h for control cells not exposed to TCE. Doubling time for a control culture continually exposed to TCE over a 9-h period was 5.0 h. In control experiments conducted with perchloroethylene to determine if inactivation could be attributed to loss of membrane integrity, no significant solvation of bacterial membranes was observed. TCE labeling experiments showed that after 10-min incubations with [^{14}C]TCE followed by TCE removal and cell fractionation, 68% of recovered radioactivity was present in protein, 20% in small molecules, 9% in RNA, 2% in DNA, and 1% in lipids. In experiments with the ammonia-oxidizing bacterium, *Nitrosomonas europaea*, Rasche et al. (1991) categorized several chlorinated aliphatic compounds into three groups according to their susceptibility to oxidation by AMO and, for those oxidized by the enzyme, their apparent ability to bring about culture inactivation. Experiments examining cellular recovery capabilities showed that recovery from TCE-associated inactivation required *de novo* protein synthesis.

Better understanding of relevant metabolic processes and development of appropriate mathematical models are needed because of the potential importance of cometabolism-based approaches for bioremediation. For example, the extent to which culture inactivation results from partial loss of key enzyme activity within live cells as compared to the extent to which it results from cell death is important since injured cells potentially can recover by resynthesizing key macromolecules, but dead cells obviously cannot. Knowledge of how or if sublethal injury may be controlled and/or natural cellular recovery processes facilitated would be helpful. We have sought to better characterize the nature of the inactivation process and the potential for cellular recovery by examining physiological effects of exposure to TCE and other chlorinated solvents in *N. europaea*.

N. europaea are obligate autotrophs that oxidize ammonia (NH_3) to hydroxylamine (NH_2OH) and then to nitrite (NO_2^-). The first reaction, catalyzed by the membrane-bound enzyme, AMO, is a net reductant-consuming step that requires two electrons. The second reaction, catalyzed by hydroxylamine oxidoreductase (HAO) located in the periplasm, generates four electrons. Two of these can enter the electron transport chain and support electron transport phosphorylation, whereas the other two are required to support AMO catalysis. At least two inferences may be drawn from these observations:

1. Because substrate (NH_2OH) is passed from AMO to HAO and electrons are passed back from HAO to AMO, the two enzymes may be closely associated in the cell, perhaps at or near the periplasmic surface of the cell membrane and perhaps as a complex.
2. Because they are obligate autotrophs, *N. europaea* grow relatively slowly (doubling time is about 8 h under maximum-growth-rate conditions).

Because of their low growth rate and the fact that AMO possesses a wide substrate range, *N. europaea* are useful for studying cometabolic enzyme kinetics over relatively long time periods (i.e., several hours) without significant cell growth. This paper summarizes an approach for examining cometabolic enzyme kinetics, results from experiments with TCE, and a model that accounts for changes in culture specific activity associated with inactivation and recovery. These topics are presented in detail elsewhere (Ely et al. 1995a,b).

EXPERIMENTAL PROCEDURES

Quasi-Steady-State Reactor Experiments

Cells were grown axenically and harvested as previously described (Ely et al. 1995b). The general approach used in reactor experiments was (1) to establish a quasi-steady-state condition of ammonia oxidation in a small (1.7 L), sealed bioreactor by feeding ammonia with a syringe pump at a constant rate far below the maximum substrate utilization rate, (2) to perturb the system by adding a pulse injection of TCE, and (3) to observe the residual TCE concentration and the development of TCE-associated effects (i.e., inhibition, inactivation, and recovery) on bacterial activity with time. Experimental data included nitrite concentration, TCE concentration, O_2 uptake rate, and quantity of ammonia fed (to allow an ammonia mass balance) versus time, and initial and final protein concentrations (to assess growth).

Hydroxylamine Oxidoreductase Experiments

Because *N. europaea* grow relatively slowly under favorable conditions and presumably grow even more slowly when stressed or injured, traditional methods of assessing cell viability based on growth can be misleading. Instead, to evaluate whether decreases in activity resulted from changes in specific enzyme activity or from partial death of the culture, we separately examined AMO- and HAO-dependent activities prior to and during TCE exposure. Because of its anticipated close association with AMO and its relatively large size (200 kDa), if a reactive, short-lived TCE oxidation product was to diffuse away from the AMO active site and cause damage to other cellular constituents, it seems reasonable that HAO would be a likely target. The integrity of HAO (and other, HAO-dependent, electron transport proteins to the terminal electron acceptor) can be evaluated separately by specifically blocking AMO activity, providing an alternative HAO substrate, and measuring HAO-dependent O_2 uptake rates. If HAO-dependent activity remains essentially unchanged after TCE exposure, insignificant damage to HAO and other electron transport proteins may be inferred, from which it may be inferred that TCE-associated protein damage is restricted primarily to AMO and that recovery is possible. In our experiments, hydrazine was provided as an alternative substrate for HAO after specifically blocking AMO activity with allyl thiourea (ATU). Details are described elsewhere (Rasche et al. 1991; Ely et al. 1995b).

MODEL DEVELOPMENT

Results from TCE degradation experiments conducted with the quasi-steady-state reactor system led to several observations: (1) inhibition and inactivation effects could be observed concurrently in real time; (2) inhibition of AMO was much more significant initially than inactivation, although inactivation became more significant than inhibition in the long term; (3) recovery capability was apparent even while TCE was present in the reactor and TCE degradation was taking place; (4) a conceptual and mathematical model was needed; and (5) previously published models did not adequately described our experimental results. Consequently, a model was developed by considering interactions between the enzyme (AMO), growth substrate (NH_3), and non-growth substrate (TCE), using the following simplified reaction equations:

$$E + S \underset{k_{-1}}{\overset{k_1}{\rightleftharpoons}} ES \xrightarrow{k_2} E + P_1 + \text{some amount of new enzyme, } E_{new} \qquad (1)$$

$$E + I \underset{k_{-3}}{\overset{k_3}{\rightleftharpoons}} EI \xrightarrow{k_4} E\left(1 - \frac{E'}{E}\right) + E' + P_2 \qquad (2)$$

$$ES + I \underset{k_{-5}}{\overset{k_5}{\longleftrightarrow}} ESI \qquad (3)$$

where E represents AMO, S is NH_3, P_1 is NO_2^-, I is nongrowth substrate (TCE, in this case), and P_2 is the product(s) of nongrowth substrate conversion. These equations differ from traditional enzyme reaction equations in two ways: (1) they indicate that some amount of enzyme (E') is lost (i.e., inactivated) as the nongrowth substrate, I, is converted; and (2) they include a term (E_{new}) to account for respondent synthesis of some amount of new enzyme (recovery). E' is assumed equal to some undefined function, f, of the amount of nongrowth substrate oxidized to product(s), P_2, i.e., $E' = f(P_2)$. In addition, because energy is required to support enzyme synthesis, E_{new} is assumed equal to some function, g, of the amount of growth substrate oxidized to final product(s), P_1, i.e., $E_{new} = g(P_1)$.

The traditional Michaelis-Menten modeling approach assumes thermodynamic equilibrium between the enzyme-substrate and enzyme-inhibitor association/dissociation reactions, i.e., that [E][S]/[ES] is equal to a constant (K_s) and that [E][I]/[EI] is equal to a constant (K_I) (Fersht 1985). Because this equilibrium condition will occur only if $k_2 << k_{-1}$ and $k_4 << k_{-3}$, Briggs and Haldane proposed a more general approach. However, their approach relies on the assumed existence of a steady-state condition, i.e., that the concentrations of enzyme-substrate complex and enzyme-inhibitor complex are constant ($^{d[ES]}/_{dt} = 0$ and $^{d[EI]}/_{dt} = 0$). Clearly, in a situation in which enzyme is being destroyed by inactivation and new enzyme is being synthesized, such an assumption may not hold. Nevertheless, if the net rate of enzyme loss or creation is slow compared to enzyme-substrate and enzyme-inhibitor association/dissociation rates, the process may be modeled adequately as pseudo-steady-state. In other words, it may be assumed that ($^{d[ES]}/_{dt} \approx 0$ and $^{d[EI]}/_{dt} \approx 0$). The error introduced in making this assumption requires experimental evaluation.

A mathematical model was derived from Equations 1, 2, and 3, based on the following assumptions: (1) the process may be modeled as pseudo-steady-state; (2) enzyme inactivation is proportional to nongrowth substrate conversion; and (3) new enzyme synthesis is proportional to the oxidation of growth substrate to final product. Excluding cell growth, the model may be expressed as a system of two ordinary, nonlinear differential equations:

$$v = \frac{1}{X}\frac{dP_1}{dt} = \frac{(k - k_{inact}P_2 + k_{rec}P_1)S}{K_m\left(1 + \frac{I}{K_{I1}}\right) + S\left(1 + \frac{I}{K_{I2}}\right)} \qquad (4)$$

$$v_I = \frac{1}{X}\frac{dI}{dt} = \frac{\frac{k_1}{k}(k - k_{inact}P_2 + k_{rec}P_1)I}{K_{I1}\left[\left(1 + \frac{S}{K_m}\right)\left(1 + \frac{I}{K_{I2}}\right)\right] + I} \qquad (5)$$

(Another equation may be added to account for cell growth, if desired.) In these equations, k_{inact} and k_{rec} are assumed constant; k_{inact} relates decreases in culture activity to nongrowth substrate conversion while k_{rec} relates increases in culture activity to growth substrate oxidation to final product. All other terms are as typically defined and used. The model may be solved by numerical integration and nonlinear regression/optimization. Model development, analyses, solution, and Monte Carlo testing are described more fully in Ely et al. (1995a).

RESULTS

Quasi-steady-state reactor experiments were conducted with several combinations of culture specific activity (0.20 to 0.54 μmol NH_3/min-mg protein), biomass concentration (57.8 to 189 mg protein/L), and initial TCE liquid-phase concentrations (4.5 μM (590 μg/L) to 26.8 μM (3.5 mg/L)). Figure 1 shows nitrite concentration (P_1), TCE concentration (I), and O_2 uptake rate data (a measure of dP_1/dt) obtained in a representative experiment, along with best-fit model curves. Since only five of the six model parameters are independent (Ely et al. 1995a), K_m was constrained to 40 ± 4 μM NH_3, consistent with reported values (Suzuki et al. 1974; Hyman and Wood 1983), to solve the model and obtain parameter estimates.

TCE was shown to be a potent, competitive inhibitor of AMO activity, with an estimated K_I (identified as K_{I1} in the model equations) of 10.7 ± 6.3 μM TCE (mean ± standard deviation for 13 experiments), though it was oxidized quite slowly (k_I = 11.0 ± 5.3 nmol TCE/min-mg protein) compared to ammonia oxidation (k = 830 ± 150 nmol NH_3/min-mg protein). Inactivation appeared proportional to TCE oxidation (k_{inact} = 0.069 ± 0.021 μmol NO_2^--L/μmol TCE-min-mg protein), as had been assumed, while k_{rec} estimates were more variable (0.26 ± 0.23 μmol NO_2^--L/mmol NO_2^--min-mg protein). Initial and final protein measurements showed no significant cell growth to occur. Comparison with Monte Carlo results (Ely et al. 1995a) suggested that most of the variability in parameter estimates resulted from data variability. Results from experiments examining the effects of TCE on AMO- and HAO-dependent activities indicated that no significant damage occurred to HAO or other electron transport proteins at TCE concentrations up to about 27 μM (3.5 mg/L), even though AMO was substantially inactivated. In a separate analysis, it was estimated that the assumption of a pseudo-steady-state condition made in deriving the model introduced insignificant error in these experiments (0.000002 to 0.0002%).

DISCUSSION

Distinct inhibition, inactivation, and recovery effects were observed in these experiments, and model parameters were estimated. Precise k_{rec} estimates

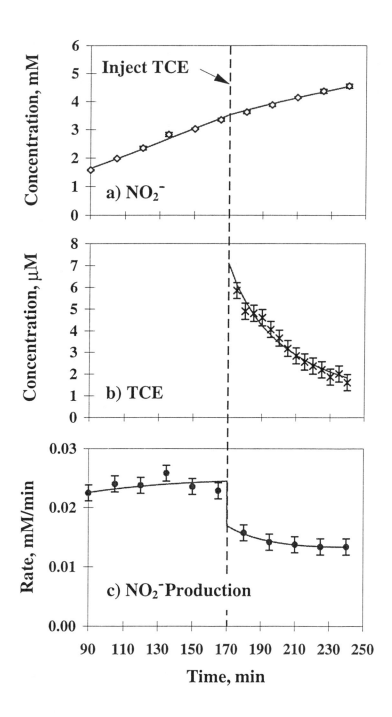

FIGURE 1. Observed nitrite concentration (◇), TCE concentration (×), and O_2 uptake rate (converted to NO_2^- production rate using observed stoichiometry) (●), with error bars, and best-fit model curves for a representative experiment. TCE was injected at t = 170 min, as indicated.

were difficult to obtain (see sensitivity analysis in Ely et al. 1995a), but a statistically significant recovery effect (as measured by perturbations in χ^2 values) was observed in 8 of 12 experiments in which recovery was possible. (Recovery was precluded in one of the 13 experiments by adding rifampicin, a transcription inhibitor.) Because of the difficulty in obtaining precise estimates of k_{rec}, failure of the model to detect statistically significant recovery in 4 of the 12 experiments does not necessarily indicate that recovery did not occur. Data variability prevented detection of recovery in some Monte Carlo analyses, even though recovery was included in the Monte Carlo data (Ely et al. 1995a). Results suggested that k_{rec} may tend to be higher at higher TCE dosages. We have speculated (Ely et al. 1995b) that this could be due to the creation of a higher impetus for recovery when inhibition and/or inactivation are greater. Additional and/or longer term experiments are needed to clarify this point.

Limitation of TCE-related damage primarily to AMO-dependent activity and actualization of recovery in these experiments are significant and suggest the following observations: (1) with the exception of AMO-dependent activity, electron transport proteins could function to provide a trans-membrane proton gradient; (2) the cell membrane was sufficiently intact to maintain the proton gradient; (3) ADP phosphorylation systems were adequately operational; and (4) protein synthesis and supporting systems (i.e., CO_2 fixation, DNA, mRNA, tRNA, etc.) were functional. Based on these observations, we conclude that decreases in culture activity resulted primarily from changes in specific activity, i.e., active enzyme level within live cells, rather than from cell death. Bacterial capability to recover from injury suggests that sustainable approaches for cometabolic degradation of chlorinated solvents may be achievable if conditions are controlled within acceptable limits.

REFERENCES

Alvarez-Cohen, L., and P. L. McCarty. 1991. "Product toxicity and cometabolic inhibition modeling of chloroform and trichloroethylene transformation by methanotrophic resting cells." *Appl. Environ. Microbiol.* 57(4):1031-1037.

Criddle, C. S. 1993. "The kinetics of cometabolism." *Biotechnol. Bioeng.* 41:1048-1056.

Ely, R. L., K. J. Williamson, R. B. Guenther, M. R. Hyman, and D. J. Arp. 1995a. "A cometabolic kinetics model incorporating enzyme inhibition, inactivation, and recovery. 1. Model development, analysis, and testing." *Biotechnol. Bioeng.* 46(3):218-231.

Ely, R. L., M. R. Hyman, D. J. Arp, R. B. Guenther, and K. J. Williamson. 1995b. "A cometabolic kinetics model incorporating enzyme inhibition, inactivation, and recovery. 2. Trichloroethylene degradation experiments." *Biotechnol. Bioeng.* 46(3):232-245.

Fersht, A. 1985. *Enzyme structure and mechanism.* 2nd ed. W. H. Freeman, New York, NY.

Fox, B. G., J. G. Borneman, L. P. Wackett, and J. D. Lipscomb. 1990. "Haloalkane oxidation by the soluble methane monooxygenase from *Methylosinus trichosporium* OB3b: mechanistic and environmental implications." *Biochem.* 29(27):6419-6427.

Hyman, M. R., and P. M. Wood. 1983. "Methane oxidation by *Nitrosomonas europaea*." *Biochem. J.* 212:31-37.

Oldenhuis, R., J. Y. Oedzes, J. J. van der Waarde, and D. B. Janssen. 1991. "Kinetics of chlorinated hydrocarbon degradation by *Methylosinus trichosporium* OB3b and toxicity of trichloroethylene." *Appl. Environ. Microbiol.* 57(1):7-14.

Rasche, M. E., M. R. Hyman, and D. J. Arp. 1991. "Factors limiting aliphatic chlorocarbon degradation by *Nitrosomonas europaea*: cometabolic inactivation of ammonia monooxygenase and substrate specificity." *Appl. Environ. Microbiol.* 57(10):2986-2994.

Suzuki, I., U. Dular, and S. C. Kwok. 1974. "Ammonia or ammonium ion as substrate for oxidation by *Nitrosomonas europaea* cells and extracts." *J. Bacteriol.* 120(1):556-558.

Wackett, L., and S. Householder. 1989. "Toxicity of trichloroethylene to *Pseudomonas putida* F1 is mediated by toluene dioxygenase." *Appl. Environ. Microbiol.* 55(10):2723-2725.

TCE Degradation by Toluene/Benzene Monooxygenase of *Pseudomonas aeruginosa* JI104 and *Escherichia coli* Recombinant

Jun-ichi Koizumi and Atsushi Kitayama

ABSTRACT

Pseudomonas aeruginosa JI104 incorporates more than three degradation pathways for aromatic compounds such as benzene, toluene, and xylene. A dioxygenase and two monooxygenases were cloned in *Escherichia coli* XL1-Blue. The dioxygenase yielding *cis*-toluene dihydrodiol and one of the monooxygenases producing *o*-cresol from toluene did not exhibit conspicuous activity in trichloroethylene (TCE) oxygenation, although DNA sequencing proved that the former enzyme was an isozyme of toluene dioxygenase of the known TCE decomposer *P. putida* F1. The other toluene/benzene monooxygenase that could generate *o*-, *m*-, and *p*-cresol simultaneously from toluene showed TCE oxygenation activity resulting in TCE decomposition in *E. coli*. The activity was inhibited competitively by toluene, ethylbenzene, and *o*- and *m*-xylene: their inhibition constants were greater than those of propylbenzene and *p*-xylene. When the *E. coli* recombinant harboring the monooxygenase was induced by isopropyl β-D-thiogalactopyranoside (IPTG) and incubated in the absence of toluene, TCE degradation activity decreased during incubation, compared to that with toluene. Toluene probably controlled the lifetime of the enzyme.

INTRODUCTION

Trichloroethylene (TCE) is a volatile chlorinated compound and a suspected carcinogen. Almost all of the prefecture governments in Japan have reported TCE contamination in soil and groundwater. For about the last 3 years, they have considered bioreclamation technologies as a potential method for remediating TCE contamination, if the control techniques of microbes can completely degrade TCE. This paper describes the TCE-degrading activities of *Pseudomonas aeruginosa* JI104 and the *Escherichia coli* recombinant harboring the gene responsible for TCE oxygenation.

EXPERIMENTAL PROCEDURES AND MATERIALS

Strains and Plasmids

Pseudomonas aeruginosa JI104 isolated from a soil in a gasworks was used. *E. coli* XL1-Blue (Δ*lac*), *endA1, gyrA96, hsdR17* (r_k^-, m_k^-), *recA1, relA1, supE44, thi-1*, [F', *lacI*^{q}, *lacZ*ΔM15, *proAB*, Tn10 (tet^r)]) and HB101 (F$^-$, *hsd20* (r_B^-, m_B^-), *recA13, ara-14, proA2, lacY1, galK2, rpsL20* (Sm^r), *xyl-5, mtl-1, supE44*, (λ$^-$, *mcrA*$^+$, *mcrB*$^-$) were employed as hosts, and pUC19 and Bluescript plasmid were used as cloning vectors.

TCE Degradation

P. aeruginosa JI104 was grown on a mineral salts medium containing 0.2% (w/v) L-arginine and 0.05 (w/v) glucose at 30°C, whereas *E. coli* was grown on LB medium at 37°C. To induce the oxygenases, toluene or IPTG was used. Cells were harvested by centrifugation at 15,000 × g for 5 min, and resuspended in mineral salts medium to a turbidimetric density at 600 nm (OD_{600}) of 1.0 to 6.5. A 2-mL cell suspension was dispensed into a 13.4-mL glass vial that was crimp-sealed with Teflon™-coated rubber septum. An appropriate volume of water saturated with TCE was added into a vial (e.g., 18.2 μL for 20 μg-TCE/vial). Toluene was supplemented at the final concentration of 2 mM in the suspension when required. These vials were incubated with shaking at 25°C. At appropriate intervals, 25-μL headspace samples were taken with a gastight syringe and analyzed on a gas chromatograph (GC) equipped with a flame ionization detector (FID). GC-FID also was used for the metabolites from various aromatic compounds.

RESULTS AND DISCUSSION

TCE Degradation by *P. aeruginosa* JI104

P. aeruginosa JI104 was able to degrade TCE (Fig. 1). When toluene was cosupplemented in the vial (Fig. 1), around 3 to 4 h of lag time for TCE degradation was observed before the toluene was consumed to an undetectable level. Toluene acted as a competitive inhibitor for TCE degradation. The effect of toluene as a competitive inhibitor was corroborated with the resupplement experiment of toluene in the vial after all the toluene had been degraded (data not shown). Although it might be misinterpreted from Fig. 1 that the initial degradation rate was influenced by the TCE concentration because of the relative value on the vertical axis, the initial TCE decomposition by JI104 was a zero-order reaction, i.e., the initial velocity of TCE degradation was independent of TCE concentration and was ca. 2 μg/mL·OD_{600}·h. Without the supplement of toluene to the incubation vial, TCE degradation leveled off even though the cells had been induced by toluene (data not shown).

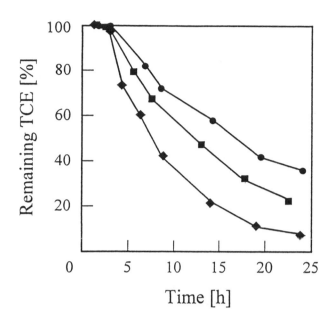

FIGURE 1. TCE degradation by *P. aeruginosa* JI104. TCE charged was 20 (●), 40 (■), or 60 (◆) µg/vial.

Recombinant Plasmids and *E. coli* Harboring the Recombinant Plasmids

Through cloning experiments it was found that *P. aeruginosa* JI104 incorporated multiple catabolic pathway for aromatic compounds (Kitayama et al. 1995, personal communication). Three different fragments expressing the catabolization activity in *E. coli* were obtained as recombinant plasmids named pIB1101, pBsR201, and pBP1. Incubated with toluene, the recombinant cells with pBsR201 and pBP1 generated *o*-cresol and *cis*-toluene dihydrodiol, respectively; the latter recombinant also oxygenated biphenyl (Fig. 2). It was interesting that the cells transformed with pIB1101 produced *o*-, *m*-, and *p*-cresol from toluene (Fig. 2 and Fig. 3B) to exhibit low regiospecificity (Kitayama et al. 1995, personal communication). The low regiospecificity was also observed for *o*-xylene (Fig. 3C), ethylbenzene (Fig. 3F), and *n*-propylbenzene (Fig. 3G) (Kitayama et al. 1995, personal communication).

TCE Degradation by *E. coli* Harboring the Recombinant Plasmids

Among three recombinants, only *E. coli* harboring pIB1101 could degrade TCE (Fig. 4). The remaining two were not effective for TCE degradation. The TCE-degrading rate of *E. coli* with pIB1101 depended on the cell density. A proportional increase in the initial rate of degradation of TCE was observed throughout the range studied here, and TCE was decomposed to 10%

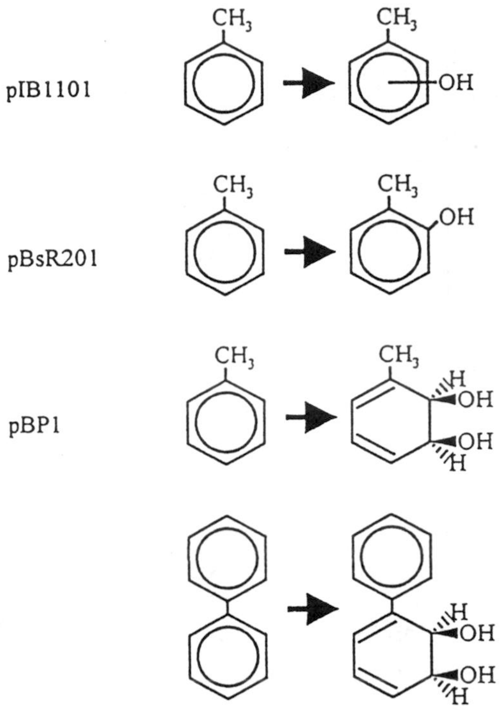

FIGURE 2. Products from toluene and/or biphenyl by *E. coli* recombinants with pIB1101, pBsR201, and pBP1.

remaining by 5 h when the cell density was 6.5 of OD at 600 nm. All of the aromatic compounds examined in Fig. 3 inhibited TCE decomposition by the transformant with pIB1101, and the inhibition constants of ethylbenzene, *o*-xylene, and *m*-xylene were greater than those of *p*-xylene and *n*-propylbenzene (data not shown).

The specificity of enzyme reaction (Fig. 2) implied that pBsR201 and pBP1 encoded the isozymes of oxygenases of the known TCE decomposers, *P. cepacia* G4 and *P. putida* F1, respectively. However, neither of the transformants could degrade TCE. DNA sequencing showed that the dioxygenase in pBP1 was exactly the same as in *bphA* of *P. putida* KF707 (data not shown), and there remains a question why the dioxygenase cloned in pBP1 was able to catabolize toluene. Another DNA sequencing revealed that the monooxygenase in pIB1101 was encoded in an operon consisting of 6 open reading frames (data not shown). On the 4th frame, two putative ribosome-binding sequences were found (data not shown).

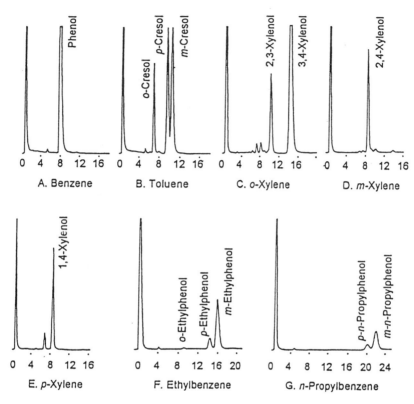

FIGURE 3. Gas chromatography of the products formed from aromatic hydrocarbons by *E. coli* recombinant with pIB1101.

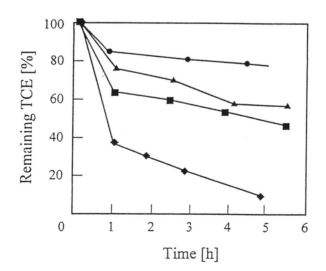

FIGURE 4. TCE degradation by *P. aeruginosa* JI104. Optical density of cells at 600 nm was 1.0 (●), 2.0 (▲), 3.0 (■), or 6.5 (◆).

Cometabolic Biodegradation of Trichloroethylene in a Biofilm Reactor

Jean-Pierre Arcangeli, Erik Arvin, and Hanne Møller Jensen

ABSTRACT

Cometabolic degradation of trichloroethylene (TCE) in an aerobic biofilm system with toluene as primary substrate was investigated. TCE degradation rate was first-order, giving an average first-order surface removal rate constant, $k_{1,a}$, of 0.26 m/d. TCE was probably degraded by a toluene-induced enzyme. However, if toluene was provided in high concentrations, degradation of TCE was inhibited. Furthermore, it appeared that TCE inhibited toluene degradation. This inhibition increased with the TCE concentration in the reactor, but it decreased with an increasing toluene concentration. We conclude that these interactions could be the result of a competitive inhibition between TCE and toluene. Practically, this shows that degradation of TCE can be maximized if an optimum concentration of toluene is provided. An example presented in this paper reveals that the optimum toluene concentration was in the range of 200 to 500 µg/L for a TCE inlet concentration of 135 µg/L. Under these optimal conditions, the TCE degradation rate was 0.045 g m^{-2} d^{-1}, leading to a first-order surface removal rate constant of 0.4 m/d and a transformation yield of 0.05 g TCE/g toluene degraded.

INTRODUCTION

Chlorinated aliphatic hydrocarbons (CAHs) are probably the most serious groundwater contaminants today. The compounds migrate quickly through soils, and many are persistent under aerobic conditions. In laboratory experiments and in pilot-scale field studies, aerobic cometabolic degradation of CAHs has been achieved, and has been shown to depend on the supplement of a primary substrate like methane, phenol, or toluene for growth of bacteria. Although the biodegradability of CAH compounds is well established with suspended culture, comparatively little is known about the kinetics of biodegradation in biofilm systems. This paper discusses the cometabolic degradation of trichloroethylene (TCE) with a mixed culture of toluene-oxidizing bacteria in a biofilm system.

MATERIALS AND METHODS

Experimental Setup

The experimental system is shown in Figure 1. The bioreactor used is a so-called biodrum reactor consisting of a drum rotating inside another one (Kristensen & Jansen 1980). The biofilm grew both on the stator and the rotator. The rotation (200 rpm), as well as the recycling systems, nearly ensure total mixing of the bulk liquid and therefore provide a relatively uniform biofilm growth in the reactor. The characteristics of the biofilm reactor and the operating conditions for the experiments are described in Table 1. The inoculation procedure and the composition of the mineral medium are reported in Arcangeli and Arvin (1992).

Analytical Techniques

Analysis of TCE and toluene were conducted isothermally at 50°C on a DANI 8520 gas chromatograph equipped with a flame ionization detector for

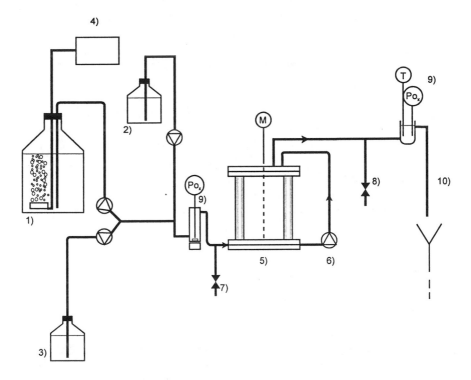

FIGURE 1. Experimental setup: (1) mineral medium, (2) toluene stock solution, (3) TCE stock solution, (4) aerator (oxygen supply), (5) biofilm reactor, (6) recirculation pump, (7) inlet sampling port, (8) outlet sampling port, (9) temperature and oxygen control, (10) outlet.

TABLE 1. Operational data for the biofilm reactor during the biofilm growth.

	Unit	Value
Surface area	m^2	0.16
Reactor volume	m^3	0.96×10^{-3}
Average flow of water	L/h	4.5
Residence time	min.	12.8
Toluene infl. concentration	mg/L	2.5
Oxygen in the reactor	mg/L	4 to 5
Nitrate infl.	mg N/L	40
Phosphate infl.	mg P/L	46.5
Temperature	°C	20 to 21
Alkalinity	meq/L	1.6
pH in the reactor		6.9 to 7.1

the analysis of toluene and with an electron capture detector for the analysis of chlorinated compounds. The GC column was a 30 m J&W DB5 capillary column, i.d. 0.53 mm, and the carrier gas used was nitrogen. Extraction of toluene and TCE was carried out in a 10-mL volumetric flask with 500 µL of pentane; the solution was shaken vigorously for 2 min; 1 µL of pentane was injected into the GC using heptane (6 mg/L) as an internal standard for toluene, and dibromochloromethane (0.28 mg/L) as an internal standard for TCE. Lithium was analyzed on an atomic absorption spectrophotometer Perkin Elmer 370. Biofilm parameters (thickness and dry weight density) were measured according to the procedure reported by Arcangeli and Arvin (1992).

Experimental Strategy

The reactor was inoculated with a mixed culture obtained from a creosote-polluted sandy aquifer (Frederiksborg, Denmark). Biofilms were grown aerobically with toluene as a sole carbon source (Run 1 to 4). For different levels of biofilm thickness, degradation experiments were performed with TCE and toluene. Results related to the biofilm growth and to the kinetics of toluene biodegradation are reported in Arcangeli and Arvin (1992).

The first type of degradation study was a series of injection experiments. They consisted of injecting certain amounts of TCE, together with a conservative tracer (LiCl), in the reactor while continuously feeding with toluene. Because the reactor was completely mixed, the time response to a pulse increase of the concentration of lithium and TCE in the input at time zero leads to an exponential decrease of the lithium and TCE concentration versus time in the outlet of the reactor. For a TCE concentration ranging from 0.05 to 2 mg/L, TCE was biodegraded according to a first-order reaction (Jensen 1994). Consequently, a mass balance for TCE and for the conservative tracer, lithium, gives the following equations.

$$\frac{S_{TCE}}{S_{TCE,o}} = EXP\left[-(\frac{Q}{V} + k_{1,a}\frac{A}{V})\,t\right] \qquad \text{for TCE} \qquad (1)$$

$$\frac{S_{Li}}{S_{Li,o}} = EXP\left[-\frac{Q}{V}t\right] \qquad \text{for lithium} \qquad (2)$$

with

$$k_{1,a} = k_{1,f}\,L\,\varepsilon = \frac{k_x}{K_s}\,X_f\,L\,\varepsilon \qquad (3)$$

and

$$\varepsilon = \frac{Tanh\,\alpha}{\alpha} \qquad (4) \qquad\qquad \text{where} \quad \alpha = \sqrt{\frac{k_{1,a}\,L}{\varepsilon\,D}} \qquad (5)$$

Where A   = surface area (m^2)
       D   = diffusion coefficient of the substrate (d^{-1})
       $k_{1,a}$ = first-order surface removal rate constants (m/d)
       $k_{1,f}$ = first-order intrinsic reaction rate constants in the biofilm (d^{-1})
       k_x = maximum substrate utilization rate (d^{-1})
       K_s = Monod's constant (half-saturation constant) (mg/L)
       L   = biofilm thickness (m)
       Q   = liquid flow (m^3/d)
     $S_{TCE,o}$ = TCE concentration at time zero (mg/L)
     $S_{Li,o}$ = lithium concentration at time zero (mg/L)
     S_{TCE} = TCE concentration (mg/L)
     S_{Li} = lithium concentration (mg/L)
       t   = time (day)
       V   = volume of the reactor (m^3)
       X_f = biofilm dry weight density (g m^{-3})
       α = biofilm constant (–)
       ε = first-order efficiency (–).

Therefore, if plotted in dimensionless units, the difference between the TCE and the lithium concentration in the reactor is a measure of the TCE degradation. For further details about the theoretical background, refer to Harremoës (1978).

The second type of degradation study was performed for another biofilm growth (inoculated with the same culture). It consisted of supplying the reactor continuously with toluene and TCE to investigate the influence of the toluene concentration on the TCE removal rate and the inhibition effect of TCE on the toluene removal rate.

RESULTS

Kinetic data related to the injection experiments are compiled in Table 2. A control test (run 4) was performed to evaluate the loss of TCE due to absorption in the reactor and in the biomass. The results of run 4 suggested that for the "injection experiments," absorption could be neglected.

Kinetics of TCE Biodegradation

An average first-order surface removal rate constant, $k_{1,a}$, of 0.26 ± 0.07 m/d was found for runs 3a-f with a toluene concentration in the reactor of approximately 1 mg/L. However, if the toluene concentration was lowered (runs 1a-c, 2), the first-order TCE degradation rate constant, $k_{1,a}$, decreased with an increasing TCE concentration (Figure 2a and 2b). The effect of toluene on TCE biodegradation is obvious in Figure 3. In this experiment both toluene and TCE were supplied continuously into the reactor. The toluene concentration in the reactor was varied in the range from 0 to 12 mg/L whereas the TCE inlet concentration was kept constant at 135 µg/L. The TCE degradation was measured for

TABLE 2. Kinetic data for the cometabolic biodegradation of TCE with toluene as a primary substrate, under aerobic conditions.

Run #	$S_{TCE,o}$ (mg/L)	S_{tol} (mg/L)	$k_{1,a}$ (m/d)	k_1' (L mg^{-1} d^{-1})	L (µm)	X_f (g m^{-3})	T (°C)
1a	0.05	0.07	0.859	0.643	373	20,900	25
1b	0.1	0.07	0.486	0.207	375	20,900	25
1c	0.3	0.07	0.269	0.069	373	20,900	25
2	2	0.32	0.045	0.0064	200	38,600	20
3a	0.05	0.91	0.242	0.088	520	12,000	20
3b	0.1	0.96	0.392	0.219	520	12,000	20
3c	0.5	0.97	0.252	0.094	520	12,000	20
3d	1	0.97	0.2	0.062	520	12,000	20
3e	1.5	0.99	0.211	0.068	520	12,000	20
3f	2	1.01	0.244	0.089	520	12,000	20
4 (Control)	1.8	0	0.008	$8.4 \ 10^{-4}$	860	12,000	20

Nomenclature: k_1': first-order specific transformation rate ($= k_X/K_S$) (L/mg dry weight d^{-1}); S_{tol}: toluene concentration (mg/L); T: temperature (°C).

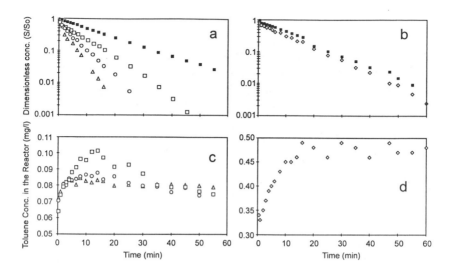

FIGURE 2. Example of injection experiments: (■) lithium. (Δ) Run 1-a: $S_{TCE,o}$ = 0.05 mg/L; (○) Run 1-b: $S_{TCE,o}$ = 0.1 mg/L. (□) Run 1-c: $S_{TCE,o}$ = 0.3 mg/L. (◇) Run 2: $S_{TCE,o}$ = 2 mg/L.

different toluene concentrations in the reactor. Figure 3 shows that, if no toluene was supplied in the reactor, the TCE was not degraded. On the other hand, if the toluene was provided in concentrations that were too high, the degradation of TCE was inhibited. In this experiment, a maximum TCE removal rate of 0.045 g m^{-2} d^{-1} was obtained, with a toluene concentration in the reactor ranging from 0.2 to 0.5 mg/L. This leads to a first-order rate constant, $k_{1,a}$, for TCE of 0.4 m/d.

TCE Inhibition

The presence of TCE in the reactor seemed to inhibit toluene degradation. This inhibition was observed in runs 1a-c and 2 where the toluene concentration in the reactor was lower than in any of the other degradation tests listed in Table 2. Furthermore, the inhibition appeared to be temporary if small concentrations of TCE were tested (run 1a-c; Figure 2c). However, if high concentrations of TCE were investigated (run 2; Figure 2b-d), the biofilm stopped to grow and the toluene removal dropped from 95 to 55%, indicating an irreversible destruction of the biofilm. This inhibition phenomenon was further investigated in a reactor fed with toluene and TCE continuously. The TCE concentration was increased in the reactor, while the toluene concentration was maintained constant in the reactor inlet. The toluene removal rate decreased even at very low TCE concentrations (below 50 μg/L). However, this inhibition appeared to be temporary, because the toluene removal rate came back to its initial level as the TCE supply in the reactor was stopped (data not shown).

DISCUSSION

Wackett and Gibson (1988) reported a first-order specific transformation rate, k_1', of 0.032 L/mg protein d^{-1} at 30°C for *Pseudomonas putida* F1. Assuming a conversion factor of 0.5 mg protein/mg cell dry weight, this kinetic parameter is in the same order of magnitude as those listed in Table 2. However, the rate of degradation of TCE is much lower compared with the rate achieved with the methanotrophic bacterium *Methylosinus trichosporium* OB3b, for which Oldenhuis et al. (1989) reported a k_1' of 6.2 L/mg protein d^{-1} at 30°C.

Data compiled in Table 2 showed inconsistencies among first-order rate constants, $k_{1,a}$. For example, in runs 3a-f, the $k_{1,a}$ values were within the same order of magnitude; whereas in run 1a-c, an inverse relationship was observed between the degradation and the concentration of TCE. On the other hand, the $k_{1,a}$ values observed in run 1a-b were higher than those found in run 3a-b, indicating that the $k_{1,a}$ values increased with a decreasing toluene concentration.

These inconsistencies revealed a cross-inhibition phenomenon between toluene and TCE, where the presence of toluene affected the kinetics of TCE degradation and vice versa. These interactions may be explained by competitive inhibition, if it is assumed that TCE and toluene are degraded by the same enzyme and that this enzyme has a lower affinity for TCE than for toluene. The competitive inhibition affects the transformation yield, i.e., the quantity of TCE degraded per unit of degraded toluene. For each data point shown in Figure 3, the transformation yield was calculated. The value was about 0.05 g TCE/g toluene at the maximum TCE degradation rate when the concentration of toluene ranged from 0.2 to 0.5 mg/L. However, as the toluene concentration in the

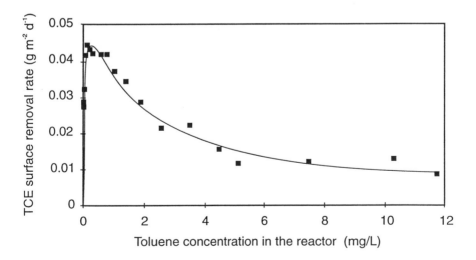

FIGURE 3. Effect of toluene concentration on TCE removal. The TCE inlet concentration was 135 µg/L. Biofilm average thickness: 538 µm; biofilm dry weight density: 24,400 g/m^3.

reactor increased, the transformation yield decreased dramatically to a value of 0.006 g TCE/g toluene for a toluene concentration above 5 mg/L.

Similar interactions have been observed for the aerobic biodegradation of benzene, toluene and p-xylene (Iversen 1992; Chang et al. 1993), for the biodegradation of toluene and o-xylene under denitrifying conditions (Arcangeli & Arvin 1995), and for the biotransformation of TCE by methanotrophic bacteria (Broholm et al. 1992; Semprini et al. 1991). Consequently, the knowledge of mechanisms controlling substrate interactions and the kinetics are of practical interest for the design of treatment processes for groundwater remediation, because the biodegradation of some pollutants can be maximized using the optimum concentration of primary substrate.

CONCLUSION

The degradation of TCE by a mixed toluene-oxidizing culture was controlled by a first-order reaction within a TCE concentration range from 0.05 to 2 mg/L. An average first-order rate constant, $k_{1,a}$, of 0.26 ± 0.07 m/d was found. However, the interaction phenomena observed between TCE and toluene revealed that TCE degradation could be maximized if an optimum concentration of toluene was maintained in the reactor. This optimum concentration of toluene ranged from 0.2 to 0.5 mg/L for a TCE inlet concentration of 135 µg/L. In this case, the TCE degradation rate was 0.045 g/m^2 d^{-1}, giving a first-order rate constant, $k_{1,a}$, of 0.4 m/d and a transformation yield of 0.05 g TCE/g toluene degraded.

ACKNOWLEDGMENT

This study was funded by the Commission of the European Communities. Contract EV5V-CT92-0239; BIODEC project.

REFERENCES

Arcangeli, J. P., and E. Arvin. 1992. "Toluene biodegradation and biofilm growth in an aerobic fixed film reactor." *Appl. Microbiol. Biotechnol.* 37:510-517.
Arcangeli, J. P., and E. Arvin. 1995. "Cometabolic transformation of o-xylene in a biofilm system under denitrifying conditions." *Biodegradation.* 6:19-27.
Broholm K., T. H. Christensen, and B. K. Jensen. 1992. "Modelling TCE degradation by a mixed culture of methane-oxidizing bacteria." *Wat. Res.* 26:1177-1185.
Chang, M. K., V. C. Voice, and C. S. Criddle. 1993. "Kinetics of competitive inhibition and cometabolism in the biodegradation of benzene, toluene, and p-xylene by two *Pseudomonas* isolates." *Biotech. and Bioeng.* 41:1057-1065.
Harremoës, P. 1978. "Biofilm kinetics." In R. Mitchell (Ed.), *Water Pollution Microbiology*, Vol. 2, pp. 82-109. John Wiley, New York, NY.

Jensen, H. M. 1992. "Kinetics of toluene and benzene biodegradation. Statistical analysis." Masters thesis report (In Danish). The Institute of Mathematical Statistics and Operations Research, The Technical University of Denmark, DK 2800 Lyngby, Denmark.

Jensen, H. M. 1994. "Cometabolic degradation of chlorinated aliphatic hydrocarbons." Ph.D. Thesis. Department of Environmental Engineering, The Technical University of Denmark, DK 2800 Lyngby, Denmark.

Kristensen, G. H., and J.l.C. Jansen. 1980. *Fixed film kinetics. Description of laboratory equipment.* Department of Environmental Engineering, Technical University of Denmark, DK 2800 Lyngby, Denmark.

Oldenhuis, R., R.L.j.M. Vink, D. B. Janssen, and B. Witholt. 1989. "Degradation of chlorinated aliphatic hydrocarbons by *Methylosinus trichosporium* OB3b expressing soluble methane monooxygenase." *Appl. Environ. Microbiol.* 55:2819-2826.

Semprini, L., G. D. Hopkins, P. V. Roberts, D. Grbić-Galić, and P. L. McCarty. 1991. "A field evaluation of *in-situ* biodegradation of chlorinated ethenes: Part 3, Studies of competitive inhibition." *Ground Water.* 29:239-250.

Wackett, L. P., and D. T. Gibson. 1988. "Degradation of trichloroethylene by toluene dioxygenase in whole-cell studies with *Pseudomonas putida* F1." *Appl. Environ. Microbiol.* 54:1703-1708.

Biofouling Effects on In Situ TCE Bioremediation by Phenol Utilizers

Thomas R. MacDonald and Peter K. Kitanidis

ABSTRACT

In situ bioremediation involves stimulating the growth of bacteria within the contaminated region of an aquifer to break down the contaminants. This large bacteria population often can have the unwanted effect of clogging the porous media. The clogging reduces the porosity and the hydraulic conductivity of the soil. These hydrodynamic changes can affect the flow of groundwater used to deliver nutrients to the bacteria in the contaminated region of the aquifer. We developed a mathematical model to study the impact of biofouling on a recirculation well flow system used to mix nutrients with contaminated groundwater. The insights gained from this examination can aid in designing a system to minimize biofouling problems.

INTRODUCTION

Enhanced in situ bioremediation of contaminated aquifers relies on stimulating the growth of bacteria to increase the rate at which the contaminant is destroyed. A larger population can degrade more of the contaminants and thereby clean up the aquifer in a shorter time. The bacteria population is increased by delivering the necessary nutrients through some type of well system, in this case, a recirculation well. The enlarged bacteria population, although necessary for enhanced in situ bioremediation, can have potentially deleterious side effects, such as clogging of the aquifer that reduces both the hydraulic conductivity and the porosity.

In applications of enhanced in situ remediation, it is necessary to create a flow system to effectively mix nutrient-rich water with the contaminated groundwater. One such method involves the use of a recirculation well that consists of a well with two screen sections and a pump located in between. Water flows into one of the screens, nutrients are added, and the enriched groundwater is pumped back into the aquifer through the second screen (see

Figure 1). Other investigators have also begun examining flow from such a well (Herrling et al. 1991).

PROCEDURE

The flow field created by the recirculation well is calculated with the perturbation boundary element method (PBEM). This method can accurately determine the flow field in heterogeneous media (Cheng 1984), including highly heterogeneous media if the PBEM includes the use of Pade approximants or continued fractions (Sato 1992). The formation is confined and initially homogeneous. To focus only on the effects of biofouling, heterogeneity is eliminated except for that created by the growth of biomass. Any background regional groundwater flow was eliminated because, in most applications, the flow from the recirculation well tends to overwhelm any practical regional head gradients, except of course far away from the well. This allows the simulations to have axisymmetry. The axisymmetric flow field calculated with the PBEM was divided into 20 different stream tubes to represent the flow from the recirculation well. Each stream tube represents 5% of the total flow.

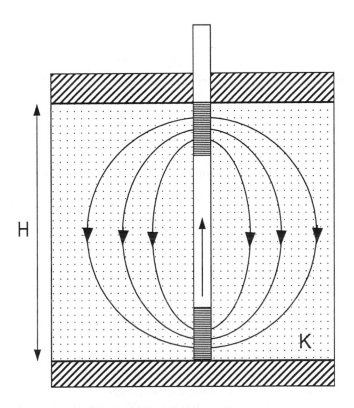

FIGURE 1. Schematic of the recirculation well.

The coupled, nonlinear advection-dispersion-reaction equations for the biomass, oxygen, phenol, and TCE described by the dispersed growth model are solved along the one-dimensional flowpaths of each stream tube. There is no lateral communication between stream tubes. We feel that this is a reasonable approximation because realistic values of the longitudinal dispersivity are extremely small and transverse dispersivity values, which govern lateral communication, are even smaller (Lang et al. 1994). Rate-limited sorption of the TCE is also included in the model.

The growth of the bacteria, modeled using dual Monod kinetics, and the decay of dead bacteria determine the consumption of oxygen. The consumption of phenol and the degradation of TCE is governed by Monod kinetics with competitive inhibition. Values for the various constants needed for these equations are taken from analysis of data from a field-scale experiment (Semprini et al. 1993). Reductions in the hydraulic conductivity were calculated based on a relationship similar in form to those found in column studies (Taylor et al. 1990) and with a maximum reduction in hydraulic conductivity of one order of magnitude (Kalish et al. 1964).

The system has been nondimensionalized to make the results more generally applicable. Lengths are nondimensionalized by the aquifer thickness, H, and times are nondimensionalized by the aquifer thickness divided by the initial hydraulic conductivity before biofouling, K_0. Two simulations were performed to assess the effects of biofouling. The first is the base case in which biofouling is neglected, the second includes biofouling. The details of the simulations can be found in MacDonald (1995).

RESULTS

Our simulations provide some indications of the effects biofouling can have on in situ bioremediation schemes. The simulations were run until the biomass concentration predicted by the dispersed growth model completely clogged the formation. The stream tubes and the flowrates were recalculated periodically during the simulation as clogging occurred. The pumping rate of the well was determined by maintaining a constant pumping horsepower. Therefore, the flowrate decreased and the head difference applied between the two well screens increased as the bacteria began clogging the formation. Figure 2 shows the decrease in the flowrate with time due to biofouling. The reduction in flowrate reduces the mass flux of nutrients entering the formation, thereby reducing the growth rate of the biomass.

Figure 3 shows the change in the breakthrough curve for a conservative tracer. A decrease in porosity due to biofouling tends to reduce breakthrough time, but this effect is offset by the reduction in both the pumping rate and the hydraulic conductivity, resulting in longer breakthrough times. The increase in breakthrough time has important implications for in situ bioremediation. The increased residence time means that contaminants remain in the aquifer longer

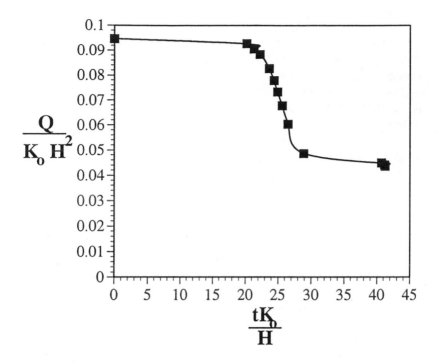

FIGURE 2. Decrease in flowrate with time due to biofouling.

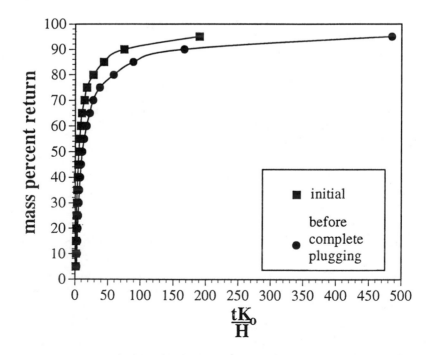

FIGURE 3. The effect of biofouling on breakthrough of a conservative tracer.

before entering the recirculation well to be reinjected into the biomass-rich area near the injection screen.

The distribution of biomass in the aquifer changes if the effects of biofouling are included. Figure 4 shows the difference in biomass concentration ($[X_a]$(neglect biofouling)-$[X_a]$(account for biofouling)) between simulations that do not account for changes in porosity and hydraulic conductivity due to biofouling and those that do at a time shortly before complete clogging of the formation occurred. The most obvious feature is the difference near the injection well screen where concentrations are much higher for the case neglecting biofouling. The case that accounts for biofouling has slightly larger concentrations farther from the screen than the case that neglects biofouling. This effect is a result of several factors, the most important being that nutrients recirculate faster without biofouling. This faster nutrient recirculation almost eliminates the benefits of a pulsed nutrient addition scheme. The pulsing works slightly better for the biofouling case because the longer residence times result in fewer nutrients recirculating.

The remediation of TCE also changes (Figure 5). Next to the injection well screen, the TCE concentration is slightly lower if biofouling is neglected. Just

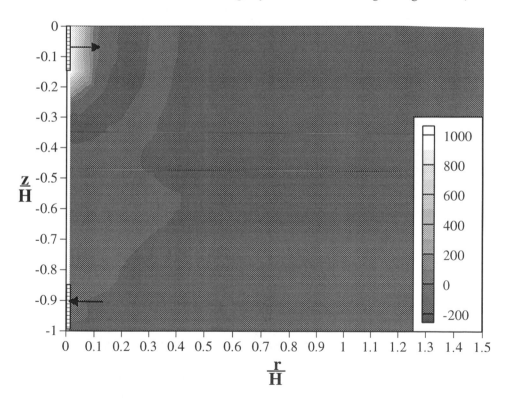

FIGURE 4. The difference in biomass concentrations (neglect biofouling– account for biofouling) to demonstrate effect of biofouling. Concentrations in mg/L.

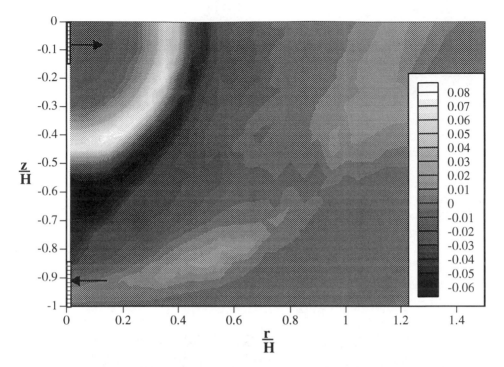

FIGURE 5. **The difference in TCE concentrations (neglect biofouling–account for biofouling) to demonstrate the effect of biofouling. Concentrations in mg/L.**

beyond this is a region in which the TCE concentration is higher in the absence of biofouling. Just beyond that region is another zone in which the TCE concentration is lower without biofouling. These differences are in part due to the difference in biomass distribution and in part due to the change in flowrates. The lower flowrates caused by biofouling affect not only the required time for TCE to be recirculated, but also the time during which the contaminants are in contact with the biomass and the biomass concentrations.

DISCUSSION

This study indicates that biofouling may reduce the effectiveness of in situ bioremediation projects. We determined that unless biomass growth is controlled by mass transfer limitations due to biofilm kinetics or is redistributed by shearing mechanisms, biofouling can significantly decrease the pumping rate of the well, thereby reducing the nutrient delivery to the aquifer. An examination of the importance of biofilm kinetics and shearing mechanisms can be found in MacDonald (1995). The change in flow properties had a significant impact on the distribution and concentration of both the biomass and the contaminant, TCE. Thus, the poɩ ɪtial for biofouling should be considered in the design of engineered in situ biʋremediation schemes, and strategies should be devised to repress biofouling.

The results of this study are preliminary. There is uncertainty about how the hydraulic conductivity and the porosity are affected by biomass growth. It is also uncertain how the kinetic parameters in the Monod equation should change as the biomass grows thicker on the soil particles. The rapid plugging of the formation suggested by the dispersed growth model needs to be examined. Biofilm kinetics or shear forces will have to be incorporated into the model at some point to address this issue (see MacDonald 1995). Even with these uncertainties, we believe that our results provide insight into the potential impact of biofouling on in situ bioremediation schemes.

ACKNOWLEDGMENTS

This work is funded by the Office of Technology Development, within the U.S. Department of Energy's Office of Environmental Management, under the In Situ Remediation Integrated Project.

REFERENCES

Cheng, A. H.-D. 1984. "Darcy's Flow with Variable Permeability—A Boundary Integral Solution." *WRR* 20(7): 980-984.

Herrling, B., J. Stamm, and W. Buermann. 1991. "Hydraulic Circulation System for *In Situ* bioreclamation and/or *In Situ* Remediation of Strippable Contamination." In R. E. Hinchee and R. F. Olfenbuttel (Ed.), *In Situ Reclamation Applications and Investigations for Hydrocarbon and Contaminated Site Remediation.* Butterworth-Heinemann, Stoneham, MA. pp. 173-195.

Kalish, P. J., J. A. Stewart, W. F. Rogers, and E. O. Bennett. 1964. "The Effect of Bacteria on Sandstone Permeability." *Journal of Pet. Tech.* 16: 805-814.

Lang, M. M., L. Semprini, and P. V. Roberts. 1993. "In-situ bioremediation using a recirculation well." In C. Y. Kuo (Ed.), *ASCE Engineering Hydrology*, pp. 880-885. San Francisco, CA.

MacDonald, T. R. 1995. "Flow from a Recirculation Well for Enhanced In Situ Bioremediation." Ph.D. Thesis, Stanford University, Stanford, CA.

Sato, K. 1992. "Accelerated Perturbation Boundary Element Model for Flow Problems in Heterogeneous Reservoirs." Ph.D. Thesis, Stanford University, Stanford, CA.

Semprini, L., G. D. Hopkins, and P. L. McCarty. 1994. "A field and modeling comparison of in situ transformation of trichloroethylene by methane utilizers and phenol utilizers." *Bioremediation of Chlorinated and Polycyclic Aromatic Hydrocarbon Compounds.* Lewis Publishers, Chelsea, MI. pp. 248-254.

Taylor, S. W., P. C. D. Milly, and P. R. Jaffe. 1990. "Biofilm Growth and the Related Changes in the Physical Properties of a Porous Medium. 2. Permeability." *Water Resources Research* 26(9): 2161-2169.

Bioremediation of Chlorinated Benzene Compounds

Philip C. Peck, Stuart H. Rhodes, and Turlough F. Guerin

ABSTRACT

In early 1994, investigations at a pharmaceutical manufacturing site revealed extensive areas of soil contaminated with chlorinated benzenes. The soil was a heavy clay and contained chlorobenzene (CB), 1,2-dichlorobenzene (referred to in this paper as DCB), and small amounts of 1,3- and 1,4-dichlorobenzene and other solvents. The soil was bioremediated in a pilot-scale treatment using an ex situ process with various inorganic and organic amendments. Approximately 90% of the DCB mass present in the soil was removed over a period of 2 to 3 weeks. Up to 100-fold increases in both total heterotrophs and specific degraders were observed. Residual concentrations of chlorinated benzenes were generally below detection limits (0.2 mg/kg). Adding organic matter did not appear to significantly enhance the treatment efficiency. Mass balance calculations applied to the treatment indicated that less than 5% of the chlorinated benzenes were removed by volatilization. Evidence was obtained that approximately 90% of the DCB was removed by biodegradation in these pilot-scale trials. Laboratory shake flask trials were conducted which confirmed that the soils in the pilot-scale treatment contained microorganisms capable of mineralizing CB and DCB.

INTRODUCTION

An examination of the literature regarding the bioremediation of chlorinated benzenes revealed that there was sufficient previous evidence to warrant large-scale evaluation of aerobic biodegradation processes (Litchfield et al. 1991; Shields et al. 1994). Because a known DCB-degrading culture was not available, the possible bioremediation processes for this material ranged from simple biostimulation (aeration and nutrient supplementation) to controlled composting. Due to the texture and structure of the soil to be used in the trial amendments, these processes were considered necessary to improve the soil's physical properties for bioremediation. Pilot-scale treatments were designed to test the effectiveness of stimulating native microflora by varying the amount and type of inorganic and organic amendments. The design attempted to provide favorable aerobic conditions to stimulate the removal of CB and DCB by the native microflora, and assumed the presence of such microorganisms from previous work (Van Der Meer et al. 1991).

It was recognized that, if a controlled composting approach were to be applied, at least two additional factors must be addressed: (1) increased losses of volatile organic compounds (VOCs) due to elevated soil temperatures, and (2) additional materials handling considerations associated with the organic material required for the composting process. Thus, the pilot-scale process compared the removal rates of CB and DCB by the two major mechanisms, i.e., biodegradation and volatilization. The process employed three static pile soil treatments that were amended with varying amounts of inorganic nutrients and organic supplements.

A laboratory-scale, liquid-culture trial was conducted in conjunction with the pilot-scale trial in order to conclusively establish the presence of, and to identify where possible, any microorganisms in the soil from the site that were capable of using chlorinated benzenes for growth. The laboratory trial had the additional goal of isolating cultures of any microorganisms capable of degrading chlorinated benzenes for possible bioaugmentation of future soil treatments.

LABORATORY PROCEDURES

Setup of Liquid Culture Assays

CB and DCB (100 mg/L) were added separately to 1 L of mineral salts, chloride-free medium in 2-L Erlenmeyer flasks. Colonies were obtained from plating out soil samples from the site on agar plates with substrate supplied in the vapor phase. A consortium (CBI2-6) of isolates was obtained from the colonies on CB and DCB plates, and grown in CB/DCB mineral salts medium. The flasks were shaken at 28°C. Chloride released from the microbial degradation of CB and DCB in the liquid culture assays was measured using ion chromatography. Tubes containing NaOH solution were used to capture evolved carbon dioxide, which was then determined by titration.

PILOT-SCALE PROCEDURES

Design of Facility

The pilot-scale facility was designed to allow assessment of future full-scale options for soil remediation. The soil venting system consisted of a vacuum pump and manifold system connected to a set of air extraction pipes at the base of each of three soil bins. Throttling of the system was provided by needle valves. Each soil bin was 3 m (long) × 2 m (wide) × 1 m (high). The bins were contained by a concrete wall and polyvinyl chloride (PVC) lined timber boxing.

Airflow through the top surface of the bins to the venting pipes at the bottom was induced by the vacuum created at the bottom of the bin. Points of air ingress to the bins were controlled by perforations in the PVC covers.

The venting system was constructed with instrumentation to allow monitoring of volumetric gas flow, soil temperatures, extracted gas temperatures, vacuums, and VOC concentrations in extracted soil gas. The flowrates were adjustable to meet the oxygen consumption and cooling requirements of each soil mixture. A volatile trap consisting of two granulated activated carbon (GAC) filters linked in series was installed at the extraction system outlet to capture airborne VOCs for the mass balance assessment.

Soil Amendments

The three soil treatments tested in the trial contained 0%, 12%, and 30%(w/w), respectively, of organic material (Table 1). The soil was amended with 50 g/m^3 of controlled release inorganic nutrients with an N:P:K ratio of 22:5.7:0, prior to mixing with organic material.

Sampling and Field Monitoring

For the analyses of VOCs and microorganisms, soil samples were removed from the mid depth of the soil bins at commissioning and on several occasions throughout the period of the trial. VOCs removed by volatilization were measured in the off-gas at individual sample ports using a Photovac Model 10-S/55 portable gas chromatograph photoionization detector (GC-PID) and a Foxboro Model OVA 108 Century organic vapor analyzer (OVA). The activated carbon trap was sampled at the conclusion of the trial to allow quantitation of volatilized DCB, CB, and other VOCs from the soil, or from the organic materials introduced as soil amendments. Initial soil DCB concentrations were as high as 6 mg/kg (average 2 mg/kg) (Table 2).

TABLE 1. Pilot-scale treatment design.

Treatment Bin	% Soil (w/w)	% Mulch (w/w)	Soil Volume (m^3)	Mulch Volume (m^3)	Est. Soil Mass (kg)
Bin 1	100%	0%	6	0	10,200
Bin 2	88%	12%	4	3.5	6,800
Bin 3	65%	35%	2	7	3,400

Assumptions: Density of soil 1,700 kg/m^3 ex situ; density of mulch materials 250 kg/m^3 loosely packed; final volume of all bins 6 m^3.

TABLE 2. Initial concentrations of 1,2-dichlorobenzene in soil.

Compost Treatment	Mean (mg/kg)	No. Sample	Minimum[a] (mg/kg)	Maximum (mg/kg)
Bin 1	1.0	4	<1	2
Bin 2[b]	3.5	4	<1	6
Bin 3[b]	<1	4	<1	<1
Total (Bins 1-3)	2.0	12	<1	6

(a) Field GC-PID-identified chlorobenzenes in all soil bins.
(b) Compost samples, adjusted to mg/kg dry soil equivalent basis.

Microbial Monitoring

Using the vapor plate method (Litchfield et al. 1991), samples from the contaminated soils were enumerated for aerobic microorganisms capable of using the chlorinated benzenes for growth (CB^+ and DCB^+). Total viable counts (TVCs) were also monitored using standard microbiological techniques. Bacterial consortia were isolated on the vapor plates containing CB and DCB.

Chemical Analyses

The methods used for quantitation of chlorinated solvents (CB and DCB) and semivolatile organic compounds (as C_8 equivalent) were based on standard procedures using gas chromatograph electron capture detection (GC-ECD) and gas chromatograph flame ionization detection (GC-FID), respectively. Portable OVAs used in the field trial were calibrated against a range of air mixtures of benzene, CB, and DCB. Calibration curves were produced for the field FID and GC-PID units for adjustment of field measurements.

Mass Balance Assessment

A mass balance approach was applied to account for all forms of VOC removal. The mass of VOCs measured in soil were compared with (1) the net mass of VOCs measured in extracted soil gas and (2) the VOCs captured and measured in the GAC volatile trap at the outlet of the system. The net difference was inferred to represent removal by biodegradation.

RESULTS AND DISCUSSION

Mineralization Studies

A microbial consortium was obtained (CBI2-6) that could use CB and DCB as the sole source of carbon and energy for growth in liquid cultures. The consortium consisted of 5 members, including strains of *Pseudomonas* and *Bacillus*.

Evidence for complete degradation (mineralization) of CB and DCB was obtained based on carbon dioxide production and chloride ion release in the liquid culture shake flasks. Yields of carbon dioxide for both CB and DCB were consistent with mineralization. Between 40 and 50% of the carbon added as chlorinated benzene substrate was recovered as carbon dioxide (Table 3).

Since this test is not definitive, and the result provides only indirect evidence for mineralization (for example CO_2 may also be generated by endogenous metabolism), samples of culture medium were also analyzed for chloride. The chloride results confirm that mineralization was occurring (Table 3). The chloride yields indicate that about 20% of the dichlorobenzene and 40% of the chlorobenzene were mineralized in these tests. Based on carbon dioxide and chloride ion release, it was likely that a consortium present in the soil was capable of mineralizing both chlorinated substrates. The consortium was found to also use phenol and 1,4-dichlorobenzene, but not 1,3-dichlorobenzene.

Pilot-Scale Treatment

The initial concentrations of the chlorinated benzenes in the soil of the pilot-scale trial ranged from <0.2 mg/kg to 6 mg/kg. The final concentrations of DCB in the soil were below the target concentration of 0.2 mg/kg. Removal of the chlorinated benzenes by volatilization was measured and found to be less than 5%, compared with biodegradation of almost 90%. The results of field studies on CB/DCB degradation are shown in Table 4.

TABLE 3. Mineralization of chlorinated benzenes by consortium CBI2-6.

Parameter	Chlorobenzene	Dichlorobenzene
Initial Substrate Supplied	1.09 mmol	1.15 mmol
Theoretical CO_2 Yield	6.54 mmol	6.90 mmol
Observed CO_2 Yield Conversion	3.55 mmol 54%	2.78 mmol 40%
Theoretical Chloride Yield	1.09 mmol	2.30 mmol
Observed Chloride Yield Conversion	0.48 mmol 44%	0.56 mmol 24%

TABLE 4. Mass balance of DCB from the pilot-scale bioremediation.

DCB	DCB Mass (g)
In initial soil	35.80
In final soil	3.45
Volatile loss measured in off-gas	0.90
Volatile loss measured in GAC filter	0.60
Biodegraded (total mass)	31.45 - 31.75
(%)	88 - 89

Four soil samples collected prior to the field trial contained TVCs of 2×10^5 to 2×10^6 per g soil, and up to 2×10^5 CB and DCB degraders per g soil. The initial TVCs for the three treatments ranged from approximately 8×10^5 to 6×10^7 per g soil. CB/DCB degrader populations responded to biostimulation, and after 15 days operation increased to 10^7 per g soil. CB/DCB populations then declined to 4×10^6 per g soil by the end of the trial.

CONCLUSIONS

We have reported on the findings from laboratory and pilot-scale trials conducted to determine the fate of chlorinated benzenes during bioremediation of soils from a pharmaceutical manufacturing site. Chlorinated benzene degrading microorganisms were found in the soil from the site and were shown to be active in the bioremediation treatment. It is apparent from the pilot-scale trial that simple biostimulation, with inorganic nutrients and low levels of organic material (mulch), was as effective as soil composting in the removal of DCB and CB. The numbers of chlorinated benzene degrading microorganisms in the soil increased significantly (up to 100-fold) in response to stimulation in all of the treatments.

The studies showed that the chlorinated benzenes at this site could be removed from the contaminated soil using bioremediation over a period of 2 to 5 weeks. The study also provided strong evidence that if parameters are optimized for aerobic biological activity, and losses due to volatilization are minimized, the predominant removal mechanism will be biodegradation (Guerin 1994).

REFERENCES

Guerin, T.F. 1994. "Distinguishing Between Biological and Non-biological Losses of Recalcitrant Organic Compounds: Implications for Bioremediation Trials." In *Second National Hazardous Waste Convention*, pp. 561-568, Australian Waste Water Association, Melbourne.

Litchfield, C.D., L.A. Belcher, and W.E. Bessicks. 1991. "Microbial Degradation of Chlorobenzene in Groundwater." In *Biodeterioration and Biodegradation*, Vol. 8, pp. 578-579, Elsevier, Ontario.

Shields, M.S., N.J. Reagin, R.R. Gerger, C. Somerville, R. Shaubut, R. Campbell, and J. Hu-Primmer. 1994. In R.E. Hinchee, A. Leeson, L. Semprini, and S.K. Ong (Eds.), *Bioremediation of Chlorinated and Polycyclic Aromatic Hydrocarbon Compounds*, pp. 50-65. Lewis Publishers, Boca Raton, FL.

Van Der Meer, J.R., A.R.W. Van Neerven, E.J. De Vries, W.M. De Vos, and A.J.B. Zehnder. 1991. "Cloning and Characterisation of Plasmid-Encoded Genes for the Degradation of 1,2-Dichloro-, 1,4-Dichloro-, and 1,2,4-Trichlorobenzene of *Pseudomonas sp.* Strain P51." *Journal of Bacteriology* 173: 6-15.

TCE Degradation by Methanotrophic Bacteria in a Water-Saturated Sand Column

Françoise Fayolle, Françoise Le Roux,
Chloé Treboul, and Daniel Ballerini

ABSTRACT

Trichloroethylene (TCE) degradation in a polluted aquifer was simulated using water-saturated sand columns with alternative injection of aqueous TCE/salt solution and CH_4/air mixture. Experiments were performed with two columns. The first under abiotic conditions to determine the TCE stripped fraction and the second seeded with a methanotrophic strain to quantify TCE biodegradation. Preliminary tests were performed in flasks to optimize CH_4/air injection. Stripping of TCE increased with increasing influent TCE concentration and residence time inside the column. TCE losses in gaseous effluent varied between 34% and 67% of the TCE injected. Under nonlimiting oxygen and mineral nutrient conditions, 50% of the TCE was biodegraded immediately after seeding the column, this value finally stabilizing at 20 to 30% of residual TCE after stripping.

INTRODUCTION

Methanotrophic bacteria are known to degrade chlorinated aliphatics such as trichloroethylene (TCE) in aqueous solution and under aerobic conditions (Henry et al. 1986, Little et al. 1988). The first step of TCE degradation is an epoxidation of TCE by the methane monooxygenase (MMO) of these strains, which is induced during growth on methane. When aquifers are contaminated by TCE, methanotrophic bacteria can be used in bioremediation processes as an alternative to air-stripping where the contaminated gaseous effluent is depolluted by adsorption on a carbon filter. In situ bioremediation experiments on TCE-contaminated aquifers have been performed to determine the feasibility of the process (Semprini et al. 1990, Lombard et al. 1994). It has been shown that injection of methane through soil columns results in degradation of TCE (Wilson et al. 1985). In this study, a methanotrophic strain isolated from pond mud was used in a water-saturated sand column to simulate a TCE-polluted aquifer in order to estimate the ratio of stripped to biodegraded TCE as a function of TCE influent concentration, TCE residence time, and oxygen supplementation.

EXPERIMENTAL PROCEDURES AND MATERIALS

Organism Growth Conditions

Strain C methanotrophic was isolated from pond mud by subculturing at 30°C on a mineral salt medium (Whittenbury et al. 1970) in sealed flasks using a (50/50) CH_4/air mixture as carbon and oxygen sources.

Column Setup

The bioreactor (Figure 1) consisted of a 5-cm-internal diameter × 30-cm- long glass column packed with sand (column A). The porous volume (V_p) was 240 mL. A mineral salt solution containing TCE and a CH_4/airstream (0.3 L/L.h) were injected concurrently at the column base. The system was operated at 30°C. The TCE concentration in the feeding medium was adjusted by periodically coupling the gaseous phase of the flask to the gaseous phase of a concentrated TCE aqueous solution. Teflon™ tubing was used to avoid TCE adsorption. The influent TCE solution and effluent stream were sampled in flasks that were quickly sealed after sampling. Residual TCE was extracted with pentane (1/1, v/v). Sampling was performed daily.

Strain C was grown in a 2-L flask with CH_4/air (50/50), with the resulting culture introduced onto the sand column by percolation. A similar column (column B) was set up without the methanotrophic strain and with nitrogen instead of CH_4/air as the injected gas to measure the TCE-stripped fraction.

Analytical Procedure

TCE was analyzed using a VARIAN 3000 gas chromatograph equipped with an electron capture detector on a 0.32mm × 30m DB 624 column. The temperature of the column initially was 20°C, then increased at 25°C.min^{-1} up to 250°C. The carrier gas was helium (2.5 mL/min). Pentane extracts were injected with an internal standard (1,2-dibromoethane) in the column.

RESULTS

Optimization of Methane Injection Frequence

Experiments were carried out in flasks with resting cells of strain C with and without CH_4. The specific activities of TCE degradation were 2.1 and 3 nmoles TCE degraded/min.mg protein, respectively. This result shows that the TCE degradation activity of strain C is not significantly inhibited by the presence of CH_4.

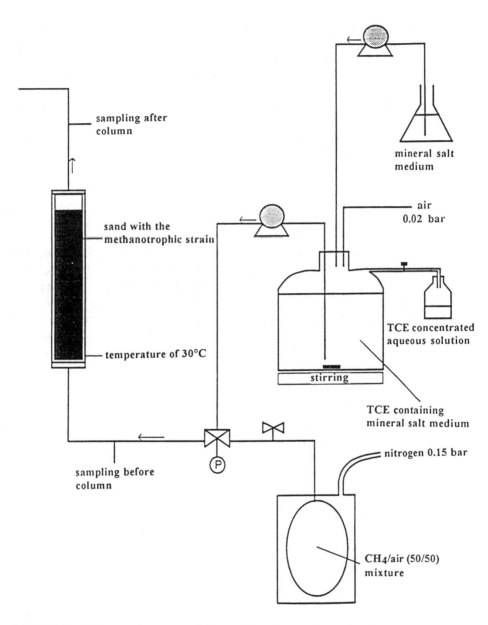

FIGURE 1. Schematic setup of the water-saturated sand column.

To determine how long TCE degradation persists in the absence of CH_4, an additional experiment was performed in flasks with resting cells of strain C with successive additions of TCE (20 μM) made every 2 hours. The result, presented in Figure 2, shows that TCE degradation activity completely disappears after 4 hours in the absence of CH_4. To confirm the necessity of CH_4

injection for maintaining MMO activity, batch experiments were performed on the water-saturated sand columns (A and B). TCE solution (four V_p) was injected to obtain the same concentration of TCE in the input and output liquid stream, approximately 20 μM. During the first 3 hours, no CH_4/air was injected, CH_4/air was then injected for 1 hour at a rate of 0.3 L/L.h. TCE concentrations were measured by sampling at 0, 3, 4, and 7 hours (Figure 3). The degradation rate of TCE before injection of CH_4/air was 0.3 μmoles/L.h, increasing to 1.1 μmoles/L.h after injection of the gas mixture. From these results, the optimum CH_4/air injection frequency was selected to be 1 hour for every 4-hour period.

Biodegradation and Stripping of TCE on Water-Saturated Sand Column

Following introduction of strain C onto the column, equal TCE concentrations in the influent and effluent streams were obtained by circulating 4 Vp of TCE solution. The TCE concentration in the influent stream was approximately 110 μM, with a TCE residence time of 48 h (dilution rate of 0.021 h^{-1}). Residual TCE concentrations were measured daily on the two columns in the effluent stream during a 14-day period with optimum CH_4/air injection (Figure 4). Under these conditions, 60% of the influent TCE was stripped on column

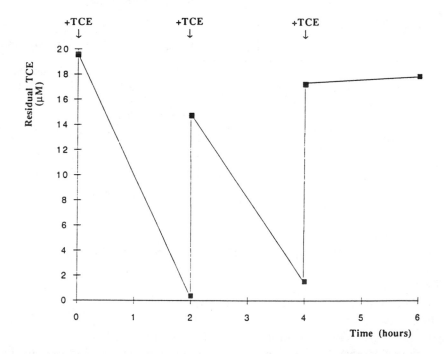

FIGURE 2. Effect of repeated additions of TCE on the TCE degradation rate in the absence of methane.

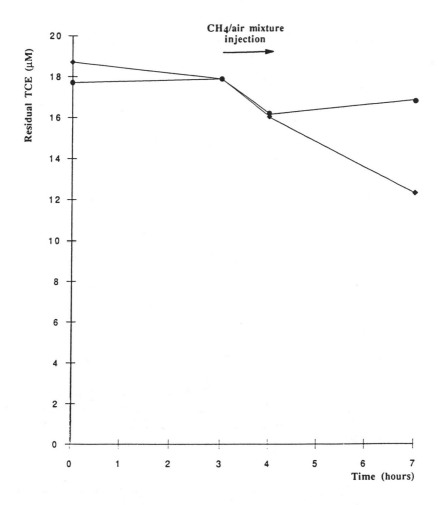

FIGURE 3. Effect of CH$_4$/air injection in batch water-sand columns. (●, abiotic column; ◆, column with methanotropic strain)

B. The percentage of TCE biodegraded was estimated from the difference in the effluent streams of the two columns as 50% of the residual TCE after stripping immediately following seeding, stabilizing at 30% after 2 days.

Changing the TCE residence time to 24 h (dilution rate of 0.042 h^{-1}) resulted in a small decrease in the TCE-stripped fraction to an average value of 49% of the influent TCE and no significant change of the biodegraded fraction (average value of 28% of residual TCE after stripping). By decreasing the TCE influent concentration to 24 μM and with the same gas injection frequency and rate, stripped and biodegraded TCE fractions were measured at two TCE residence times, 24 and 48 h. The percentage of stripped TCE was (respectively) 34% and 67%; and the percentage of biodegraded TCE was (respectively) 20%

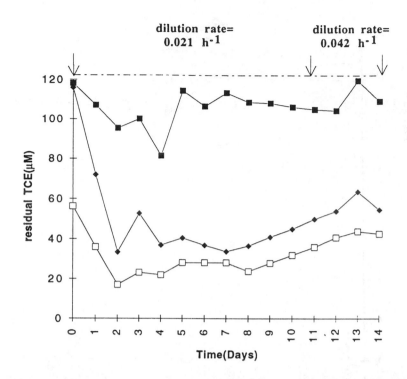

FIGURE 4. TCE degradation and stripping in column at different dilution rates with a high influent TCE concentration (■, influent TCE concetration; ◆, TCE concentration after stripping; □, TCE concentration after biodegradation).

and 28%. Rates of TCE degradation have been calculated at different TCE feeding rates (Table 1).

The replacement of air by oxygen was also tested and no change of the biodegraded TCE part was observed, indicating that oxygen was not a limiting factor, nor was the nitrogen as measured by residual nitrate in the effluent stream.

DISCUSSION

The presence of methane does not significantly inhibit the TCE degradation ability of strain C. The optimum frequency and rate of CH_4/air mixture injection in to observe TCE biodegradation by strain C immobilized on sand were 1 h/4 h period and 0.3 L/L.h, respectively.

The TCE stripped fraction increased with increasing influent TCE concentration and with increasing residence time. When the influent TCE concentra-

TABLE 1. TCE biodegradation rates at different TCE feeding rates.

TCE feeding rate (μmoles/L.h)	TCE degradation rate (μmoles/L.h)
4.6	0.65
2.3	0.27
1.0	0.13
0.5	0.05

tion was high (110 μM), the fraction of stripped TCE was 50% or more under the described operating conditions.

In the range of the TCE influent concentration studied, the biodegradation rate increased with the TCE concentration and TCE feeding rate (Table 1). Immediately after seeding, the biodegradation rate was 2.3 μmoles/L.h. This high value could be explained by the high activity of the strain seeded on the column following cultivation in the presence of methane. During continuous TCE feeding, preferential routes for methane diffusion are created, reducing the efficiency of the contact between the cells and the gaseous phase, thereby reducing the TCE biodegradation ability of strain C.

REFERENCES

Henry, S.M., and D. Grbic-Galic. 1986. "Aerobic Degradation of Trichlorethylene by Methanotrophic Isolated from a Contaminated Aquifer."Abstr. *Q-64,Proceedings of Annu. Meet. Am. Soc. Microbiol.* p. 294.

Little, C.D., A.V. Palumbo, S.E. Herbes, M.E. Lidstrom, R.L. Tyndall, and P.J. Gilmer. 1988. "Trichlorethylene Biodegradation by a Methane-Oxidizing Bacterium." *Appl. Environ. Microbiol. 54:*951-956.

Lombard, K.H., J.W. Borthen, and T.C. Hazen. 1994. "The Design and Management of System Components for In Situ Methanotrophic Bioremediation of Chlorinated Hydrocarbons at the Savannah River Site." In R.E. Hinchee (Ed.), *Air Sparging for Site Remediation,* pp. 81-96. CRC Press, Boca Raton, FL.

Semprini, L., P.V. Roberts, G.D. Hopkins, and P.L. Mac Carthy. 1990. "A Field Evaluation of In Situ Biodegradation of Chlorinated Ethenes: Part 2, Results of Biostimulation and Biotransformation Experiments." *Ground Water 28:* 715-727.

Whittenbury, R., K.C. Phillips, and J.F. Wilkinson. 1970. "Enrichment, Isolation and Some Properties of Methane-Utilizing Bacteria." *J. Gen. Microbiol. 24:* 225-233.

Wilson, J.T., and B.H. Wilson. 1985. "Biotransformation of Trichlorethylene in Soil." *Appl. Environ. Microbiol. 49:* 242-243.

Degradation of Chlorinated Aliphatic Compounds by Methane and Phenol-Oxidizing Bacteria

Angela R. Bielefeldt, H. David Stensel, and Stuart E. Strand

ABSTRACT

This paper compares the ability of methanotrophic and phenol-oxidizing enrichments to degrade chlorinated aliphatic compounds (CACs), with a focus on the important application issues of degradation rates, competitive inhibition, and intermediate toxicity. Bioremediation implications are discussed.

INTRODUCTION

CACs, such as trichloroethylene (TCE), *cis*-1,2-dichloroethylene (*c*-DCE), chloroform (CF), and 1,1,1-trichloroethane (TCA), are common groundwater contaminants. Aerobic cometabolic degradation of these compounds has been fortuitously accomplished by nonspecific oxygenase enzymes produced by methane, phenol, propane, and ammonia-oxidizing bacteria. The reaction also requires nicotinamide adenine dinucleotide (NADH) and oxygen.

CAC degradation by methanotrophs has been frequently studied with either mixed or pure cultures (Alvarez-Cohen & McCarty 1991a, Fogel et al. 1986, Fox et al. 1990, Oldenhuis et al. 1989, Strand et al. 1990). The methanotrophs express a soluble methane monooxygenase (sMMO) enzyme under copper-limiting conditions and have degraded TCE, *c*-DCE, *t*-DCE, 11-DCE, CF, TCA, 1,1-dichloroethane (11-DCA), 12-DCA, and vinyl chloride in various studies (Tsien et al. 1989, Oldenhuis et al. 1989, Strand et al. 1990). Phenol or toluene oxidizers produce mono- or dioxygenase enzymes and have a more limited range of CACs that can be degraded. These include TCE, *c*-DCE, *t*-DCE, 11-DCE, and DCA (Shields and Reagin 1992, Bielefeldt 1994).

In situ studies by Hopkins et al. (1993) showed more efficient TCE removal by phenol-oxidizers than methanotrophs. Such results suggested that the CAC-degrading capabilities of phenol-oxidizers should be further explored and compared to that found for methanotrophic enrichments. The purpose of this paper is to compare the CAC degradative abilities of methanotrophic and phenol-oxidizing bacteria relative to important factors affecting remediation designs.

FACTORS AFFECTING BIOREMEDIATION DESIGNS

Important factors related to the engineering application of cometabolic organisms for CAC degradation include CAC degradation rates, competitive inhibition, and intermediate toxicity. Because the CACs and growth substrates are degraded by the same oxygenase enzyme, competition can occur to reduce the degradation rates of both compounds when simultaneously present. A similar competitive effect also occurs between different CACs.

Intermediate toxicity occurs when a product of CAC degradation exerts toxic effects on the cells. For example, methanotrophs convert TCE to TCE-epoxide, which is known to react with cell protein (Fox et al. 1990). As the amount of TCE degraded increases, the cell activity declines and finally cell viability is lost; this has been termed the transformation capacity (Tc), which is the mg of TCE degraded per mg cells inactivated (Alvarez-Cohen and McCarty 1991a). Intermediate toxicity poses a significant treatment limitation with methanotrophs and has affected reactor designs and TCE and methane feeding strategies (Alvarez-Cohen and McCarty 1991b, Stensel et al. 1992). Intermediate toxicity has also been observed for phenol degraders (Wackett et al. 1989, Chang and Alvarez-Cohen 1994).

Batch CAC degradation tests in the absence of the bacteria's cometabolic growth substrate have been used to observe the occurrence of intermediate toxicity (Alvarez-Cohen and McCarty 1991a, Stensel et al. 1992). In such tests, the decline in CAC degradation rates may be due to the depletion of energy reserves that supplied NADH, as well as intermediate toxicity. These two effects must be separated to determine the impacts of intermediate toxicity.

EXPERIMENTAL METHODS

CAC degradation by both methanotrophs and phenol-oxidizing cultures has been conducted at the University of Washington. Methanotrophic enrichments were grown in closed, suspended growth reactors, with 5-day solids retention time (SRT) and hydraulic (HRT) retention time. The reactor design was described previously (Strand et al. 1990). Gas-phase methane was maintained at approximately 5% partial pressure. The inorganic media were copper-free.

The phenol-oxidizing enrichments were grown at 20°C in continuously stirred 4-L flasks, stoppered with sterilized glass wool, and fed phenol as a sole carbon source. The average specific phenol loading rate was 0.36 g/g volatile suspended solids (VSS)-d, and SRT and HRT were 6 days. A filamentous bacteria dominated the cultures that had been enriched under nitrogen-limited conditions (100:2.2 COD:N ratio). Methods are given in more detail in Bielefeldt et al. (1995).

CAC degradation was determined in batch tests, typically conducted in 160 ml serum bottles incubated on a rotary shaker at 20°C. Liquid samples were analyzed by gas chromatography, and depletion of compounds from the gas-phase in the bottles calculated from Henry's equilibrium partitioning coefficients.

DEGRADATION RATES

Table 1 summarizes endogenous TCE degradation rates for our filamentous phenol-oxidizing and methanotrophic mixed cultures, along with some literature values. Pure culture results are included, but long-term activity in the environment and reactors has not been successfully demonstrated. Rates vary widely between culture used. The methanotrophs show both the highest and lowest rates. Degradation rates by phenol-oxidizers are within the range of rates by methanotrophs.

The endogenous *c*-DCE degradation rates with the filamentous phenol-oxidizing culture ranged from 0.27 to 1.5 g DCE/g VSS-d. Few DCE degradation rates have been published for other cultures. Fogel et al. (1986) reported a *c*-DCE degradation rate of 0.013 g DCE/g VSS-d for a mixed methanotrophic culture (compared to a TCE degradation rate of 0.004 g TCE/g VSS-d). Other mixed methanotrophic cultures studied by us have shown very slow or

TABLE 1. Endogenous TCE degradation kinetics for cometabolic organisms.

Culture	Max. Specific TCE degradation rate, g/g-d	TCE Ks, mg/L	Temp. of tests, °C	Reference
Filamentous phenol-oxidizers	0.10–0.25 (0.18 ± 0.03)	ND	20	Bielefeldt (1994)
Phenol-oxidizer *P. cepacia* G4	0.38–0.75 (0.55)	0.4	26	Folsom (1990)
Phenol-oxidizers	0.21	2.04	20	Chang & Alvarez-Cohen (1994)
UW Mixed methanotrophs	0.003–0.03	ND	20	Treat (1993)
Mixed methanotrophs	1.03	3.84	20	Chang & Alvarez-Cohen (1994)
Mixed methanotrophs	0.6	NR	21	Alvarez-Cohen & McCarty (1991a)
Methylosinus trichosporium OB3b	1.6–3.8	NR	30	Tsien et al. (1989)

Ks = half-saturation concentration.
() = average, ND = not determined, NR = not reported.

negligible DCE degradation rates. Hopkins et al. (1993) reported greater c-DCE transformation than TCE transformation in situ for both methane- and phenol-stimulated plots.

COMPETITIVE INHIBITION

Reports of the severity of the effect of competitive inhibition of TCE degradation rates by methane have varied (Alvarez-Cohen & McCarty 1991b, Strand et al. 1990). Strand et al. (1990) reported no consistent effect of increased dissolved methane concentration on TCE degradation rate. However, TCA degradation rates decreased significantly at higher dissolved methane concentrations.

In work with the filamentous phenol-oxidizer, phenol addition actually increased TCE degradation rates in batch tests. Figure 1 shows that the addition of phenol increased initial TCE degradation rates as much as twofold over endogenous degradation rates. The higher rate continued after phenol in the bottles was completely degraded. A similar result was found with c-DCE degradation. Increased CAC degradation rates are attributed to higher levels of cometabolic enzymes being induced in the cells due to exposure to higher phenol concentrations. The cells were exposed to low phenol concentrations during growth, less than 0.5 mg/L, but a concentration of 10 to 20 mg/L phenol was added at the start of batch tests. Enzyme competition was evident from the decreased phenol degradation rates with increasing TCE concentrations (Bielefeldt 1994). A similar effect of TCE on phenol degradation rate was noted

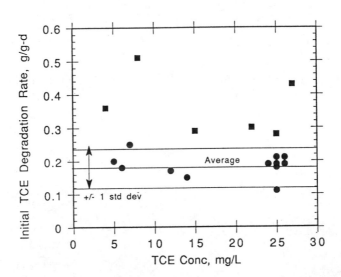

FIGURE 1. Batch test zero–order specific TCE degradation rates versus initial TCE concentration, with and without phenol addition. Symbols: ● Endogenous conditions; ■ With phenol addition.

by Folsom et al. (1990). However, competitive inhibition during CAC degradation appears to have been more than overcome by the level of enzyme induction in the cells during phenol addition.

INTERMEDIATE TOXICITY TESTS

In work by Stensel et al. (1992), TCE degradation in batch tests without methane declined after 5 h. In these tests the possible affects of NADH limitation, loss of enzyme activity, and intermediate toxicity cannot be separated. In contrast, TCE degradation was maintained for over 30 h in a similar test with methane present and with repeated respikes of TCE. The added methane should have overcome any NADH limitations, so the eventual decline in TCE degradation rate after 40 h is attributed to intermediate toxicity effects.

An experimental procedure was used with the filamentous phenol-oxidizing culture to separate the effects of electron donor depletion and intermediate toxicity. As shown in Figure 2, batch bottles were spiked with TCE at 0 and at 24 h. A second set of bottles were pre-aerated for 24 h and then spiked with

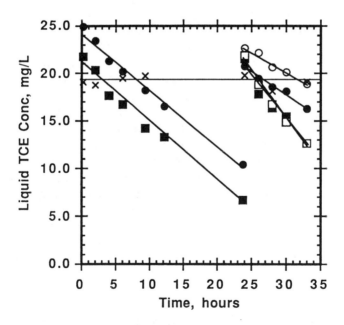

FIGURE 2. TCE degradation by filamentous phenol-oxidizer (155 mg/L VSS) after pre-aeration and TCE exposure, showing lack of intermediate toxicity and enhanced degradation with phenol addition. Symbols: ● TCE spiked at 0 and 24 hours; ■ TCE spiked at 0, TCE and phenol spiked at 24 hours; O TCE spiked at 24 hours after pre-aeration; □ TCE and phenol spiked at 24 hours after pre-aeration; × Control.

TCE. Degradation rates in these bottles were compared to the bottles initially spiked with TCE and respiked later. If the bottles with prior TCE exposure had lower subsequent degradation rates than the pre-aerated bottles (which were treated identically with the exception of TCE addition), intermediate toxicity effects could be concluded. In addition, one of the pre-aerated and TCE respiked bottles also was spiked with phenol at 24 h.

The endogenous specific TCE degradation rate declined after 24 h, from 0.20 g/g VSS-d initially to 0.12 and 0.09 g/g VSS-d with prior TCE degradation and with pre-aeration only, respectively. The decline was not due to intermediate toxicity, because the culture which degraded TCE over the first 24 h did not have a lower TCE degradation rate at 24 h than the culture that was only aerated for 24 h prior to TCE degradation (the rates are within the standard deviation for this test procedure). The decline in rate from 0 h to 24 h may be due to energy limitation or loss of enzyme activity. In addition, no intermediate toxicity was observed in bottles where 20 mg/L phenol was added at 24 h, which showed similar degradation rates both with prior TCE degradation and aeration alone (0.36 and 0.29 g/g VSS-d, respectively). Similar results were obtained for DCE degradation, with 17 mg/L DCE and 8 h endogenous time (Bielefeldt 1994).

A TCE transformation quantity (Tq) was calculated for tests with the filamentous phenol-oxidizer for comparison to reported transformation capacities (Tc) of other cometabolic bacteria. Unlike Tc, Tq represents the specific amount of TCE degraded by the cells during endogenous batch testing, where no loss of cell viability occurred. To determine Tq , the total mass of TCE degraded in the batch test was divided by the total cell mass present. The Tq for the filamentous phenol-oxidizer ranged from 0.31 to 0.51 g TCE/g VSS (the Tq for the test shown in Figure 2 was 0.33 g TCE/g VSS). Higher values were possible because the cells still possessed the ability to degrade TCE, but had merely degraded all the TCE available in the batch test. Chang and Alvarez-Cohen (1994) reported a much lower Tc value (0.031 g TCE/g cells) for a mixed culture of phenol-oxidizers. Reported methanotrophic Tc values range from 0.014 to 0.10 g TCE/g VSS (Alvarez-Cohen & McCarty 1991a; Treat 1993; Oldenhuis et al. 1989; Tsien et al. 1989; Chang & Alvarez-Cohen 1994). The filamentous phenol-oxidizers did not show intermediate toxicity and their Tq values are well above the Tc values shown for other cometabolic bacteria exhibiting intermediate toxicity effects.

BIOREMEDIATION IMPLICATIONS

Phenol-oxidizers can be considered as an alternative to methane-oxidizers for cometabolic degradation of chlorinated ethenes. Although feeding phenol or toluene to subsurface environments may raise some regulatory issues, there would be fewer barriers for use of phenol-oxidizers in ex situ applications. Phenol or toluene can be used more completely than methane to grow the cometabolic bacteria, due to the low solubility of methane and its partitioning between gas and liquid phases. In addition, the explosive issues related to methane use are avoided when growing phenol-oxidizers. Degradation rates with phenol-oxidizers are competitive with the rates for methanotrophs. Competitive inhibition between methane and the CAC to be degraded is of sufficient concern for methanotrophs to affect reactor designs and in situ feeding strategies. Similarly, in situ experiments by Hopkins et al. (1993) with pulse injection of phenol showed that the degradation of TCE and *c*-DCE was competitively inhibited by phenol addition. Competitive inhibition effects have not yet been fully characterized for the filamentous phenol-oxidizing bacteria reported here; however, the initial addition of 10 to 20 mg/L phenol in batch tests increased the CAC degradation rate. Intermediate toxicity raises serious concerns about maintaining the viability of methanotrophic bacteria during CAC degradation. This problem also was apparent in reports of TCE degradation with phenol-oxidizers (Wackett & Householder 1989). The filamentous phenol-oxidizing bacteria did not exhibit intermediate toxicity effects from TCE and *c*-DCE degradation in our tests; this could be due to a degradative pathway by which nontoxic intermediates are formed or to higher resistance of the bacteria to potentially toxic intermediate products. However, the high Tq already apparently makes use of the filamentous phenol-oxidizer a potentially attractive alternative for use in aerobic cometabolic CAC degradation.

REFERENCES

Alvarez-Cohen, L. and P.L. McCarty. 1991a. "Effects of Toxicity, Aeration, and Reductant Supply on Trichloroethylene Transformation by a Mixed Methanotrophic Culture." *Appl. Environ. Microbiol.* 57(1): 228–235.

Alvarez-Cohen, L. and P.L. McCarty. 1991b. "Product Toxicity and Cometabolic Competitive Inhibition Modeling of Chloroform and Trichloroethylene Transformation by Methanotrophic Resting Cells." *Appl. Environ. Microbiol.* 57(4): 1031–1037.

Bielefeldt, A.R. 1994. "Cometabolic Degradation of Chlorinated Aliphatics Using a Phenol-Degrading Enrichment." M.S. Thesis, University of Washington, Seattle, WA.

Bielefeldt, A.R., H.D. Stensel, and S.E. Strand. 1995. "Cometabolic Degradation of TCE and DCE Without Intermediate Toxicity." *ASCE Journ. of Environ. Engr.* (November issue, in press).

Chang, H.L. and L. Alvarez-Cohen. 1994. "Modeling Product Toxicity and Competitive Inhibition of Cometabolic Degradation of Chlorinated Organics." Presented at Water Environment Federation 67th Annual Conference.

Fogel, M.M., A.R. Taddeo, and S. Fogel. 1986. "Biodegradation of Chlorinated Ethenes by a Methane-Utilizing Mixed Culture." *Appl. Environ. Microbiol.* 51:720–724.

Folsom, B.R., P.J. Chapman, and P.H. Pritchard. 1990. "Phenol and Trichloroethylene Degradation by *Pseudomonas cepacia* G4: Kinetics and Interactions between Substrates." *Appl. Environ. Microbiol.* 56(5): 1279–1285.

Fox, B.G., J.G. Borneman, L.P. Wackett, and J.D. Lipscomb. 1990. "Haloalkene Oxidation by the Soluble Methane Monooxygenase from *Methylosinus trichosporium* OB3b: Mechanistic and Environmental Implications." *Biochemistry* 29, 6419–6427.

Hopkins, G.D., L. Semprini, and P.L. McCarty. 1993. "Microcosm and In Situ Field Studies of Enhanced Biotransformation of Trichloroethylene by Phenol-Utilizing Microorganisms." *Appl. Environ. Microbiol. 59*(7): 2227–2285.

Oldenhuis, R., R.L.J.M. Vink, D.B. Janssen, and B. Witholt. 1989. "Degradation of Chlorinated Aliphatic Hydrocarbons by *Methylosinus trichosporium* OB3b Expressing Soluble Methane Monooxygenase." *Appl. Environ. Microbiol. 55*(11): 2819–2826.

Shields, M.S. and M.J. Reagin. 1992. "Selection of a *Pseudomonas cepacia* Strain Constitutive for the Degradation of Trichloroethylene." *Appl. Environ. Microbiol. 58*: 3977–3983.

Stensel, H.D., S.C. Richards, T.P. Treat, and S.E. Strand. 1992. "Toxicity Effects of Methanotrophic Biotransformation of Trichloroethylene." WEF Conference.

Strand, S.E., M.D. Bjelland, and H.D. Stensel. 1990. "Kinetics of chlorinated hydrocarbon degradation by suspended cultures of methane-oxidizing bacteria." *Research Journal WPCF 62*(2): 124–129.

Treat, T.P. 1993. M.S. Thesis. University of Washington, Seattle.

Tsien, H.C., G.A. Brusseau, R.S. Hanson, and L.P. Wackett. 1989. "Biodegradation of Trichloroethylene by *Methylosinus trichosporium* OB3b." *Appl. Environ. Microbiol. 55*(12): 3155–3161.

Wackett, L.P. and S.R. Householder. 1989. "Toxicity of Trichloroethylene to *Pseudomonas putida* F1 Is Mediated by Toluene Dioxygenase." *Appl. Environ. Microbiol. 55*: 2723–2725.

Cometabolism of Chloroethene Mixtures by Biofilms Grown on Phenol

Robert L. Segar, Jr. and Gerald E. Speitel, Jr.

ABSTRACT

Cometabolic biological treatment of water contaminated with chlorinated solvents is attractive because the contaminants are destroyed. An aerobic, sequencing biofilm reactor using phenol as the growth and cometabolism-inducing substrate was developed to treat water contaminated with trichloroethene (TCE) to a level below the U.S. Environmental Protection Agency drinking water standard of 5 μg/L (Speitel et al. 1994). Cometabolism of chloroethene mixtures was of interest in this research because natural anaerobic dehalogenation of tetrachloroethene (PCE) and TCE produce groundwater contaminant mixtures that include dichloroethenes (DCE) and vinyl chloride (VC). Recently, research efforts to stimulate reductive dehalogenation of PCE and TCE have been promising, but the complete dehalogenation of lesser-chlorinated ethenes to ethene remains, in practice, a relatively slow process. Combinations of anaerobic dehalogenation and aerobic cometabolic oxidation processes ultimately may be the best means to completely treat the water (or remediate the aquifer) in a reasonable period of time. This paper addresses the ability of aerobic biofilms grown with phenol to cometabolize TCE and chloroethene mixtures, either in situ, by pump-and-treat, or by a combination thereof. Also, the effect of two reactor packings on biofilm production and endogenous TCE removal was evaluated in an effort to improve reactor performance.

EXPERIMENTAL PROCEDURES AND MATERIALS

Four identical bench-scale biofilm reactors were fed an aerated nutrient solution amended with phenol to grow biofilms for endogenous cometabolism of TCE and other chloroethenes. During the growth period, effluent was recycled from the reactor outlets through an air-sparged vessel, and then it was pumped to the reactor inlets. Phenol, microbial inoculum, and nutrients were added continuously to the aeration vessel. The buffered, inorganic nutrient water used for reactor feeds had total dissolved solids of 550 mg/L, phosphates of 39 mg-P/L, and nitrates of 3.5 mg-N/L, and the pH was between 6.5 to 7.0. Further details of the growth conditions are given in Table 1.

The phenol-utilizing consortium used in this work was enriched over a period of several years by periodically inoculating a chemostat with phenol-utilizing cultures obtained from activated sludge, lake water, soils, and commercial microbial products (Munox-501™, Microlife Technics, Inc.). *Pseudomonas putida* was identified as the primary phenol-utilizing organism in the chemostat when this study was conducted. It was isolated by the spread-plate technique using phenol + mineral salts agar and was identified by conducting substrate utilization screens (API-NFT™ system) with pure colonies.

The four packed-bed reactors were glass liquid chromatography columns (Ace Glass, Inc.) filled with either 4-mm glass beads or 4-mm stainless steel (SS)

TABLE 1. Reactor description and experiment conditions for the continuous-flow biofilm reactors.

Reactor Characteristics		
Configuration	Upflow, single-pass	
Type	Packed-bed, unsparged	
Material	Glass with Teflon™ fittings	
Internal Diameter	2.75 cm	
Length	45 cm	
Packing Characteristics		
Type	Glass beads	SS packing
Diameter	3.8 cm	37 cm
Specific Surface Area[a]	9.2 cm^{-1}	18.9 cm^{-1}
Porosity	0.42	0.94
Growth Conditions		
Inlet Phenol Concentration	9.9 mg/L	
Inlet Oxygen Concentration	7.7 mg/L	
Empty Bed Contact Time	2.0 min	2.0 min
Packed Bed Contact Time	0.84 min	1.9 min
Removal Conditions		
Inlet TCE Concentration	90 μg/L	
Inlet Oxygen Concentration	8.0 mg/L	
Empty Bed Contact Time	14.3 min	14.3 min
Packed Bed Contact Time	6.0 min	13.4 min

(a) Reactor volume basis.

distillation column packing (Pro-pak, Scientific Development Co.) to give two identical reactors per packing type. The SS packing is a perforated foil formed into a semicylindrical shape. It was selected for comparison to glass beads to illustrate the potential performance improvements that engineered packings may offer for cometabolism-based treatment processes. At the laboratory scale, the SS-packing provided twice the surface area as glass beads of equivalent diameter and yielded a 2.2-fold increase in packed-bed contact time (PBCT) over glass beads at the same empty-bed contact time (EBCT). Specifications of the reactor design and packings are found in Table 1. Biofilm development on the packings was followed visually and by monitoring the removal of phenol and oxygen across the reactors. Phenol concentrations were determined by the 4-aminoantipyrene technique (APHA 1989), and dissolved oxygen concentrations were measured with a membrane electrode (YSI Model 57).

After a 48-h growth period, two of the reactors were dismantled to determine their biomass concentration and to provide biofilm suspensions for batch chloroethene degradation studies. Biomass levels as volatile suspended solids (VSS) (APHA 1989) were determined for the inlet and outlet reactor half-sections. The feeds to the remaining two reactors were switched from phenol-amended to a continuous feed of TCE-amended nutrient water. TCE removal by endogenous cometabolism was monitored by sampling the reactor inlet and outlet over a 96-h period. Phenol was withheld during this period to avoid inhibiting TCE degradation by enzyme competition.

Parallel batch studies were conducted with the biofilm suspensions to obtain chloroethene degradation rates and observe the behavior of chloroethene mixtures. Typically, these batch studies consisted of four sets of simultaneous incubations: TCE only, TCE with another chloroethene, the chloroethene only, and biomass-free controls to determine the initial concentrations and monitor abiotic losses. Initial chloroethene concentrations ranged from 0.008 to 1.67 mg/L. Reactions were halted by injecting acid followed by extraction with pentane. Chloroethene concentrations in pentane extracts were analyzed with a gas chromatograph (Hewlett Packard 5890A) equipped with an electron-capture detector and megabore column (J&W Scientific DB-624).

RESULTS

Biofilm growth in the reactors was visible after 18 h and penetration of the biofilm into the reactor bed ceased after 30 h. The phenol usage rates shown in Figure 1 indicated that the SS packing reactors achieved their maximum rate of phenol usage approximately 6 h earlier than the reactors with glass beads. However, the average phenol usage rate near the end of the growth period was approximately the same for all the reactors, ranging from 6.2 to 6.6 g/L/day. Effluent oxygen concentrations from Table 2 indicate that most of the oxygen had been consumed by the biofilm, but more than 5 mg/L of phenol remained.

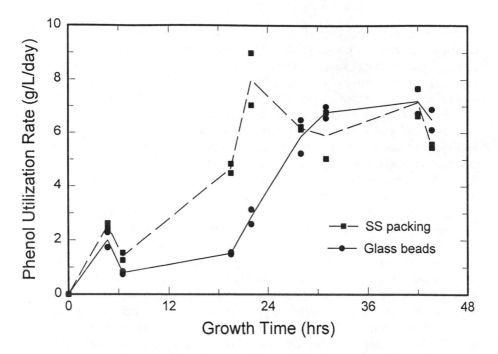

FIGURE 1. Specific phenol utilization rate (reactor volume basis) during growth of packed-bed biofilm reactors. Data are from two identical reactors filled with glass-beads and from two identical reactors filled with SS packing. Average values of the duplicate data are indicated by the plotted lines.

Biomass concentrations were approximately the same in both reactors, and less than 10% of the biomass resided in the outlet half-section of the reactors. Thus, the phenol usage rate and amount of biomass in the reactor was limited by the supply of oxygen.

The reactor containing SS packing gave higher TCE removal and removed more TCE over the period of endogenous cometabolism. During the first 12 h of TCE degradation, TCE was not detected in the effluent of the SS packing reactor. Figure 2 shows the steady decline in TCE removal over the 96-h period without phenol. Biomass concentrations at the end of the TCE removal period were used to determine that the reactor packed with glass beads lost about 42% of its initial biomass versus a 6% biomass loss for the SS packing reactor. The total mass of TCE removed and the average of initial and terminal biomass concentrations were used to calculate a specific TCE capacity of 23 mg/gVSS for the SS packing reactor and 18.3 mg/gVSS for the reactor containing glass beads.

In the batch studies, PCE was not cometabolized at a measurable rate by the biofilm suspensions, and it did not interfere with TCE degradation. As shown in Figure 3, both PCE and TCE were adsorbed by the biofilm. Adsorption was apparent because the initial concentrations measured in the

TABLE 2. Summary of biofilm reactor growth on phenol and endogenous TCE removal.

Reactor Media Type	Glass	SS packing	Unit
Outlet Phenol Concentration[a]	5.6	5.3	mg/L
Outlet Oxygen Concentration[a]	0.4	0.2	mg/L
Specific Phenol Removal Rate[a]	6.2	6.6	g/L/day
Oxygen:Phenol Mass Ratio[a]	1.70	1.63	
Biomass Concentration[b]	1.37	1.23	g/L
Specific Phenol Uptake Rate	4.5	5.4	g/g VSS/day
Initial TCE removal (hour 12)	72	>99	%
Endogenous biomass loss[c]	42	6	%
Specific TCE capacity[d]	18.3	23.3	mg/g VSS

(a) Average value near end of growth period when uptake was steady, reactor volume basis.
(b) At end of growth period; reactor volume basis.
(c) Includes endogenous decay and shearing losses over 96-h period.
(d) Based on the average biomass concentration over the TCE degradation period.

incubations were substantially less than the initial concentrations (C_o) measured in the controls. Competitive adsorption was indicated by the decrease in adsorption of TCE in the presence of PCE. Similarly, this behavior was observed for 1,1-DCE and TCE. The competitive interaction was less pronounced for *trans*-1,2-DCE and much less for *cis*-1,2-DCE.

Table 3 summarizes the relative rates of degradation, adsorption, and competitive interactions of the chloroethenes. Pseudo-first-order degradation rates were determined by kinetic modeling that accounted for instantaneous, irreversible adsorption. All isomers of DCE were cometabolized in the batch incubations. The 1,1-DCE was degraded at an initial rate similar to TCE, but the degradation rate rapidly decreased over the incubation period. Degradation of 1,1-DCE interfered with TCE degradation when both substrates were present. The *trans*-1,2-DCE was degraded more slowly than TCE, but the *cis*-1,2-DCE

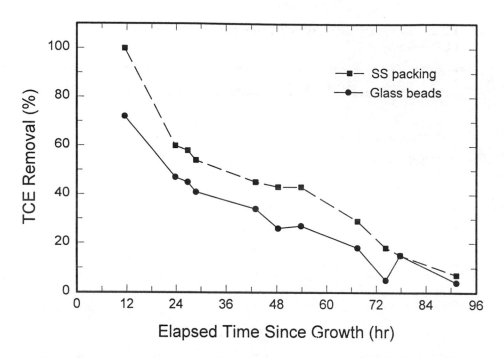

FIGURE 2. Continuous-flow TCE removal during endogenous decay of
packed-bed biofilm reactors. The average inlet TCE concentration was
90 μg/L and the EBCT was 14.3 minutes.

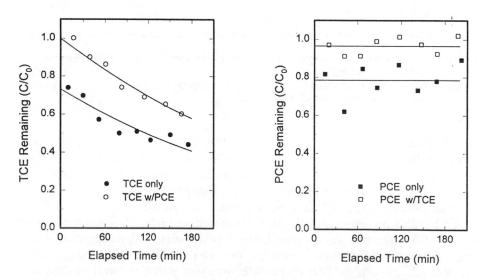

FIGURE 3. Batch TCE and PCE incubations with reactor biofilm suspensions.
Left: Effect of PCE on TCE removal.   Right: Effect of TCE on PCE
removal. The initial concentrations were TCE = 139 μg/L; PCE = 8 μg/L;
VSS = 47 mg/L.

TABLE 3. Summarized results of multisubstrate batch incubations for the chloroethenes.

Chloroethene	Rate Constant Compared to TCE[a]	Inhibition by TCE	TCE Inhibition by Chloroethene	Adsorptive Capacity Compared to TCE[b]
PCE	<5%	None	None	Similar
1,1-DCE	Similar, but toxic	Moderate	Strong	Less
cis-1,2-DCE	>500%	Moderate	Weak	Much Less
trans-1,2-DCE	~50%	Moderate	Weak	Less

(a) Based on pseudo-first-order rate constants, which for TCE averaged 07 L/g VSS/day at 0.14 mg/L.
(b) For TCE, the adsorptive capacity was 0.47 mg/g VSS at 0.14 mg/L and 4.13 mg/g VSS at 1.46 mg/L.

was degraded much more rapidly than TCE. Degradation of *cis*-1,2-DCE in the absence and presence of TCE is shown in Figure 4. The pseudo-first-order rate constant for *cis*-1,2-DCE decreased from 423 L/gVSS/day in the absence of TCE to 123 L/gVSS/day with TCE present at roughly the same concentration as *cis*-1,2-DCE. A lag in the accumulation of a *cis*-1,2-DCE metabolite was observed in the incubation with TCE.

DISCUSSION

The removal of TCE obtained in endogenous biofilm reactors depends primarily on the PBCT, the biomass concentration, and the specific degradation rate of the microorganisms. During reactor growth, the concentration of phenol or oxygen in the feed and the EBCT determines the amount of active biomass that accumulates in the reactor. In this experiment, the type of packing did not affect the concentration of biomass in the reactors because the oxygen was depleted within the reactor. Thus, the latter one-half of the reactor had little biomass and probably did not contribute much to TCE removal. The SS packing reactor obtained higher TCE removals than the reactor with glass beads primarily because it had a greater PBCT due to the high porosity of the SS packing. A secondary benefit was a slightly greater TCE cometabolizing capacity. The similar decline in loss of TCE removal over time in both reactors suggests that activity loss was primarily caused by enzyme activity loss, not by the endogenous decay of biomass. Subsequent experiments (data not shown) were conducted to increase the biomass levels in the outlet half-section of the reactors by decreasing the EBCT to 0.5 min (thus increasing the oxygen and phenol loading). Additional biomass was accumulated in these experiments, but it was evident that penetration of phenol and oxygen (and biofilm growth) into porous beds beyond a depth of 50 cm requires an extraordinarily high

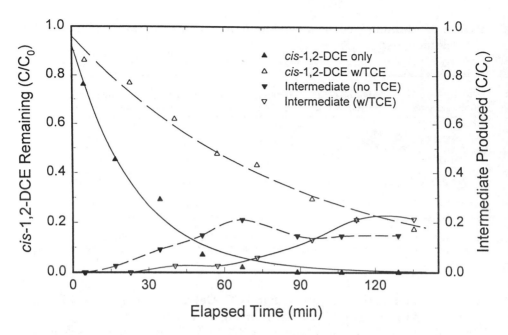

FIGURE 4. Batch degradation of *cis*-1,2-DCE by reactor biofilm suspensions. The initial concentrations were *cis*-1,2-DCE = 1.2 mg/L; TCE = 1.7 mg/L; VSS = 140 mg/L. The concentration of the *cis*-1,2-DCE metabolite was quantified from its GC peak area as an equivalent response of *cis*-1,2-DCE.

velocity or periodic feeding as used for in situ bioremediation by Semprini et al. (1994).

Cometabolism rates in multisubstrate systems may be lower than in sole-substrate systems due to enzyme competition or transformation product toxicity. The 1,1-DCE was inhibitory to TCE degradation, and its degradation was detrimental to the biofilm organisms. Palumbo et al. (1992) has reported inhibitory effects of 1,1-DCE to cometabolizing microorganisms grown with methane. The GC analysis of pentane extracts from 1,1-DCE incubations never revealed metabolite peaks, whereas *cis*- or *trans*-1,2-DCE incubations consistently provided evidence of relatively stable intermediates (see Figure 4). Presumably, 1,1-DCE is transformed by oxygenation into a reactive compound causing much greater toxicity than the other DCE isomer or TCE metabolites cause.

In the multisubstrate batch incubations, the rate of degradation for TCE and the DCEs was less than in the sole-substrate incubations, indicating competitive inhibition occurs between these chloroethenes. PCE had no effect on TCE degradation, but appeared to compete with the adsorption of TCE. Adsorption of TCE has been reported by Smets and Rittmann (1990) for bacteria and algae with adsorption capacities comparable to those obtained in this work. At low concentrations of chloroethenes, adsorption can cause an overprediction of degradation rate if not properly addressed during data analysis. The *cis*-1,2-DCE was rapidly degraded, and its presence poses a minimal problem for the degradation of mixtures. Generally, degradation rates in the batch studies were higher than those reported by Semprini et al. (1994) for in situ bioremediation, possibly reflecting the better control over growth conditions obtained in biofilm reactors than in the subsurface. Whether phenol-utilizers are used in situ or in ex situ bioreactors, the means of delivery for oxygen and phenol, the resulting biomass distribution, and the cometabolism kinetics of multisubstrate interactions will require consideration in the design of a successful bioremediation scheme.

ACKNOWLEDGMENTS

This research was funded by the Gulf Coast Hazardous Substance Research Center and by the Abel Wolman Doctoral Fellowship for R.L.S. from the American Water Works Association.

REFERENCES

American Public Health Association. 1989. *Standard Methods for the Examination of Water and Wastewater*. 17th ed., Washington, D.C.

Palumbo, A.V., W. Eng, P.A. Boerman, G.W. Strandberg, T.L. Donaldson, and S.E. Herbes. 1992. "Effects of Diverse Organic Contaminants on Trichloroethylene Degradation by Methanotrophic Bacteria and Methane-Utilizing Consortia." In R.E. Hinchee and R.F. Olfenbuttel (Eds.), *On-Site Bioreclamation: Processes for Xenobiotic and Hydrocarbon Treatment*, pp. 77-91. Butterworth-Heinemann Publishers, Stoneham, MA.

Semprini, L., G.D. Hopkins, and P.L. McCarty. 1994. "A Field and Modeling Comparison of In Situ Transformation of Trichloroethylene by Methane Utilizers and Phenol Utilizers." In R.E. Hinchee, A. Leeson, L. Semprini, and S.K. Ong (Eds.), *Bioremediation of Chlorinated and Polycyclic Aromatic Hydrocarbon Compounds*, pp. 248-254. Lewis Publishers, Boca Raton, FL.

Smets, B.F., and B. E. Rittmann. 1990. "Sorption Equilibria for Trichloroethene on Algae." *Water Research* 24(3): 355-360.

Speitel, G.E. Jr., R.L. Segar Jr., and S.L. De Wys. 1994. "Trichloroethylene Cometabolism by Phenol-Degrading Bacteria in Sequencing Biofilm Reactors." In R.E. Hinchee, A. Leeson, L. Semprini, and S.K. Ong (Eds.), *Bioremediation of Chlorinated and Polycyclic Aromatic Hydrocarbon Compounds*, pp. 333-338. Lewis Publishers, Boca Raton, FL.

Evaluation of Trichloroethylene and *cis*-1,2-Dichloroethylene Bioremediation in Groundwater

Laurie T. LaPat-Polasko, Natalie C. Lazarr, and Jennifer L. Rock

ABSTRACT

To evaluate the potential for bioremediation of trichloroethylene (TCE) and *cis*-1,2-dichloroethylene (*cis*-1,2-DCE)-contaminated groundwater, a bench-scale study was conducted using soil and groundwater collected from the Landfill 4 site at Offutt Air Force Base, Nebraska. Numerous studies have demonstrated that aerobic microorganisms are capable of degrading TCE and other chlorinated organics by cometabolizing various aromatic compounds such as toluene, phenol, or specific amino acids (Semprini et al. 1994, Speitel et al. 1994, Folsom et al. 1990, Nelson et al. 1986, United States Patent No. 4,925,802). The Offutt soil column study was designed to evaluate aerobic cometabolism of the contaminants through the addition of phenol and tyrosine to groundwater during soil column exchanges. Reductive dehalogenation was evaluated by promoting anaerobic conditions.

EXPERIMENTAL PROCEDURES AND MATERIALS

Table 1 identifies the amendments to each of six semicontinuous soil columns and native groundwater and soil contaminant concentrations. Hydrogen peroxide (H_2O_2) was amended as an oxygen source, ammonium phosphate served as a nutrient source, and sodium sulfite was amended to promote reduced (anaerobic) conditions. Both phenol and tyrosine were amended as potential cometabolic inducers. Soil columns were prepared in low-pressure glass chromatography columns 100 cm in height and 2.5 cm in diameter (BioRad), as described previously (LaPat-Polasko et al. 1994). Each column was packed with 696 g (received weight) of soil. The sterile soil column, prepared to evaluate abiotic losses, contained autoclaved soil (6 successive days with sterilization verification) and was exchanged with filter-sterilized groundwater (0.45 μm). All columns were exchanged in an upflow direction with a syringe pump on either 7-day, 3-day, or 1-day intervals. Columns 2, 3, 5, and 6 were discontinued after the second 3-day exchange

TABLE 1. Column identification key. Amendments were prepared in Offutt Landfill 4 groundwater which contained TCE, *cis*-1,2-DCE, and vinyl chloride at average concentrations of approximately 640 μg/L, 560 μg/L, and 650 μg/L, respectively. Offutt soil initially contained TCE and DCE at concentrations of approximately 30 and 900 μg/kg, respectively.

Column Number	Name	Amendments	Concentration
1	Sterile Column	Hydrogen peroxide Ammonium phosphate	50 mg/L 0.15 mg/L
2	Reduced Column	Sodium sulfite	2,000 mg/L
3	Hydrogen Peroxide Column	Hydrogen peroxide	50 mg/L
4	Nutrient Column	Hydrogen peroxide Ammonium phosphate	50 mg/L 0.15 mg/L
5	Phenol Column	Hydrogen peroxide Ammonium phosphate Phenol	50 mg/L 0.15 mg/L 100 μM
6	Tyrosine Column	Hydrogen peroxide Ammonium phosphate Tyrosine	50 mg/L 0.15 mg/L 50 μM

period. During the exchange process, approximately 100 mL of exchange fluid was pumped through each column at an average rate of 0.2 mL/min. The effluent fraction used for volatile organic compound (VOC) analysis included the first 5 mL to 25 mL exiting the column. Volatile organic compound (VOC) analyses were performed using U.S. Environmental Protection Agency (EPA) Methods 601 (liquids) and 8010 (solids). Aerobic by-products and inducer compound concentrations were not monitored. Dissolved oxygen concentrations were determined using an Orion Model 97-08 oxygen electrode.

RESULTS

Percent Reductions in TCE Concentrations

Figure 1 shows the percent TCE removed versus time in groundwater effluents collected from Columns 1, 5, and 6. These results indicate that during the first 9 weeks of the study, percent TCE removals in the Sterile Column were less than approximately 20% and decreased to less than 5% during the latter

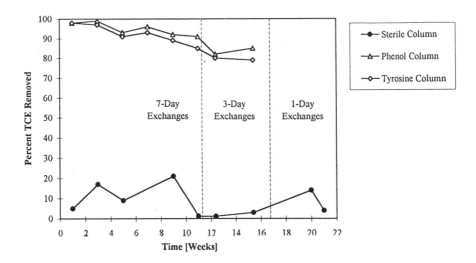

FIGURE 1. Percent TCE removed versus time for Column 1 (the Sterile Column), Column 5 (the Phenol Column), and Column 6 (the Tyrosine Column).

part of the study. Figure 1 also shows that percent TCE removals in the two cometabolic inducer-amended columns (Columns 5 and 6) were similar and ranged from approximately 80 to 90% TCE removed.

Figure 2 shows the percent TCE removed versus time in the groundwater effluents collected from Columns 1, 3, 4, and 5. Percent TCE removals in the Hydrogen Peroxide and Nutrient Columns were as high as the percent TCE removals in the column amended with the cometabolic inducer, phenol. The results shown in Figure 2 also indicate that percent TCE removals in columns amended with nutrients and an oxygen source were comparable to columns that did not receive a nutrient amendment.

Figure 3 shows the percent TCE removed versus time in groundwater effluents collected from Columns 1 and 2. The results shown in Figure 3 suggest that the groundwater and/or soil contains an anaerobic microbial population capable of significant TCE degradation. Percent TCE removals in the Reduced Column appeared to decrease progressively throughout the study.

Percent Reductions in *cis*-1,2-DCE Concentrations

Figure 4 shows the percent *cis*-1,2-DCE removed versus time in groundwater effluents collected from Columns 1, 3, and 4. Figure 4 indicates that the indigenous microbial population was capable of significantly degrading *cis*-1,2-DCE when the time period between exchanges was at least 3 days. However, when the exchange interval was reduced to a 1-day period, percent *cis*-1,2-DCE removals decreased to approximately 30%.

Figure 5 shows the percent *cis*-1,2-DCE removed versus time in groundwater effluents collected from Columns 1 and 2. The Reduced Column showed

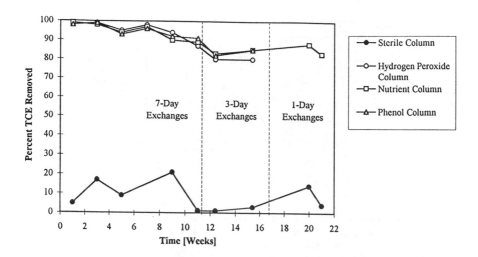

FIGURE 2. **Percent TCE removed versus time for Column 1 (the Sterile Column), Column 3 (the Hydrogen Peroxide Column), Column 4 (the Nutrient Column), and Column 5 (the Phenol Column).**

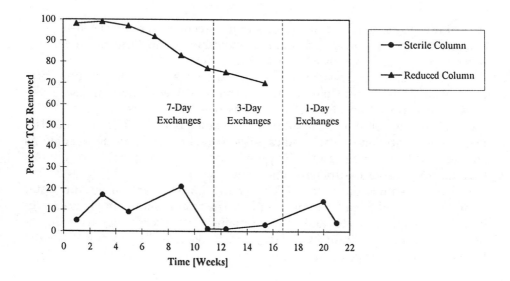

FIGURE 3. **Percent TCE removed versus time for Column 1 (the Sterile Column) and Column 2 (the Reduced Column).**

high percent *cis*-1,2-DCE removals during the first two 7-day exchange intervals. However, after these initial exchanges, the percent *cis*-1,2-DCE removed decreased to approximately 50% for the remaining 7-day exchange intervals and to approximately 30% for the 3-day exchange intervals.

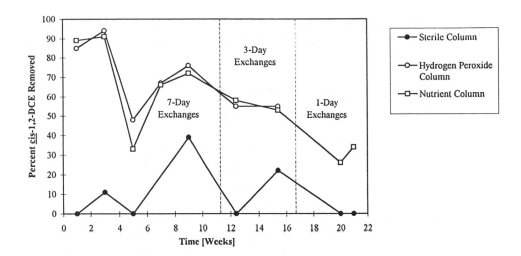

FIGURE 4. Percent *cis*-1,2-DCE removed versus time for Column 1 (the Sterile Column), Column 3 (the Hydrogen Peroxide Column), and Column 4 (the Nutrient Column).

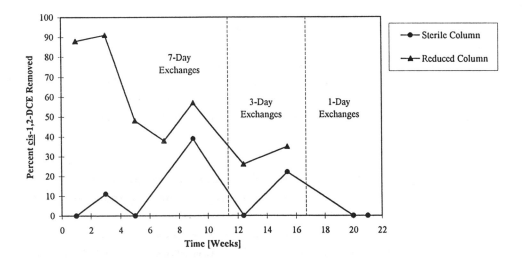

FIGURE 5. Percent *cis*-1,2-DCE removed versus time for Column 1 (the Sterile Column), and Column 2 (the Reduced Column).

Dissolved Oxygen Concentrations

Dissolved oxygen concentrations in effluents collected from columns amended with hydrogen peroxide ranged from 3.1 mg/L to 8.3 mg/L throughout the study period. In the reduced column, effluent dissolved oxygen concentrations ranged initially from 2.8 mg/L to 4.1 mg/L during the first 4 weeks of the study and decreased to 0.6 mg/L (on average) for all subsequent exchanges.

TCE and *cis*-1,2-DCE Mass Balances

Table 2 shows the results of a mass balance analysis based on soil and groundwater analytical results. Cumulative mass adsorbed to soil was determined for each column by analyzing soil column contents following the final exchange period. Contaminant mass removed due to other abiotic removal mechanisms (e.g. chemical interactions and volatilization into pore spaces) in biotic columns was assumed by subtracting the mass adsorbed onto soil in the sterile column from the cumulative mass applied to the sterile column. Contaminant mass removed due to biodegradation was calculated by subtracting the mass adsorbed onto column soil and the assumed mass removed via other abiotic mechanisms from the cumulative mass applied to each column. In general, contaminant removal due to biodegradation was similar regardless of incubation conditions. TCE removal due to biodegradation appeared to be greater than *cis*-1,2-DCE removal due to biodegradation.

DISCUSSION

Based on the results of the 21-week bench-scale study, the indigenous microbial population from the Offutt Landfill 4 site was capable of degrading TCE and *cis*-1,2-DCE under both aerobic and reduced bulk conditions. Significant TCE percent removals were achieved regardless of nutrient or cometabolic inducer amendment.

TABLE 2. Results of TCE and *cis*-1,2-DCE mass balance analysis. Cumulative mass removed is expressed as a percentage of the total mass applied to the individual column.

Column Number	Cumulative Mass TCE Removed			Cumulative Mass *cis*-1,2-DCE Removed		
	Adsorption to Soil	Other Abiotic Mechanisms	Bio-degradation	Adsorption to Soil	Other Abiotic Mechanisms	Bio-degradation
1	<4%	<18%	0%	<3%	<52%	0%
2	11%	18%	60%	4%	52%	21%
3	14%	18%	61%	3%	52%	29%
4	17%	18%	55%	3%	52%	24%
5	10%	18%	66%	4%	52%	36%
6	10%	18%	63%	3%	52%	29%

Under bulk aerobic conditions, percent TCE removals were as high in the noncometabolic inducer-amended columns as in those columns that were amended with an inducer. These results indicate that if TCE was degraded aerobically, the groundwater and/or soil may already contain a natural cometabolic inducer. To evaluate the potential existence of a natural cometabolic inducer in Offutt Landfill 4 groundwater, Oak Ridge National Laboratory conducted analyses of site groundwater using the *Pseudomonas putida*, B-2 lux bioreporter (Applegate et al. unpublished data). Results indicated that the lux bioreporter was not induced by selected site groundwater samples.

Documented anaerobic biotransformation of TCE under bulk aerobic conditions has been reported previously (Enzien et al. 1994) and may be responsible for the microbial transformation of TCE described herein. However, no accumulation of reductive dehalogenation by-products were observed during the study. Based on the results of an independent evaluation investigating the interaction between sodium sulfite (the amended reductant) and vinyl chloride, significant chemical degradation occurred (LaPat-Polasko unpublished data).

REFERENCES

Enzien, M. V., F. Picardal, T. C. Hazen, R. G. Arnold, and C. B. Fliermans. 1994. "Reductive Dechlorination of Trichloroethylene and Tetrachloroethylene Under Aerobic Conditions in a Sediment Column." *Appl. and Environ. Microbiol.* 60:2200–2204.

Folsom, B. R., P. J. Chapman, and P. H. Pritchard. 1990. "Phenol and Trichloroethylene Degradation by *Pseudomonas cepacia* G4." *Appl. and Environ. Microbiol.* 56: 1279–1285.

LaPat-Polasko, L. T., G. A. Fisher, and V. H. Bess. 1994. "Laboratory Evaluation of Aerobic Transformation of Trichloroethylene in Groundwater Using Soil Column Studies." In R. E. Hinchee, A. Leeson, L. Semprini, and S. K. Ong (Eds.), *Bioremediation of Chlorinated and Polycyclic Aromatic Hydrocarbon Compounds*, pp. 349–353. Lewis Publishers, Ann Arbor, MI.

Nelson, M. J. K., S. O. Montgomery, E. J. O'Neill, and P. W. Pritchard. 1986. "Aerobic Metabolism of Trichloroethylene by a Bacterial Isolate." *Appl. and Environ. Microbiol.* 51: 383–384.

Semprini, L., G. D. Hopkins, and P. L. McCarty. 1994. "A Field and Modeling Comparison of In Situ Transformation of Trichloroethylene by Methane Utilizers and Phenol Utilizers." In R.E. Hinchee, A. Leeson, L. Semprini, and S.K. Ong (Eds.), *Bioremediation of Chlorinated and Polycyclic Aromatic Hydrocarbon Compounds.* Lewis Publishers, Ann Arbor, MI, 248–254.

Speitel, G. E., R. L. Segar, Jr., and S. L. DeWys. 1994. "Trichloroethylene Cometabolism by Phenol-Degrading Bacteria in Sequencing Biofilm Reactors." In R. E. Hinchee, A. Leeson, L. Semprini, and S. K. Ong (Eds.), *Bioremediation of Chlorinated and Polycyclic Aromatic Hydrocarbon Compounds*, pp. 333–338. Lewis Publishers, Ann Arbor, MI.

Subsurface Microbial Communities and Degradative Capacities During Trichloroethylene Bioremediation

Susan M. Pfiffner, David B. Ringelberg, David B. Hedrick, Tommy J. Phelps, and Anthony V. Palumbo

ABSTRACT

Subsurface amendments of air, methane, and nutrients were investigated for the in situ stimulation of trichloroethylene-degrading microorganisms at the U.S. DOE, Savannah River Integrated Demonstration. Amendments were injected into a lower horizontal well coupled with vacuum extraction from the vadose zone horizontal well. The amendments were sequenced to give increasingly more aggressive treatments. Microbial populations and degradative capacities were monitored in groundwaters sampled bimonthly from 12 vertical wells. Data are presented from two representative wells. The distance from the injection well clearly had an effect on the response of microbial populations in the monitoring wells. Major changes in microbial populations and activities occurred in response to 1% methane and the nutrient injections. Methanotrophs increased in and adjacent to the treatment zone during 1% methane injection, decreased during 4% methane injection, and increased with nutrient supplements. The addition of triethyl phosphate (TEP) and nitrous oxide (NO) to pulsed methane delivery resulted in dramatic stimulation of trichloroethylene (TCE)- and tetrachloroethylene (PCE)-degrading potentials observed from groundwater enrichments. In general, wells located between the horizontal wells (e.g., in MHT-6c) were affected more rapidly and to a greater extent than those outside the horizontal wells (e.g., MHT-10c).

INTRODUCTION

Diverse microbial communities with varied metabolic capabilities are present in subsurface environments (Balkwill 1989, Chapelle & Lovley 1990), including those capable of degrading chlorinated hydrocarbons such as TCE

(Phelps et al. 1989a). Microbial biomass, community structure, and biodegradative activities are limited by properties of the subsurface environment such as moisture, pH, and the availability of carbon, nutrients, and electron donors/acceptors. Laboratory studies have shown that the addition of methane to bioreactors and cultures does enhance the degradation of TCE by methanotrophs (Wilson & Wilson 1985, Little et al. 1988, Niedzielski et al. 1990, Lackey et al. 1993). To determine if cometabolic degradation of TCE could be enhanced by stimulation of the natural microbial community, DOE instituted an in situ remediation demonstration (Lombard et al. 1994) at the Westinghouse Savannah River Site (WSRS). The goal of this demonstration was to determine if in situ bioremediation can be enhanced by appropriate modifications of the environment.

Effective in situ bioremediation strategies require an understanding of the effects of pollutants and remediation techniques on subsurface microbial communities. Therefore, detailed characterization of the microbial community at a remediation site is important. The objective of this research was to apply microbiological and biochemical measurement techniques to monitor changes in the subsurface groundwater microbial communities and degradative capacities during in situ bioremediation. This paper examines some of the resulting changes in microbial populations near and away from the zone of maximum influence by comparing data from two representative wells. This comparison provides a preliminary evaluation of the effects of the injection regimes.

MATERIALS AND METHODS

The in situ bioremediation demonstration at the WSRS targeted an area of subsurface and groundwater contaminated with TCE and PCE from an abandoned process sewer line. A detailed description of the Integrated Demonstration is available elsewhere (Lombard et al. 1994). Briefly, a series of operational campaigns (Table 1) (Palumbo et al. 1995) was designed and implemented. Horizontal wells were used to deliver nutrients to the subsurface, and twelve vertical wells were used for monitoring. Data from two representative wells, MHT-6c (Fig. 1) located between the injection and extraction wells and well MHT-10c (Fig. 1) outside the injection and extraction wells, are the focus of this paper. Well MHT-6c was in the zone of maximum effect of added nutrients, while at MHT-10c methane concentrations were generally lower (Fig. 2). Groundwater was sampled twice monthly for microbiological studies according to documented WSRS well sampling protocols (Lombard et al. 1994). Measurements were initiated on site or samples were stored on dry ice until processed.

Biomass was assayed using ester-linked phospholipid fatty acids (PLFA) representative of the living biomass (White et al. 1979) and by most probable number (MPN) enumeration of the methanotrophs. PLFA was recovered from 1 to 1.5 L groundwater samples filtered through 0.2-μm pore size inorganic filters

TABLE 1. Operational campaigns.

Treatment	Date	Days[a]	Abbreviation
Post Air-Stripping			S
Vacuum Extraction	2/26/92	0	V
Air Injection	3/18/92	21	A
1% Methane Injection	4/20/92	54	1
4% Methane Injection	8/5/92	161	4
Pulsed Methane/Air	10/23/92	240	P
Nutrient Addition	1/18/93	327	N
Posttreatment	4/20/93	419	E

(a) Days are numbered sequentially from the initiation of bioremediation operations.

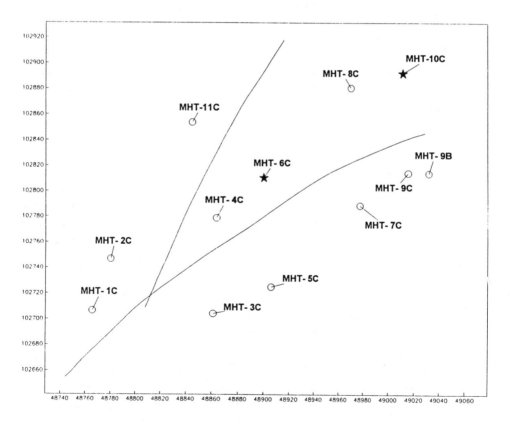

FIGURE 1. Map of demonstration site showing the horizontal wells (lines) and the monitoring wells (circles or stars). Coordinates are in feet (1 ft = 0.3 m) and wells MHT-6c and MHT-10c are indicated by stars.

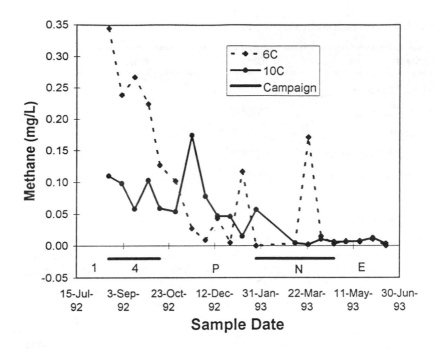

FIGURE 2. Dissolved methane concentrations in groundwater from wells MHT-6c and MHT-10c with operating conditions delineated by solid lines at the bottom of the figure. See Figure 3 for abbreviations.

(Alltech). PLFA was quantitatively extracted from the frozen (–50°C) filters by a single phase chloroform-methanol-buffer method (White et al. 1979). The extract was fractionated into specific lipid classes by silicic acid column chromatography and trans-esterified into esters in a mild-alkaline methanolic solution (Guckert et al. 1985). MPN enumerations of the methanotrophs were based on turbidity in a phosphate-buffered (pH 7.1) mineral salt medium (Palumbo et al. 1995). After sample inoculation, methane was added at 5% of headspace.

Biodegradative capacities were assessed by mineralization experiments that used 10 mL groundwater and 0.5 μCi of carrier-free [1,2-^{14}C]-TCE and -PCE as previously described (Palumbo et al. 1995, Phelps et al. 1989b). Double strength methanotrophic medium was used, except that phosphate was kept at 2 mM. All mineralizations were incubated for 30 days at room temperature, inhibited with 0.4 mL 2M NaOH, and refrigerated until analyzed. Radioactive ^{14}CO$_2$ was determined by gas chromatography-gas proportional counting (Phelps et al. 1989b) with a Shimadzu 8A gas chromatograph equipped with a thermal conductivity detector and Packard 894 gas proportional counter. Tubes were acidified with 0.5 mL of 6M hydrochloric acid 1 hour before analysis. Data are expressed as percent mineralization to CO$_2$.

One-way analysis of variance (ANOVA) was performed on the data using the least significant difference method at a confidence interval of 95%. The statistical software used was Statgraphics® Plus by Manugistics, Inc.

RESULTS

The well (MHT-6c) between the horizontal wells clearly exhibited a more pronounced biomass response to the amendments than did the well farther away from the horizontal wells (Fig. 3). In well MHT-6c, total biomass, measured as PLFA, generally increased from the beginning of the methane additions through the nutrient addition treatment, and decreased to near background levels during the posttreatment phase (Fig. 3A). Total biomass was stimulated to a much lower extent in well MHT-10c and did not decrease to background levels during the posttreatment phase (Fig. 3A). The highest total biomass in well MHT-6c was approximately four times the highest biomass detected in well MHT-10c. ANOVA analysis for MHT-10c showed no statistical difference at 95% significance level, while MHT-6c showed the pulsed treatment to be significantly different from the 1% methane and the posttreatment periods.

As with total biomass, both wells exhibited increased numbers of methanotrophs during methane and methane plus nutrient injection (Fig. 3B). Methanotrophs in well MHT-6c greatly increased in the second half of the 1% methane treatment, dropped during the 4% methane treatment, rose again during the pulsed phase, and peaked during the nutrient treatment (Fig. 3B). Methanotroph numbers also increased in well MHT-10c, but at a much lower level than in MHT-6c. The highest methanotroph density detected in well MHT-6c was five times the maximum detected in well MHT-10c (Fig. 3B). ANOVA analysis of methanotroph enumerations on well MHT-6c showed statistically significant difference (95% confidence level) between the premethane treatments and treatments with methane additions. Furthermore, 1% methane addition differed significantly from pulsed and posttreatment regimes. Well MHT-10c showed a similar pattern to MHT-6c with the exception that air injection was not significantly different from 1% methane injection. The 1% methane treatment was significantly different from the pulsed methane in air treatment.

Changes in TCE and PCE degradation potentials (as measured by ^{14}C-contaminant mineralization) in wells MHT-6c and 10c were quite different during the various treatments (Fig. 4). During the post air stripping treatment, TCE and PCE degradation potentials were below detectable limits in well MHT-6c. In well MHT-6c, TCE degradation potential was highly variable over the entire course of treatments (Fig. 4A), whereas PCE degradation potential increased during vacuum extraction and then decreased over the air and 1% methane treatments. PCE degradation potential was not detectable during the

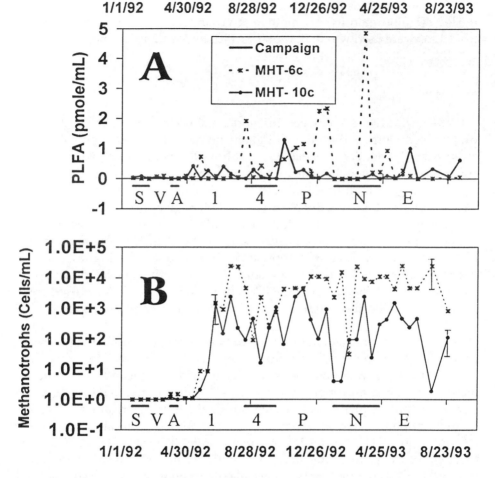

FIGURE 3. Changes in total biomass, as measured by PLFA (A) and total number of methanotrophs (B) during the WSRS demonstration. The 95% confidence level for MPN enumeration is depicted by error bars on three representative values. Different operating campaigns are delineated by solid bars at the bottom of the figure and include post air-stripping (S), vacuum extraction (V), air injection (A), 1% methane injection (1), 4% methane injection (4), pulsed methane/air injection (P), nutrient addition (N) treatments, and period after end of treatment (E).

4% methane and the first half of the pulsed treatments, then peaked again during the nutrient addition and posttreatment period (Fig. 4B).

The pattern in well MHT-10c was different; there was PCE biodegradation during the post air stripping period, but not during the vacuum treatment (Fig. 4B). The 1% methane, 4% methane, nutrient addition, and posttreatment regimes each exhibited PCE degradation potentials. In well MHT-10c, these

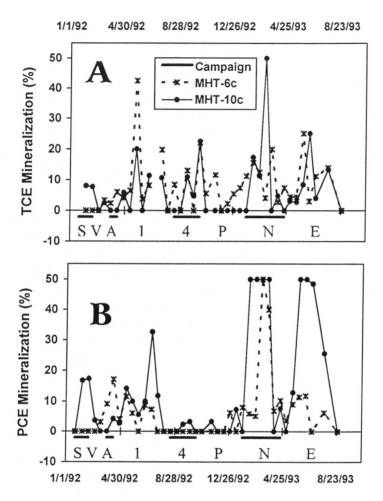

FIGURE 4. Mineralization potentials for TCE (A) and PCE (B) in wells MHT-6c and MHT-10c. Abbreviations are the same as for Figure 3.

same regimes evidenced TCE degradation potentials. Overall, the TCE and PCE degradation potentials exhibited similar patterns in well MHT-10c, with the exception of the 4% methane treatment (Fig. 4). One-way analysis of variance on TCE mineralizations showed a significant difference between the pulsed and nutrient treatments for well MHT-10c.

Statistical analysis of PCE mineralization for well MHT-6c indicated a significant difference between the nutrient treatment and all other treatments except air injection. Results of the nutrient treatment from well MHT-10c differed significantly from vacuum, air, 1% methane, 4% methane, and pulsed methane treatments. Importantly, the average and maximum TCE and PCE mineralization values for treatments increased with the addition of methane to the subsurface.

DISCUSSION

The addition of methane and nutrients to the subsurface environment at WSRS dramatically increased microbial biomass, methanotroph counts, and the toxicant biodegradative potential of the microbiota in the groundwaters from these two monitoring wells. An effect of the sampling well locations was seen in that much higher microbial biomass and methanotrophic counts were observed in well MHT-6c than MHT-10c. Well MHT-6c was located between the horizontal nutrient injection and extraction wells, while MHT-10c was outside this area and past the end of the injection well (Fig. 1). Higher densities of methanotrophs were expected in well MHT-6c due to its location in the zone of expected higher methane concentrations as was demonstrated in figure 2. This is significant because it shows that the addition of methane to the subsurface did stimulate the methanotrophs in the zone of effect.

Decreased microbial activities during the 4% methane treatment may have resulted from insufficient nutrient and mineral availability. For example, phosphate concentrations were low and nitrate decreased during 1% methane injection (data not shown). Monitoring the microbial community and the degradative potentials provide indications of stress that may adversely affect toxicant-degrading microorganisms. The impacts of the nutrient injection may have been more dramatic than evidenced by these two wells and are discussed elsewhere (Palumbo et al. 1995).

The highly heterogeneous nature of the subsurface environment can be seen in the values measured for TCE and PCE biodegradative potentials in well MHT-10c during the post air stripping treatment before methane addition, versus the below detectable limit values for MHT-6c. Significant amounts of biodegradation were observed in the early samples from MHT-10c, though the biomass values were low and methanotrophs were not detectable. Organisms capable of degrading chlorinated hydrocarbons have been found in subsurface environments (Wilson & Wilson 1985, Fliermans et al. 1988), including methanotrophs (Henry & Grbic´-Galic´ 1990) at low population densities. The very high levels of biodegradative potential observed in MHT-10c during the nutrient and immediate posttreatment regimes may have been due to a preadapted population existing at this site. Apparently, due to the noncontinuous layers of clay at the site, the zone of influence of the injection extended much farther than the original design intended, thus biostimulating the methanotrophs and increasing the degradation potentials seen in well MHT-10c.

The extent of the biological effects was clearly detected by the various microbial monitoring techniques applied. These monitoring techniques provided essential information on spatial changes in microbial community structure and degradation potentials that was used to design the nutrient injection campaign and to improve the effectiveness of the bioremediation operations.

ACKNOWLEDGMENTS

This research was funded by the Office of Technology Development within the U.S. Department of Energy's Office of Environmental Management, under the Non-Arid Sites Integrated Demonstration at WSRS. We thank Terry Hazen, Carl Fliermans, Ken Lombard, Brian Looney, and the WSRS staff for site management and sampling assistance. We also thank Mary Anne Bogle of Oak Ridge National Laboratory for assistance with the site map.

REFERENCES

Balkwill, D. L. 1989. "Numbers, Diversity and Morphological Characteristics of Aerobic, Chemoheterotrophic Bacteria in Deep Subsurface Sediments from a Site in South Carolina." *Geomicrobiol. J.* 6(1):733-752.

Chapelle, F. H., and D. R. Lovley. 1990. "Rates of Microbial Metabolism in Deep Coastal Plain Aquifers." *Appl. Environ. Microbiol.* 56(6):1865-1874.

Fliermans, C. B., T. J. Phelps, D. Ringelberg, A. T. Mikell, and D. C. White. 1988. "Mineralization of Trichloroethylene by Heterotrophic Enrichment Cultures." *Appl. Environ. Microbiol.* 54(7):1709-1716.

Guckert, J. B., C. P. Antworth, P. D. Nichols, and D. C. White. 1985. "Phospholipid, Ester-linked Fatty Acid Profiles as Reproducible Assays for Changes in Prokaryotic Community Structure of Estuarine Sediments." *FEMS Microbiology Ecology* 31:147-158.

Henry, S., and D. Grbic´-Galic´. 1990. "Effect of Mineral Media on Trichloroethylene Oxidation by Aquifer Methanotrophs." *Microb. Ecol.* 20(2):151-169.

Lackey, L. W., T. J. Phelps, P. R. Bienkowski, and D. C. White. 1993. "Biodegradation of Chlorinated Aliphatic Hydrocarbon Mixtures in a Single-pass Packed-bed Reactor." *Appl. Biochem. Bioeng.* 39-40(1):701-713.

Little, C. D., A. V. Palumbo, S. E. Herbes, M. E. Lindstrom, R. L. Tyndall, and P. J. Gilmer. 1988. "Trichloroethylene Biodegradation by Pure Cultures of a Methane-Oxidizing Bacterium." *Appl. Environ. Microbiol.* 54(4):951-956.

Lombard, K. H., J. W. Borthen, and T. C. Hazen. 1994. "The Design and Management of System Components for In Situ Methanotrophic Bioremediation of Chlorinated Hydrocarbons at the Savannah River Site." In R. E. Hinchee (Ed.), *Air Sparging for Site Remediation*, pp. 81-96. Lewis Publishers, Boca Raton, FL.

Niedzielski, J. J., T. J. Phelps, R. M. Schram, S. E. Herbes, and D. C. White. 1990. "Biodegradation of Trichloroethylene in Continuous-Recycle Expanded-bed Bioreactors." *Appl. Environ. Microbiol.* 56(6):1702-1709.

Palumbo, A. V., S. P. Scarborough, S. M. Pfiffner, and T. J. Phelps. 1995. "Influence of Nitrogen and Phosphorous on the In-Situ Bioremediation of Trichloroethylene." *Appl. Biochem. Biotech.* In Press.

Phelps, T. J., D. Ringelberg, D. Hedrick, J. Davis, C. B. Fliermans, and D.C. White. 1989a. "Microbial Activities and Biomass Associated with Subsurface Environments Contaminated with Chlorinated Hydrocarbons." *Geomicrobiology J.* 6(1):157-170.

Phelps, T. J., E. G. Raione, D. C. White, and C. B. Fliermans. 1989b. "Microbial Activities in Deep Subsurface Environments." *Geomicrobiol J.* 6(1):79-91

White, D. C., W. M. Davis, J. S. Nickels, J. D. King, and R. J. Bobbie. 1979. "Determination of the Sedimentary Microbial Biomass by Extractable Lipid Phosphate." *Oceologia* 40(1):51-62.

Wilson, J. T., and B. H. Wilson. 1985. "Biotransformation of Trichloroethylene in Soil." *Appl. Environ. Microbiol.* 49(1):242-243.

Selection of Methylotrophs for TCE Degradation in an Electroosmotic Environment

Stephen J. Vesper, Souhail Al-Abed,
Priyamvada Patnaik, Lawrence C. Murdoch,
Leland M. Vane, Jonathan G. Herrmann, and
Wendy J. Davis-Hoover

ABSTRACT

Electroosmosis has been used for more than 50 years to dewater soils for engineering purposes. This process is now being used to extract or move contaminants in soil. The conditions created by electroosmosis may be quite restrictive to the use of bioremediation due to the production of extremes in pH and high temperatures. Methylotrophs or methanotrophs have been demonstrated to degrade trichloroethylene (TCE) in situ. Both native methylotrophs and isolates from Yellowstone National Park are being tested as TCE degraders for use in bioremediation during electroosmosis. Also, a simple laboratory model of the electroosmosis process was developed.

INTRODUCTION

Trichloroethylene (TCE) is a chlorinated organic solvent that has been widely used in metal processing, electronics, dry cleaning, paint, and many other industries. The methylotrophic bacteria have been shown to cometabolize TCE. These organisms have been used as degraders of TCE in aquifers (Semprini et al. 1987), which suggests that the use of methylotrophs is a practical solution for TCE contamination.

Electroosmosis is a relatively old technique used to dewater soils (Casagrande 1949). Electroosmosis uses a direct-current electric field to induce motion in liquid and dissolved ions, which transports the contaminants in the electric field. This technique is limited to fairly soluble contaminants such as benzene, toluene, xylenes, phenol, and chlorinated solvents (Bruell et al. 1992, Acar et al. 1992, Shapiro & Probstein 1993). The goal of our study is to develop in situ technology for biodegrading TCE in an electroosmotic environment.

EXPERIMENTAL PROCEDURES AND MATERIALS

Electroosmosis Microcosms

A model electroosmosis microcosm (EM) was developed to test the bioremediation process. This design (Figure 1) allows for maintenance-free operation for at least a month. The design also allows for manipulation of various components of the EM to test concepts for field applications.

The EM is 20 cm long and 10 cm wide and is made of polyethylene phthalate (PET). The bottom of each EM is filled with 400 g which acts as a reservoir for water. On top of the sand, a mixture of 100 g of granular graphite and 50 g of granular-activated carbon is layered to make up the anode. Next, a soil slurry is made that contains 1,500 g of soil obtained from the subsurface. The soil is mixed for 1 h in a Hobart mixer with 600 mL of tap water. The whole soil slurry can be added to make a complete soil column.

If a remediation zone is to be added, only half of this mix is then gently poured on to the anode electrode. The remediation zone is created by adding

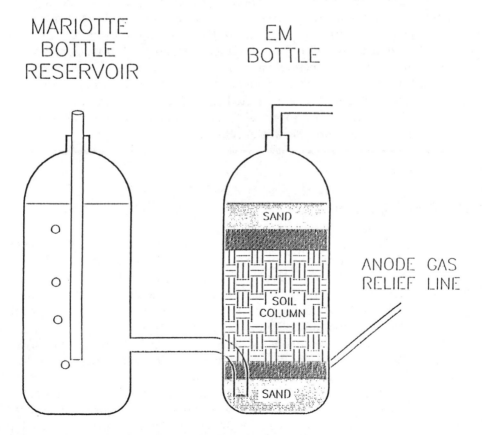

FIGURE 1. Schematic view of electroosmosis microcosm design.

150 g of granular-activated carbon on top of the first layer of soil plus 50 mL of tap water. Then the rest of the soil is added gently on top of the remediation layer. Another electrode, identical to the first, is poured on top of this soil to make the cathode. On top of the cathode is added 400 g of sand, and the rest of the chamber is filled with tap water. A Mariotte Bottle is used to control the water level. A gas relief valve is added to the anode to remove gas generated at the anode.

Tests were made using the EM. All tests were run at a constant 25 V unless otherwise noted. The current, voltage, and accumulated purged water were monitored daily. At the end of the test, the pH and temperature of the soil were taken with a probe thermometer and pH paper, respectively, every 2 cm along the soil column length.

Isolate Screening for TCE Degradation

Samples were taken at 10 major locations in Yellowstone Park and shipped back to the laboratory the next day. When the samples arrived in the laboratory, they were placed in water baths at 50, 60, 70, or 80°C, whichever temperature was closest to the temperature from which they came. Next, 5 g of sample was placed in 20 mL of "L" medium (ATCC 1984) in stoppered serum bottles and purged with 3% methane in air. The bottles were placed in water baths at the appropriate temperature. After 10 days, the bottles were subsampled by placing 1 mL of the aqueous phase on L agar plates containing 2% agar. These plates were placed in a desiccators, which were sealed, then purged with the methane in air mixture for 30 min. The desiccators were placed in incubators at the appropriate temperature.

As different colonies appeared, they were picked and placed in L medium or L medium diluted 1:10 with water and returned to stopped serum bottles. These were incubated in the methane in air mixture and restreaked for purity. The isolates were then streaked on two identical plates, and one plate was placed in a desiccator with the methane:air mixture and the other in an incubator without methane. The isolates that grew in the methane in air mixture, but which did not grow in the desiccator without methane, were screened for TCE degradation.

Each isolate was grown in duplicate 60-mL stoppered serum bottles for at least 1 month in 20 mL of L medium or L medium diluted 1:10 with deionized water. The medium was adjusted to pH 4, 8, or 10 with concentrated NaOH or HCl. The bottles were periodically purged with a mixture of 3% methane in air to provide a carbon source.

After a growth period of at least 1 month, each isolate was tested for TCE degradation. The methane was first removed by placing them in the laminar flow hood for 1 h. At that point, the serum bottle was resealed, and 2 μL of a TCE stock solution was added to each bottle so that the final concentration of TCE added was approximately 10 mg/L in the vapor phase. On the same day that the serum bottles received the TCE, the starting concentrations were determined for each. TCE was monitored using a gas chromatography-flame ionization detector (GC-FID). Headspace samples (0.5 mL) were removed from the serum bottles and injected into the GC. On the fifth and eighth days, the bottles were resampled to determine the TCE concentration. The concentrations of TCE obtained from the second and third analyses were subtracted from the initial value of TCE to indicate the level of TCE degradation (or disappearance).

After screening the isolates, one of interest was tested further. In this case, duplicate cultures of the isolates were grown in L medium at pH 4, 8, and 10 and at temperatures of 50, 60, 70, or 80°C. Controls were stoppered serum bottles with L medium but without microorganisms. The loss of TCE from the controls was subtracted from the bottles with the isolates to determine the amount of TCE degraded.

RESULTS

Electroosmosis Microcosms

Duplicate EMs were run for 6 h at 100 V to simulate a long field experiment. The temperature rose as high as 82°C in these chambers. However, if the EMs were run at 25 V for 1 month, the soil stayed in equilibrium with the ambient temperature, about 22°C.

Further EM experiments compared the operational characteristics of EMs run at 25 V with and without a remediation zone of activated carbon. Figure 2 (left) shows the results of the EM tests for the typical soil column with no activated carbon remediation zone. The current and monitoring voltage both started high, but dropped quickly. The purge water started flowing immediately, but slowed after about 400 h. It took nearly 600 h to move 2 pore volumes (pore volume is 969 mL). When an activated carbon layer was added as a remediation zone in the middle of the soil column (Figure 2, right), the current started out higher than in the soil-alone configuration, but the current rapidly dropped to about 35 mA. Voltage also started very high, declined, then rose again. About 3 pore volumes were moved in 300 h.

The results of the pH analysis of the soil at the end of the experiment indicated that the pH near the anode was 4, but near the cathode the pH was as highly alkaline (Figure 3). However, the pH in the remediation zone (activated carbon layer) was near neutral.

Isolate Screening

The early predictions about electroosmosis suggested that the temperatures would be too high and the pHs would be too extreme for microbial survival. Our screening program pursued microorganisms that would tolerate these conditions. The screening program to date has identified several isolates that appear to have degraded TCE in the serum bottles. One isolate, 50102, was tested further at three pHs, i.e., 4, 8, and 10, and at 4 temperatures, i.e., 50, 60,

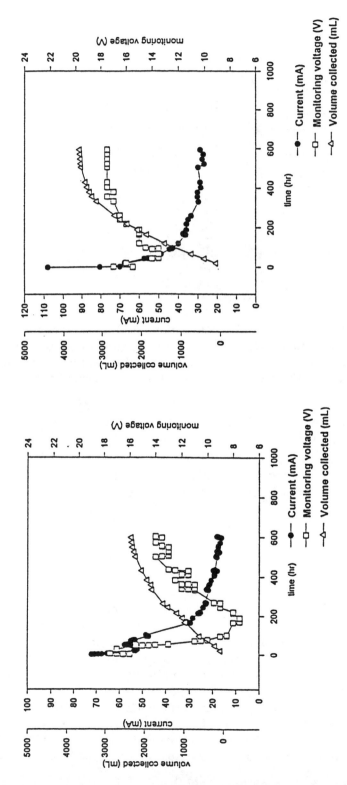

FIGURE 2. Comparison of electrical changes and purge water collected between two EMs run at 25 V, either without an activated carbon remediation zone (left) or with an activated carbon remediation zone (right).

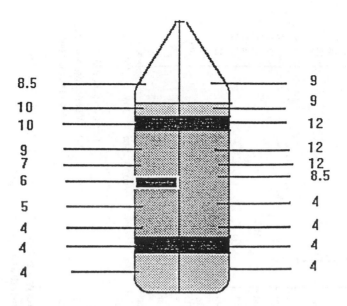

FIGURE 3. The pH changes in soil columns subjected to an electroosmotic field of 25 V for 1 month. The left side of the figure shows the results from an EM containing a granular activated carbon remediation zone. The right side shows the pH results of an EM without a remediation zone.

70, and 80°C. The results are shown in Figure 4. The left panel is pH 4, the center panel is pH 8, and the right panel is pH 10. The TCE was maximally degraded by day 8 of the exposure of TCE to isolate 50102 at 80°C at all pHs. The day 8 value (not shown) was 45% at pH 8 and 80°C, and appears to be the optimum temperature and pH tested for degradation of TCE by this isolate.

DISCUSSION

The EMs developed appear to function electrically as expected for this process. The placement of the granular activated carbon as the remediation zone actually enhances the electrical and water-moving capacity of the system. This is apparently because this layer acts as an additional electrode. This remediation layer may have one additional effect which is to maintain a more neutral pH. This suggests that the extreme pHs may not be as limiting to biodegradation as expected. We will need to examine more soils to confirm this.

The actual temperature expected in the field may not be well modeled by the EMs. Depending on the voltage applied, the temperature in the soil may rise significantly or remain unchanged. It appears that only the tests under field conditions will reveal what temperatures the microorganisms will be subjected to in the electroosmotic field.

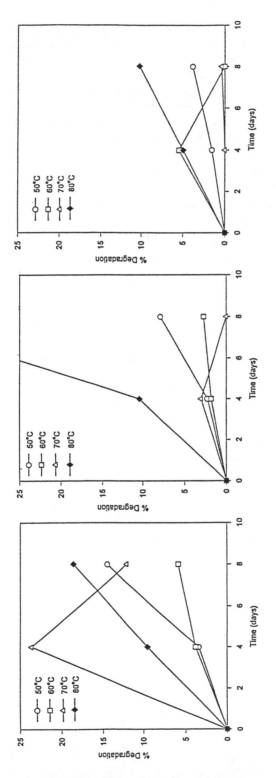

FIGURE 4. TCE degradation by isolate 50102 at four temperatures (50, 60, 70, 80°C) and at pH 4 (left), 8 (center), and 10 (right).

The selection of the TCE degrader population began with the assumption that high temperatures and extremes of pH will occur in the electroosmotic field. Thus, we believed that isolates from Yellowstone National Park might be useful. It does appear that at least one isolate obtained may have the capacity to survive the extreme pH and high temperatures predicted by some models of the electroosmosis process. This organism grew very slowly and the apparent TCE degradation was very slow. It must be noted that we have not proven that any of these isolates are methylotrophs nor that the TCE was biodegraded. These questions will await further analysis.

Since the conditions in the electroosmotic field may not be as extreme as expected, we are now examining the possibility of enhancing the native methanotroph populations in the soil. We believe that following both approaches, introduced and native populations, will allow us to achieve our goal of a viable in situ biodegradation process for TCE in the electroosmotic field.

REFERENCES

Acar, Y. B., H Li, and R. J. Gale. 1992. "Phenol Removal from Kaolinite by Electrokinetics." *Geotech. Eng. 118* (2):1837-1841.

ATCC. 1984. *Media Handbook.* American Type Culture Collection, Rockville, MD.

Bruell, C. J., B. A. Segall, and M. T. Walsh. 1992."Electro-Osmotic Removal of Gasoline Hydrocarbons and TCE from Clay." *J. Environ. Eng. 118* (4): 68-74.

Casagrande, L. 1949. "Electroosmosis in Soils." *Geotechnique 1*(3): 159-166.

Semprini, L., V. P. Roberts, G. D. Hopkins, and D. M. Mackay. 1987. *A Field Evaluation of In-Situ Biodegradation for Aquifer Restoration.* U.S. Environmental Protection Agency. EPA/600/2-87/096. Washington, D. C.

Shapiro, A. P. and R. F. Probstein. 1993. "Removal of Contaminants from Saturated Clay by Electroosmosis." *Environ. Sci. Tech. 27* (3):283-287.

Nitrogen Sources for Bioremediation of TCE in Unsaturated Porous Media

Kung-Hui Chu, Jane Vernalia, and Lisa Alvarez-Cohen

ABSTRACT

Methane-oxidizing bacteria degrade a wide range of subsurface contaminants, grow on gaseous substrates, and fix molecular nitrogen as a nitrogen source, indicating that they may be excellent candidates for vadose zone bioremediation. Experimental studies were conducted to evaluate the effect of nitrogen source on trichloroethylene (TCE) degradation by methane oxidizers in unsaturated porous media. In batch studies with methane oxidizing cells grown with nitrate (NO_3^-), ammonia (NH_4^+), or molecular nitrogen (N_2) as nitrogen sources, NH_4^+ supplied cells showed the most rapid growth and methane oxidation rates. However, N_2-fixing cells oxidized TCE as efficiently as NO_3^- and NH_4^+ supplied cells, and exhibited the highest TCE transformation capacity (T_c, mg TCE/mg cells) and the lowest product toxicity of all cells. When grown on unsaturated porous materials, the methane oxidizers were unable to develop stable populations on either glass beads or silicalite pellets, whereas stable populations of methane oxidizers grew 3 to 4 times more rapidly on diatomaceous earth pellets (DEP) than on Monterey sand (MS), regardless of nitrogen source. Continuous gasflow columns were packed with DEP, seeded with methane oxidizers, and enriched with CH_4 and water-saturated air under different nutrient conditions. Preliminary studies suggested that ≥ 95% of influent TCE (80 μg/L to 700 μg/L) could be degraded within the NO_3^--supplied columns and the N_2-fixing column for time periods of 5 to 7 days with little effect on biofilm activity. Further studies are required to evaluate long-term column performance.

INTRODUCTION

In situ bioremediation of subsurface contamination is a favorable treatment option for degradable compounds. TCE, a commonly encountered subsurface contaminant, can be aerobically degraded to carbon dioxide and nonvolatile intermediates by methane-oxidizing bacteria. Although methane oxidizers are indigenous inhabitants of most soils and water bodies, their application for in situ bioremediation of saturated aquifers may be severely limited by the low solubility of methane and oxygen in water. However, methane oxidizers may be an attractive option for bioremediation of volatile compounds within the vadose zone where delivery of gas-phase substrates is facilitated. In addition, some methane-oxidizing cultures are capable of fixing molecular nitrogen as their sole N-source (Whittenbury and Dalton 1980), eliminating the potential hindrance to microbial growth within the vadose zone due to limited transport of soluble fixed nitrogen. Although the nitrogen-fixing capability of methane-oxidizing cultures has been reported extensively, little research has been performed to assess the potential application of this characteristic to enhance in situ bioremediation. In this study, the effects of different nitrogen sources on TCE degradation by methane-oxidizing bacteria in unsaturated porous media are evaluated.

MATERIALS AND METHODS

Methane-Oxidizing Cultures

The methane-oxidizing cultures used in this study were enriched from a mixed culture grown in copper-free nitrate mineral salts media within a high-agitation (500 RPM) continuous-flow chemostat with a 5-day retention time (Chang and Alvarez-Cohen 1995). The chemostat culture was grown at 20°C with a continuous gas feed (13% methane in air) at a rate of 200 mL/min, a pH of 6.8 to 7.0, and a suspended cell density of 3.7 g/L.

Cell densities were determined by absorbance at 600 nm using a spectrophotometer and correlated with dry cell mass, which was measured from the gravimetric difference after drying at 105°C overnight and after combustion at 550°C for 2 h.

Suspended Cell Batch Studies

Experiments to evaluate the effects of nitrogen sources on the growth of methane oxidizers were conducted at 20°C in 500-mL side-armed nephelo culture flasks containing 50 mL of mineral medium and 1 mL of the chemostat culture. Three types of mineral salts media were used: ammonia mineral salts (AMS), nitrate mineral salts (NMS), and nitrogen-free mineral salts (NFMS) media as described in Chu and Alvarez-Cohen (1995). Flasks containing NFMS initially had 10% methane, 5% oxygen, and 85% nitrogen gases in the headspace, whereas the ammonia- and nitrate-supplied flasks initially contained 10% methane, 20% oxygen, and 70% nitrogen gases. Each experiment was repeated three times with similar results.

Methane and TCE oxidation were measured in duplicate initial rate experiments at 20°C as described previously (Chu and Alvarez-Cohen 1995) with 5 mL of cell suspension (0.37 to 0.87 g/L) and 0.02 mg TCE or 0.15 mg methane added to 26-mL vials sealed with Mininert valves and 20 mM of sodium formate solution added to the formate-amended vials. TCE transformation capacities were calculated from the change of total TCE mass in vials (0.5 mg total TCE added) divided by the total dry cell mass after incubation times of ≥ 24 h, a period during which TCE activity was experimentally determined to have ceased. Cell-free controls exhibited less than 5% losses during the experimental time range. All reported values were based on the average of duplicate samples.

Cell Growth on Support Media at 25% Water Saturation

Growth experiments were conducted in triplicate with 60-mL vials for each of the four support materials and three nitrogen sources. Vials contained 2 g DEP, 6 g MS, 6.5 g glass beads, or 2 g silicalite pellets to achieve similar porous media surface areas. AMS, NMS, or NFMS media and suspended cell culture were added to vials so that the total liquid added resulted in $\approx 25\%$ saturation by volume. Vials were adjusted with 5% oxygen, 10% methane, and 85% nitrogen gases in the headspace. Gas samples were taken periodically and vials were purged to restore 10% methane and 5% oxygen when levels dropped below 5% or 2%, respectively.

Column Studies

Four columns consisting of glass tubes (5 cm ID $\times$ 30 cm long) with Teflon™ end caps were packed with diatomaceous earth pellets and inoculated with nitrate-supplied methane-oxidizing cells by passing cell rich chemostat effluent through the columns for 2 hours. During column operation, water-saturated methane and air were fed to the columns at a rate of 10 mL/min. Columns were flooded with liquid nutrient solution every 2 to 3 weeks for 1 hour. The gas and media compositions of each column are listed in Table 1. For TCE degradation

TABLE 1. Gas and nutrient conditions of four unsaturated porous media columns.[a]

	Control column	Column #1	Column #2	Column #3
Media[b]	NMS	NMS	NMS	NFMS
$CH_4\%$	0	10	10	10
$O_2\%$	20	20	10	10
$N_2\%$	80	70	80	80

(a) All columns had continuous gas flowrates of 10 mL/min and gas residence times of 20 min.
(b) All columns were flooded with media every 2 to 3 weeks for 1 hour.

experiments, vapor-phase TCE was mixed with the gas mixture at concentrations of 80 to 700 μg/L prior to entry into the column. The influent and effluent gas lines were connected to a 16-port Valco valve for automated gas sampling. All tubing and fittings were constructed of stainless steel, glass, and Teflon™.

Analytical Methods

Headspace gases (CH_4, O_2, CO_2, and N_2) were measured with a gas chromatograph equipped with a CTR1 column (Alltech Co., Illinois) and a thermal conductivity detector. Headspace organics were measured using a gas chromatograph with an electron capture detector (ECD) for TCE analysis and a flame ionization detector (FID) for methane measurements.

RESULTS AND DISCUSSION

Growth of Methane Oxidizers with Different N-Sources

The effects of nitrogen sources on the microbial growth of methane oxidizers in liquid phase were evaluated in batch flasks containing methane, air, and mineral media amended with NO_3^-, NH_4^+, or no fixed nitrogen (Chu and Alvarez-Cohen 1995). Cells grew rapidly after 50 h of incubation with each N-source, with NH_4^+ and NO_3^- supplied cells growing slightly faster than N_2-fixing cells. Methane oxidation rates paralleled growth rates, with the NH_4^+-supplied cultures consuming methane (0.35 ±0.007 mg CH_4/mg cell/day) approximately 10% faster than NO_3^- supplied (0.32 ±0.001 mg CH_4/mg cell/day) and N_2-fixing cells (0.32 ±0.008 mg CH_4/mg cell/day).

As shown in Figure 1, methane-oxidizing cultures enriched with each of the three nitrogen sources were capable of degrading 1.5 mg/L TCE in the absence of methane. The TCE degradation rate by NH_4^+-supplied cells was lower than that for the other cells. The addition of formate as a source of reducing energy increased TCE degradation rates by 30 to 40% for all three cultures, with NO_3^--supplied cells exhibiting the highest rate (0.58 ± 0.03 g TCE/g cell • d), followed by N_2-fixing cells (0.54 ± 0.04 g TCE/g cell • d) and NH_4^+-supplied cells (0.32 ± 0.03 g TCE/g cell • d).

Due to TCE product toxicity, the amount of TCE that can be transformed by methane oxidizers before inactivation can be quantified as the transformation capacity (T_c). The T_c of TCE by methane oxidizers was significantly higher for N_2-fixers than for the other cells. In addition, the N_2-fixing cells exhibited the lowest product toxicity by retaining the highest methane usage after TCE degradation among the three cultures, whereas the NO_3^--supplied cells exhibited the greatest toxicity.

Growth of Methane Oxidizers on Unsaturated Porous Media

Four different types of porous media: glass beads, silicalite pellets, Monterey sand (MS), and diatomaceous earth pellets (DEP), were tested as

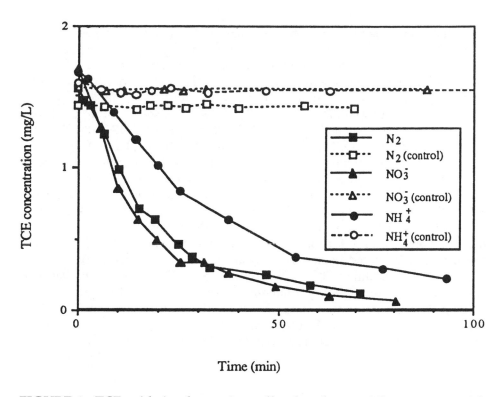

FIGURE 1. TCE oxidation by resting cells of methane oxidizers grown with three different nitrogen sources.

growth supports for methane oxidizers with nitrate mineral salts media at 25% water saturation. The amount of carbon dioxide that evolved from cells grown on each of the support media suggested that microbial activity on the DEP was significantly greater than on any of the other materials. No carbon dioxide evolution was measured in vials containing silicalite pellets after 21 days of incubation. The cell density was approximately five times greater using DEP compared to MS, and was more than 10 times greater on DEP compared to glass beads. Cell density for silicalite pellets was nondetectable.

DEP and MS at 25% water saturation were then used to evaluate the growth of methane oxidizers with various nitrogen sources (nitrate, ammonia, and molecular nitrogen). N_2-fixing cells exhibiting slightly lower cell densities than the NH_4^+ or NO_3^--supplied cells grown on either DEP or MS. However, cell densities for all cells supported on DEP were 3 to 4 times greater than for cells supported on MS, regardless of N source.

Unsaturated Porous Media Column Studies

Experiments were conducted to evaluate the growth of methane oxidizers within unsaturated DEP columns with three different gas and nutrient conditions (listed in Table 1). A microbial population developed rapidly on the surface of DEP in all three columns. CO_2, production, O_2, and CH_4 consumption indicated stable growth for most of the 300-day column operation. The biofilm that developed within Column #3, the nitrogen-fixing column, appeared slimier than the other two columns; and the CO_2 production was significantly higher than that of Columns #1 and #2. The volatile solids within the three columns were approximately 22 mg/g DEP for Column #1, 17 mg/g DEP for Column #2, and 14 mg/g DEP for Column #3, respectively. TCE tracer studies conducted with the abiotic control column indicated insignificant TCE losses within the column following initial sorptive uptake onto the DEP (Figure 2).

Continuous TCE input (ca. 0.08 mg/L) to Column #1 was applied for nearly 5 days (Figure 3). The effluent TCE concentration remained below 0.005 mg/L for the duration of the experiment. During the experiment, methane consumption and CO_2 production did not change significantly while the oxygen consumption increased (Figure 4). Similar results were observed for Column #2. Upon TCE addition (0.7 mg/L) to Column #3, ≥98% TCE removal was observed for 7 continuous days of operation, resulting in ≤0.003 mg/L TCE gas effluent. An important preliminary result of this study was that the microbial columns recovered to typical steady-state conditions (as evidenced by the consumption and production of gases) rapidly following TCE degradation.

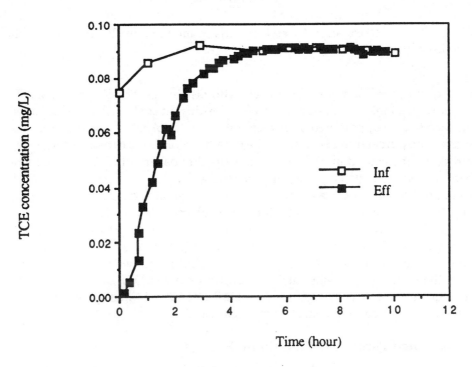

FIGURE 2. Column tracer test using TCE and an uninoculated control column.

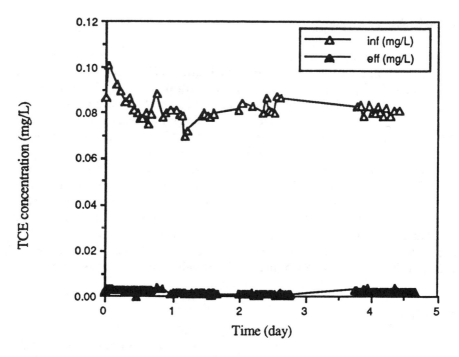

FIGURE 3. TCE oxidation within column #1, a nitrate-supplied column containing methane oxidizers.

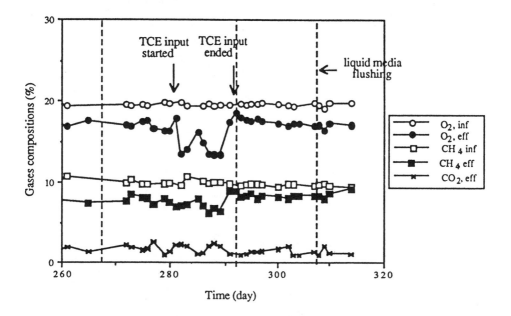

FIGURE 4. Influent and effluent gas compositions for column #1, a nitrate-supplied column containing methane oxidizers.

SUMMARY

Nitrogen-fixing methane oxidizers degraded TCE as rapidly as ammonia- or nitrate-supplied cells and exhibited the highest TCE transformation capacity and lowest TCE product toxicity of all the cultures. Methane oxidizers were unable to develop significant stable populations at 25% water saturation on glass beads or silicalite pellets, whereas stable populations grew on DEP and MS with any of the three nitrogen sources. DEP was able to support 3 to 4 times the methane oxidizer growth that MS supported.

Methane oxidizers grew rapidly within continuous gasflow columns packed with DEP and enriched with CH_4 under different nutrient conditions. Preliminary TCE degradation studies within the columns suggested that $\geq 95\%$ TCE removals could be achieved within each of the biologically active columns for time periods of 5 to 7 days with little effect on the biofilm activity. Long-term column performance and the effects of varying operational parameters such as TCE and methane loadings will be evaluated in future studies.

REFERENCES

Chang, H-L., and L. Alvarez-Cohen. 1995. "Transformation Capacities of Chlorinated Organics by Mixed Cultures Enriched on Methane, Propane, Toluene or Phenol." *Biotechnology and Bioengineering*, 45:440-449.

Chu, K. H., and L. Alvarez-Cohen. 1995. "TCE Degradation by Methane Oxidizing Cultures Grown with Various Nitrogen Sources." *Water Environment Research*, in press.

Whittenbury, R. and H. Dalton. 1980. "The Methylotrophic Bacteria." In M.P. Starr et al. (Eds.), *The Prokaryotes*, p. 894. Springer-Verlag, New York, NY.

Cost Effectiveness of In Situ Bioremediation at Savannah River

Ramiz P. Saaty, W. Eric Showalter,
and Steven R. Booth

ABSTRACT

In situ bioremediation (ISBR) is an innovative new remediation technology for the removal of chlorinated solvents from contaminated soils and groundwater. The principal contaminant at the Savannah River Integrated Demonstration (SRID) is trichloroethylene (TCE), which is a volatile organic compound (VOC). A 384-day test run at Savannah River, sponsored by the U.S. Department of Energy (U.S. DOE), Office of Technology Development (EM-50), furnished information about the performance and applications of ISBR. As tested, ISBR is based on two distinct processes occurring simultaneously; the physical process of in situ air stripping and the biological process of bioremediation. Both processes have the potential to remediate some amount of contamination. A quantity of VOCs, directly measured from the extracted airstream, was removed from the test area by the physical process of air stripping. The biological process is difficult to examine. However, the results of several tests performed at the SRID and independent numerical modeling determined that the biological process remediated an additional 40% above the physical process. Given these data, the cost effectiveness of this new technology can be evaluated.

INTRODUCTION

The purpose of this paper is to study the cost effectiveness of ISBR with horizontal wells as tested at the SRID site in Aiken, South Carolina. ISBR is an innovative new remediation technology for the removal of chlorinated solvents from contaminated soils and groundwater. The principal contaminant at the SRID is TCE. A 384-day test run at Savannah River furnished information about the performance and applications of ISBR.

- The overall cost effectiveness of ISBR is based on the cost sensitivity of the biological component; as the biological addition increases, the cost per pound of VOCs remediated decreases.
- The short-term cost of ISBR with a biological addition of 40% above the vacuum component is $21 per pound (or per kg) of VOCs remediated. The worst case scenario, ISBR + 0% addition, costs $29/lb (per 0.45 kg) of VOCs remediated, and is based solely on the vacuum component.
- The baseline pump-and-treat/soil vapor extraction system costs $31/lb (per 0.45 kg) in the short-term and has no possibility of a biological addition.
- Life-cycle analysis shows that ISBR is more cost effective than the baseline pump-and-treat/soil vapor extraction system.
- As demonstrated, ISBR has a possible savings of $1 million at the SRID site alone.

In situ bioremediation is based on two distinct processes occurring simultaneously: the physical process of in situ air stripping and the biological process of bioremediation (see Figure 1). Both processes have the potential to

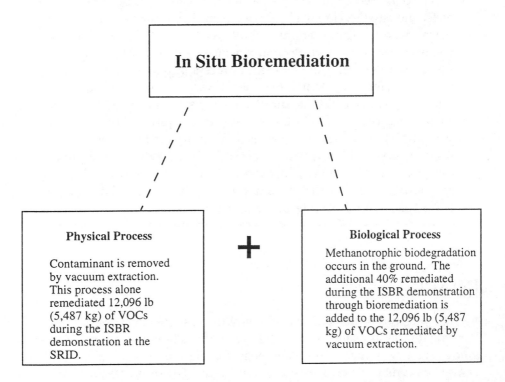

FIGURE 1. Schematic diagram of the two processes involved in in situ bioremediation.

remediate some amount of contamination. A quantity of VOCs, directly measured from the extracted airstream, was removed from the test area by the physical process of air stripping. The biological process is more difficult to examine. However, the results of several tests performed at the SRID and independent numerical modeling determined that the biological process remediated an additional 40% above the physical process. Given these data, the cost effectiveness of this new technology can be evaluated. In addition to calculating the cost effectiveness on the ISBR demonstration at the SRID, sensitivity analysis is conducted to determine how the overall cost of ISBR changes in regard to the performance of the biological component. By comparing the overall cost of this system and the price per pound of VOCs remediated against a conventional pump-and-treat/soil vapor extraction system, we can evaluate the overall cost effectiveness of the alternative technologies.

SYSTEM CAVEATS

The ISBR demonstration at the SRID was set up to address a "hot spot" of an overall larger VOC contaminant plume. The pump-and-treat/soil vapor extraction system is engineer-designed and presumed to perform optimally. Both pump-and-treat and soil vapor extraction systems have been tested at the SRID. The baseline system (a combination of pump-and-treat/soil vapor extraction apparatus) is integrated to avoid overlapping of equipment and materials, and is located in an area exactly like the ISBR demonstration in regard to all necessary site characteristics, including overall concentration of contaminants. By designing both the baseline and the innovative systems to handle equal flow and assuming equal vacuum extraction performance, a level playing field for a cost comparison is created.

ANALYSIS

The data used in these analyses have a "field demonstration" level of confidence and are based on an actual field demonstration. The performance comparison consists of Plan 1, which is based on the new ISBR technology as demonstrated at the SRID, and Plan 2, which is based on "equivalent" conventional technologies, pump-and-treat/soil vapor extraction, necessary to remediate the contamination problems addressed by ISBR. Plan 2 is constructed so that it remediates the same conditions treated by ISBR at the SRID. To be fair to both technologies, equal physical process performance is forced from both Plan 1 and Plan 2. Plan 1 and Plan 2 are compared based on what it costs to operate them over equal periods of time. Performance data indicate that the vacuum component of ISBR destroyed 12,096 lb (5,487 kg) of VOCs in 384 days, and an additional 40% above the vacuum component was destroyed by bioremediation. The vacuum component data are used in the pump-and-treat/soil vapor extraction system, assuming that the equal flowrates will remove the same quantity in an equal amount of time.

The ISBR system, as tested, uses two horizontal wells. The first well is an injection well, 300 ft (91.4 m) long and 165 ft (50.3 m) deep (about 35 ft or 10.6 m below the water table). The second well is an extraction well, 175 ft (53 m) long and 75 ft (23 m) below the surface (in the vadose zone). A concentration of methane (between 1% and 4%) and any necessary chemical nutrients (nitrogen in the form of nitrous oxide and phosphorus in the form of triethyl phosphate) are blended into the injected airstream to create a biological element for remediation. The methane provides the necessary material substrate for the indigenous microorganism to produce the enzyme methane monooxygenase which, in turn, degrades the principal contaminant, TCE. For the conventional technologies used in Plan 2, four vertical soil vapor extraction (SVE) wells are assumed to be equal in area influenced to the one horizontal extraction well of ISBR. One vertical pump-and-treat well is also used. Volatilized contaminants from both remediation systems are sent to a catalytic oxidation off-gas system where they are destroyed.

Economic comparisons for short-term costs are made by relying on actual field data and using cost sensitivity analysis; life-cycle costs are estimated in relation to possible time to achieve cleanup. The first economic comparison is a calculation of the short-term costs in relation to performance. Short-term costs are those expenses incurred during the immediate field test demonstration of the technologies compared (generally about a year). The equipment capital costs are amortized yearly over the useful life of the equipment, which is assumed to be 10 years. All short-term equipment costs are amortized at 7%, which is the interest on the loan.

For ISBR the total cost is about $354,000, with a total of 16,934 lb (7,681 kg) of VOCs destroyed by the vacuum component and biological component, giving a cost per pound of VOCs remediated of about $21. The integrated pump-and-treat/soil vapor extraction with four vertical SVE wells costs about $380,000. Assuming an equal vacuum extraction performance of 12,096 lb (5,487 kg) of VOCs removed, the integrated system has a cost of about $31 per lb (per 0.45 kg) of VOCs remediated. A ratio of ISBR to the baseline shows that ISBR is 32% less expensive than the baseline.

Next, an analysis of life-cycle cost is conducted. A real discount rate of 2.3% is used to calculate the present value. ISBR, with its combination of vacuum component and bioremediation, costs $1 million and remediates the site in only 3 years. The baseline takes 10 years to remediate the site and costs $2 million. ISBR, therefore, saves $1 million and 7 years of remediation. Even when we assume that the baseline can perform at twice the expected time and clean the site in only 5 years, it still costs $1.4 million. ISBR still beats the baseline by $400,000 and 2 years remediation time.

Where ISBR has the potential to exceed the baseline technologies is its ability to remediate a portion of the contamination in situ, thereby eliminating the need to physically remove the contaminant and process it. Because ISBR relies heavily on the biological component to achieve greater performance, sensitivity analysis is conducted to compare the cost per pound of VOCs remediated versus the performance of the biological component. Of particular interest is ISBR + 0% addition. This is a worst case scenario based on a 0% addition from the biological component. It assumes that all the necessary materials are added to stimulate the biological addition, but no additional remediation occurs. In this situation, ISBR still costs slightly less than the baseline, $29 versus $31, respectively. By adding a percent addition of pounds of VOCs destroyed by bioremediation in addition to that removed via the vacuum component, we can examine how the cost per pound changes with respect to the biological component. Six hypothetical percentages are used to account for the bioremediation levels: 0%, 20%, 40%, 50%, 70%, and 90%. Figure 2 shows the various hypothetical additions and the decrease in cost per pound of VOCs remediated.

The baseline technologies in Plan 2 have a constant price per pound of VOCs remediated of $31 because there is no biological component. As the biological addition of ISBR increases, the price per pound of VOCs decreases. So, even in the worse case scenario where no bioremediation occurs, ISBR breaks even with the baseline. There is, therefore, no cost risk to run ISBR over the baseline system. The savings, however, are quite substantial when the biological component is stimulated. In order for the biological component to occur, it is necessary to inject methane and nutrients into the system. Without this material, only the physical, vacuum component of ISBR is possible. Because the cost of the biological component is so inexpensive, ISBR only has to remediate an additional 1,570 lb (712 kg) of VOCs over the 12,096 lb (5,487 kg) of

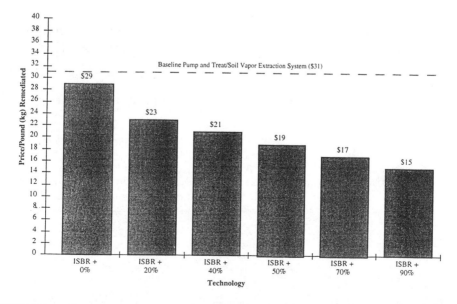

FIGURE 2. Comparison of short-term costs with various biological additions.

VOCs remediated with the vacuum component in order for the system to completely pay for the cost of the methane injection. Any additional remediation is achieved at no extra cost and increases the cost savings of ISBR over the baseline technologies.

Next, the total present value cost for operating each plan for five years, including all necessary equipment, is computed. The total equipment costs are included in the first year so that no amortization is needed. As with the short-term cost, the potential cost savings for ISBR is in its ability to remediate VOCs in addition to the physical process, thereby lowering the cost per pound and increasing the total amount remediated over equal time. The same hypothetical percent additions of 0%, 20%, 40%, 50%, 70%, and 90% are used. Table 1 shows the decrease in price per pound as bioremediation increases. The $38 per lb (per 0.45 kg) of VOCs remediated with the pump-and-treat/soil vapor extraction remains constant because there is no equivalent biological addition.

PERSPECTIVES AND COST DRIVERS

The two largest categories in regard to cost for both ISBR and the baseline system are the costs of consumables and labor. The labor and consumables are greater than 85% of the overall operating costs; therefore, if the overall remediation time of the project is shortened, the cost will drop. This is due to the nature of the labor and consumables that are incurred each day of operation. Because ISBR can significantly decrease operation time, ISBR lowers the overall cost of the remediation effort.

APPLICABILITY

ISBR can be very effective in settings where some interbedded thin and/or discontinuous clays are present. ISBR should prove even more successful than

TABLE 1: Life-cycle cost of ISBR over 5-year operation compared to percent addition (where 1 lb = 0.45 kg).

Hypothetical percent addition	Physical component from life-cycle costs (lb)	Additional pounds remediated via biological component	New total pounds VOC remediated	Price per pound VOC remediated
0%	37,375	0	37,375	$38
20%	37,375	7,475	44,850	$31
40%	37,375	14,950	52,325	$27
50%	37,375	18,688	56,063	$25
70%	37,375	26,162	63,537	$22
90%	37,375	33,638	71,013	$20

in situ air stripping alone because ISBR contains a biological component as well as the physical air-stripping process. A potential concern with the use of ISBR is the possible lateral spread of the contaminant plume. If the geology constricts vertical flow, the injection process can push the dissolved contamination concentrically from the injection point. Thus, it may be advisable in heterogeneous formations to use ISBR in conjunction with a surrounding pump-and-treat system that provides hydraulic control at the site. Note that the limitations on applicable geologic settings described above also apply to soil vapor extraction and pump-and-treat systems.

Anaerobic and Aerobic/Anaerobic Treatment for Tetrachloroethylene (PCE)

Serge R. Guiot, Xin Kuang, Chantal Beaulieu,
Alain Corriveau, and Jalal Hawari

ABSTRACT

The reductive dechlorination of tetrachloroethylene (PCE) was studied in a laboratory-scale upflow anaerobic sludge bed (UASB) reactor using sucrose, lactic acid, propionic acid, and methanol as cosubstrates. Parallel experiments were performed to compare the novel coupled anaerobic/aerobic reactor with the conventional UASB. More than 95% of PCE was transformed in both reactors. Complete dechlorination in the UASB reactor decreased with increased PCE loading, declining from 45 to 19%. Minor concentrations of trichloroethylene and of undegraded PCE were detected in the liquid effluent throughout the experiment. Dichloroethylene was the dominant metabolite at all PCE loads, while vinyl chloride was not detected in the liquid effluent. For both reactor types, increased PCE loading led to lower chemical oxygen demand (COD) removal rates caused by a decrease in the specific acetate utilization rate. This, combined with a decline of the specific total PCE dechlorination activity, may cause long-term stability problems in the UASB reactor. The coupled reactor demonstrated higher specific PCE degradation rates at all PCE loading levels and a higher specific total dechlorination rate at the highest PCE loading. These characteristics may promote long-term stability of the coupled reactor system.

INTRODUCTION

Chlorinated aliphatics are a common group of groundwater contaminants due to their widespread industrial use. In the environment, these compounds are persistent and toxic (Chaudhry & Chapalamadugu 1991). Of this group, tetrachloroethylene (perchloroethylene, PCE) is the only chloroethene isomer that persists throughout aerobic biodegradation. However, it is reductively dechlorinated under anaerobic conditions (Bouwer & McCarty 1983, Fathepure & Boyd 1988, Mohn & Tiedje 1992, de Bruin et al. 1992, Wu et al. 1993). The degradation of PCE proceeds by a sequential reductive dechlorination to TCE, dichloroethylene (DCE), vinyl chloride (VC), and even to ethylene and ethane. However, in many cases, this anaerobic dechlorination is incomplete. According to Vogel et al. (1987), the rate of transformation decreases with the decrease of chlorine substitution. In contrast to the anaerobic processes, aerobic microorganisms are efficient degraders of less-substituted compounds up to complete mineralization.

Aerobic and anaerobic environments each have limitations in their biodegrading abilities that often complement each other when they are combined (Zitomer & Speece 1993). Complete destruction of polychloroorganics can be achieved by sequencing the reductive dechlorination with the bioxidation (McCarty 1993). In the above system, the anaerobic and aerobic bacteria operate in separate units. However, knowledge of a single bacterial culture unit combining simultaneously anaerobic-aerobic degrading abilities is still limited. With this regard, the biofilm could be an appropriate matrix for integrating oxic and anaerobic niches to create a syntrophy between the oxidative and reductive catabolisms (Beunink & Rehm 1990).

This study presents attempts at exploiting the spatial structure of the anaerobic sludge granules to practically develop an aerobic and anaerobic intimate association. Anaerobic microbial granules, which include a wide variety of bacterial trophic groups, have a stratified structure, the periphery of which contains a large variety of morphotypes dominated by fermentative species, which are mostly facultative anaerobes (Guiot et al. 1992). Despite the fact that acetogens and methanogens are obligate anaerobes that are highly sensitive to traces of O_2, it is thus possible to operate these granules in highly oxic environments because the O_2 consumption by the peripheral acidogens would effectively limit the penetration of O_2 on a short distance. This shields the strict anaerobic granule core against any detrimental O_2 (Guiot et al. 1993). By supplying a small amount of O_2 to a UASB reactor and thereby supporting aerobic metabolism in an otherwise anaerobic environment, a coupled reactor allows for an aerobic microbial layer to surround the anaerobic granules. Supporting both anaerobes and aerobes in the same reactor can extend the possible applications and effectiveness of either an anaerobic or aerobic system on its own.

The objectives of our research were to compare the performance of a coupled aerobic/anaerobic reactor system with a conventional UASB system on the following basis: (1) the PCE dechlorinating efficiency, and (2) the effects of high PCE loading rates on the viability of the biological treatment system.

MATERIALS AND METHODS

Reactor Operation

The UASB reactor was a glass column (9.35 cm inside diameter, 80 cm length, effective liquid volume 4.3 L) with Viton™, Teflon™, and glass connections. The coupled reactor system was comprised of a similar 4.3 L reactor connected to a 1.3 L (5 cm × 80 cm) aeration column and a 0.4-L settler. Liquid was recycled in an upflow manner at a rate of 280 to 340 L/d. Pure oxygen was added into the column at a rate of 5 L/d by means of a porous glass 5-m diffuser. This resulted in a dissolved oxygen concentration at the base of the reactor of 0.3 to 0.4 mg/L and a redox potential (E_h) ranging from 28 to 103 mV (against 115 to 162 mV in the UASB). A schematic of the coupled reactor concept is given in Figure 1. The details of the apparatus can be found in Guiot (1994). Two feed streams were mixed prior to entering the 35°C reactors: a cosubstrate solution stored at 4°C, and a stock solution of PCE in methanol at 35°C. The composition of the cosubstrate influent was (g/L): 0.30 sucrose, 0.31 lactic acid, 0.22 propionic acid, 0.01 yeast extract, 0.26 NH_4HCO_3, 0.017 $(NH_4)_2SO_4$, 0.023 KH_2PO_4 and 0.029 K_2HPO_4. The partition ratio of PCE (gas-to-liquid concentration) in the stock bottle was 0.24, comparable to other studies (0.26, Kastner 1991).

Analytical Methods

The measurement of PCE, TCE, DCE, dichloromethane (DCM) and VC in the liquid samples was made using the U.S. EPA Method 8260. Analysis was

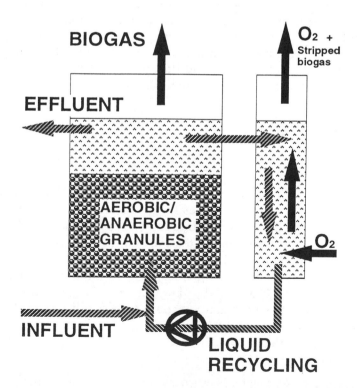

FIGURE 1. Schematic of the coupled aerobic/anaerobic system concept.

performed using a Varian 3400 (Walnut Creek, California, USA) gas chromatograph equipped with a Saturn II ion trap mass spectrometer detector, using bromofluorobenzene as a surrogate. An aliquot of 5 mL was purged and trapped by a Tekmar (Cincinnati, Ohio, USA) LSC 2000 and 2016 autosampler. The desorbed material was passed through a 60 m $\times$ 0.32 mm DB-624 capillary column (J&W Scientific, Folson, California, USA) by means of a helium carrier gas. The injector and interface temperatures were 125 and 220°C respectively. The oven temperature was maintained at an initial temperature of 40°C for 2 minutes, then raised at a rate of 4°C/min to 50°C and increased by 8.5°C/min to reach a final temperature of 225°C.

The concentrations of PCE, TCE, DCE and VC in the gas phase were determined by a Sigma 2000 gas chromatograph with an FID detector (Perkin-Elmer, Norwalk, Connecticut, USA). A 100- to 200-L gas sample was injected on a 1.8 m Carbopack B/1% SP-1000 column (Supelco, Bellafonte, Pennsylvania, USA) using nitrogen as the carrier gas. The oven temperature was programmed at 40°C for 1.5 min, 25°C/min to 150°C, and 10°C/min to 200°C for 6 min.

The inorganic chloride concentration was determined by ion chromatography using a Spectra-Physics (Fremont, California, USA) HPLC, Waters 431 conductivity detector (Millipore, Milford, Massachusetts, USA) and a 250 mm $\times$ 41 mm PRP-X100 column (Hamilton, Reno, Nevada, USA). The sample supernatant was acidified to pH 6.2 $\pm$ 0.2 with 0.3 M H_2SO_4 in order to eliminate bicarbonate interference. The detailed procedures for the routine tests [pH, E_h, volatile fatty acids (VFA), gas composition, chemical oxygen demand (COD), and volatile suspended solids (VSS)] and the specific substrate activity of the reactor biomass were described previously by Guiot et al. (1992).

RESULTS AND DISCUSSION

Startup

Prior to adding PCE, the UASB reactor, which had been inoculated with sugar and acetate-fed anaerobic granules [19 g VSS/liter of reactor (L_{rx})], was operated for 3 months at an organic loading rate (OLR) of 1 g COD/L_{rx} d. Once stable operation was achieved, the reactor was fed for 90 days at a gradually increasing PCE load to end up at 29 mg PCE/L_{rx} d, while OLR and hydraulic residence time (HRT) were around 6 g COD /L_{rx}.d and 6 h, respectively. An average removal rate of 24 mg PCE/L_{rx} d was achieved. This startup phase demonstrated that the anaerobic granules could successfully remove PCE and adapt to high PCE concentrations quickly.

PCE Dechlorination

After startup, the UASB biomass was divided into two equal amounts. One half was returned to the UASB reactor and the other half served as the inocula for the coupled reactor system. The two systems were then operated in parallel at OLR ranging from 3.0 to 7.1 g COD /L_{rx}.d and at HRT, from 5.2 to 7.1 h. PCE was loaded in both systems at three different levels: at 30 mg PCE/L_{rx} d on average for 15 days with an influent concentration of approximately 8 mg PCE/L; at 41 to 45 mg PCE/L_{rx} d for the next 15 days with a 12 mg PCE/L influent concentration; and at 58 to 70 mg PCE/L_{rx} d for 40 days with an influent concentration of approximately 17 mg PCE/L. Both the COD and the PCE loading rates of the coupled reactor were based on the total volume of liquid in the system, i.e. 6.0 L.

The PCE removal rate increased with the PCE load: on average, 29, 41, and 60 mg PCE/L_{rx} d successively. In the UASB reactor, PCE was transformed at an efficiency that exceeded 95%, irrespective of the PCE load. The complete dechlorination rate was determined by means of a molar balance of the PCE load and the residual PCE plus all the chlorinated metabolites in the aqueous and gaseous phases. Complete dechlorination efficiency decreased with increasing loading rate, from 45 to 41 to 19% for the UASB reactor. The PCE removal in the coupled reactor was similarly high (93 to 95%). Although lower on average, the complete dechlorination efficiency of the coupled reactor system was largely unaffected by the PCE loading rate, ranging from 22 to 31%. The coupled reactor system demonstrated superior complete dechlorination at the highest PCE loading rate. The treated effluent streams of both reactor systems consistently bore a significant concentration of partially dechlorinated PCE metabolites. Figure 2 shows the concentration of PCE and its metabolites for both the UASB and the coupled reactors at the three PCE influent concentrations. Measurements of inorganic chlorine, DCM, DCE, TCE, and PCE revealed a near quantitative chlorine recovery. The volatilization of all PCE and chlorinated intermediates accounted for less than 0.5% of the PCE load.

The PCE concentration detected in the effluent remained very small throughout these experiments, from 1 to 5% of the total influent chlorine load. The TCE concentration decreased from 13 to 6 to 3% of the total influent chlorine load. The dominant metabolite at all loading rates was cis-1,2-DCE, which ranged from 18 to 29% of the total influent chlorine load. In some cases, particularly in the case of the UASB reactor, DCM was detected when the influent PCE was raised to the highest concentration. We are not sure whether this is a real metabolite or a system artifact. Thus far, all literature in this field excludes the formation of DCM as a PCE metabolite (McCarty 1993). No significant VC concentrations were detected in the liquid effluent (detection limit of 1 g/L), but a small amount of VC (< 10 ppm) was found in the gas phase. Measurements of ethylene in the liquid phase were not available. However, trace amounts of ethylene were detected in the gas phase. Fathepure and Tiedje (1994) achieved similar results. This indicates that the rate-limiting step was the conversion from cis-1,2-DCE to VC. However, other investigators reported the conversion of VC to ethylene to be the rate-limiting step in the PCE-dechlorination pathway (Long et al. 1993, Chu & Jewell 1994).

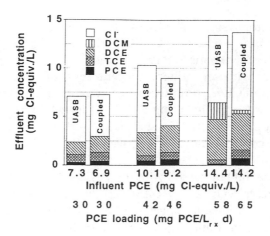

FIGURE 2. Relative concentration of the tetrachloroethylene (PCE), of its chlorinated metabolites (in equivalent-chlorines), and of inorganic chloride in the effluent as a function of the PCE loading rate and the PCE influent concentration.

Table 1 illustrates both the rates of PCE transformation and complete dechlorination per gram of biomass (mg PCE/g VSS d). The specific rate of PCE removal increased with the increase of PCE loading rate for both the UASB and the coupled reactor. This indicates that the first step of dechlorination (transformation of PCE into TCE) has never been limiting. Consistently, the lower specific rates of PCE removal for the UASB reactor were due only to the higher biomass volumic content of the UASB reactor compared to the coupled system, as the latter had a higher total liquid volume for a similar amount of biomass.

The specific rate of complete dechlorination in the UASB reactor had reached its maximum value (1.2 to 1.5 mg PCE/g VSS d) since it steadied although the PCE load increased. In contrast, in the coupled system, capacity for complete dechlorination increased as the PCE load increased, showing a maximal rate of specific complete dechlorination 2½-fold higher. In contrast to the PCE activity in the UASB reactor, the increased capacity of the coupled reactor biomass to completely dechlorinate PCE and its metabolites suggests that adaptation to even higher PCE loads may be possible.

Specific Biomass Activity

To study the effect of high concentrations of PCE on anaerobic granules, specific substrate utilization rate assays were conducted using four substrates: glucose, acetic acid, propionic acid, and hydrogen (Table 2). The specific glucose activity of the coupled system was consistently greater than that of the UASB, confirming the higher activity of the aerobically supported fermentative bacteria. This specific glucose activity decreased with elevated PCE loads and with PCE exposure time. The specific propionate and hydrogen utilization

TABLE 1. Change in the specific rates of PCE transformation and complete dechlorination as a function of the PCE loading.

Average PCE load mg PCE/L_{rx} d	PCE transformation specific rate mg PCE/g VSS d (± std. dev.)		PCE total dechlorination specific rate mg PCE/g VSS d (± std. dev.)	
	UASB reactor	Coupled reactor	UASB reactor	Coupled reactor
30	2.4 ± 0.1	4.2 ± 0.7	1.3 ± 0.35	1.3 ± 1.15
43	3.4 ± 0.2	5.8 ± 0.5	1.5 ± 0.30	1.4 ± 0.30
62	4.7 ± 0.3	9.4 ± 1.0	1.2 ± 0.82	3.1 ± 0.78

TABLE 2. Specific substrate biomass activity for UASB vs. coupled reactors.

Experiment step	Reactor	Specific Substrate Biomass Activity, g substrate/g VSS d(± std. dev.)			
		Glucose	Acetate	Propionate	Hydrogen
Startup	UASB	>1.26	0.65 ± 0.17	0.16 ± 0.04	0.37 ± 0.01
	Coupled	>2.07	0.87 ± 0.26	0.15 ± 0.01	0.56 ± 0.28
30[a] (15)[b]	UASB	1.99 ± 0.59	0.69 ± 0.21	0.13 ± 0.01	0.22 ± 0.10
	Coupled	2.95 ± 0.11	0.71 ± 0.17	0.17 ± 0.01	0.21 ± 0.01
63 (15)	UASB	0.72 ± 0.15	0.20 ± 0.04	0.09 ± 0.01	0.30 ± 0.19
	Coupled	2.67 ± 0.96	0.19 ± 0.06	0.12 ± 0.02	0.44 ± 0.07
63 (36)	UASB	0.64 ± 0.01	0.10 ± 0.03	0.11 ± 0.02	0.24 ± 0.05
	Coupled	1.28 ± 0.11	0.29 ± 0.06	0.21 ± 0.04	0.84 ± 0.31

(a)   Average PCE loading rate (mg PCE/L_{rx} d).
(b)   Operating time (day) within the experiment current step.

rates were unaffected by PCE loading rate. While not affected by the aerobic coupling, the specific acetate methanogenic activity was also lowered over exposure to the highest PCE concentrations. This observation coincided with DCE accumulation and is consistent with the lowered methane productivities and COD removal. Some studies have reported similar inhibition of methanogenesis by the high concentrations of PCE and/or its reductive products and suggest a role for other organisms in the dechlorination of PCE (DiStefano et al. 1991). Interestingly, the methanogen inhibition tended to be slightly relieved in the aerobic/anaerobic coupling, as the specific acetoclastic activity re-increased at the end of the high-PCE-load phase.

CONCLUSIONS

1. The anaerobic granules were highly resistant to oxic environments.
2. The anaerobic granules that developed in both the UASB and the coupled reactor-degraded PCE and adapted to high PCE concentrations quickly. This indicates that there was no limitation at the first step of dechlorination. DCE was the dominant metabolite at all PCE loads.
3. For both reactor types, increased PCE loading led to lower COD removal rates caused by a decrease in the specific acetate utilization rate. This, combined with a decline of the specific total PCE dechlorination activity, may cause long-term stability problems in the UASB reactor.
4. The coupled reactor demonstrated higher maximal specific rate of complete dechlorination. These characteristics may promote long-term stability of the coupled reactor system.

REFERENCES

Beunink, J., and H.J. Rehm. 1990. "Coupled reductive and oxidative degradation of 4-chloro-2-nitrophenol by a co-immobilized mixed culture system." *Appl. Microbiol. Biotechnol. 34*: 108-115.

Bouwer, E.J., and P.L. McCarty. 1983. "Transformations of 1- and 2- Carbon Halogenated Aliphatic Organic Compounds Under Methanogenic Conditions." *Appl. Environ. Microbiol. 45*(4): 1286-1294.

Chaudhry, G.R., and S. Chapalamadugu. 1991. "Biodegradation of Halogenated Organic Compounds." *Microbiol. Rev. 55*(1): 59-79.

Chu, K.H., and W.J. Jewell. 1994. "Treatment of Tetrachloroethylene with Anaerobic Attached Film Process." *J. Environ. Engin. 120* (1): 58-71.

de Bruin, W.P., M.A. Posthumus, and A.J.B. Zehnder. 1992. "Complete Biological Reductive Transformation of Tetrachloroethene to Ethane." *Appl. Environ. Microbiol. 58* (6): 1996-2000.

DiStefano, T.D., J.M. Gossett, and S.H. Zinder. 1991. "Reductive Dechlorination of High Concentration of Tetrachloroethene to Ethene by an Anaerobic Enrichment Culture in the Absence of Methanogenesis." *Appl. Environ. Microbiol. 57* (8): 2287-2292.

Fathepure, B.Z., and S.A. Boyd. 1988. "Reductive Dechlorination of Perchloroethylene and the Role of Methanogens." *FEMS Microbiol. Let. 49*: 149-156.

Fathepure, B.Z., and J.M. Tiedje. 1994. "Reductive Dechlorination of Tetrachloroethylene by a Chlorobenzoate-Enriched Biofilm Reactor." *Environ. Sci. Technol. 28* (4): 746-752.

Guiot, S.R. 1994. "Anaerobic and Aerobic Integrated System for Biotreatment of Toxic Wastes (CANOXIS)." U.S. Patent Pending. Serial No. 08/316,482. 3 October, 1994.

Guiot, S. R., J. C. Frigon, B. Darrah, M. F. Landry, and H. Macarie. 1993. "Aerobic and Anaerobic Synchronous Treatment of Toxic Wastewater." *Med. Fac. Landbouww. Univ. Gent 58*(4a): 1761-1769.

Guiot, S.R., A. Pauss, and J. W. Costerton. 1992. "A Structured Model of the Anaerobic Granule Consortium." *Wat. Sci. Technol. 25* (7): 1-10.

Kastner, M. 1991. "Reductive Dechlorination of Tri- and Tetra-chloroethylenes Depends on Transition from Aerobic to Anaerobic Conditions." *Appl. Environ. Microbiol. 57*(7): 2039-2046.

Long, J.L., H.D. Stensel, and J.F. Ferguson. 1993. "Anaerobic and Aerobic Treatment of Chlorinated Aliphatic Compounds." *J. Environ. Eng. 119* (2): 300-319.

McCarty, P.L. 1993. "In Situ Bioremediation of Chlorinated Solvents." *Current Opinion in Biotechnology.* *4* (3): 323-330.

Mohn, W.W., and J.M. Tiedje. 1992. "Microbiol Reductive Dehalogenation." *Microbiol. Rev.* *56*(3): 482-507.

Vogel, T.M., C.S. Criddle, and P.L. McCarty. 1987. "Transformations of Halogenated Aliphatic Compounds." *Environ. Sci. Technol.,* *21*: 722-736.

Wu, W.M., J. Nye, R.F. Hickey, and L. Bhatnagar. 1993. "Anaerobic Granules Developed for Reductive Dechlorination of Chlorophenols and Chlorinated Ethylene." *Proc. 48th Annual Purdue University, Industrial Waste Conference.* West Lafayette, IN, USA.

Zitomer, D.H., and R.E. Speece. 1993. "Sequential Environments for Enhanced Biotransformation of Aqueous Contaminants." *Environ. Sci. Technol.* *27*(2): 227-244.

Application of Plant Materials for the Cleanup of Wastewater

Jerzy Dec and Jean-Marc Bollag

ABSTRACT

Plant materials that contain oxidoreductases may be used as inexpensive and easy-to-apply tissues for decontamination purposes. Tissues from three plants (horseradish roots, potato tubers, and white radish roots) were minced and tested for the removal of 2,4-dichlorophenol (2,4-DCP) (up to 850 mg/L) and other chlorinated phenols from wastewater. The reactions were initiated by the addition of H_2O_2 as an electron acceptor. After a 2-hour incubation period, the percentage of pollutant removal for the different plants amounted to 100%, 97%, and 99%, respectively. Horseradish-mediated removal of 2,4-dichlorophenol from model solutions was comparable to that achieved using purified horseradish peroxidase. The proportion of plant material to polluted water was 1:10 w/v. At a concentration of 5.3 mM H_2O_2 and pH ranging from 3 to 8, more than 90% of substrate was removed; most removal (96%) occurred at pH 6. Using 10.6 mM H_2O_2 concentration, complete removal could be achieved during a 30-minute treatment. Minced horseradish roots could be reused up to 30 times.

INTRODUCTION

The ever-increasing environmental threat due to emissions of toxic pollutants by modern industry and agriculture creates an urgent need to develop alternative, more effective methods to decontaminate water and soil. Phenols and anilines are among the most common pollutants (Mills et al. 1985). If improperly handled, these contaminants can easily find their way to surface waters and soil, and from soil they may leach into groundwater.

To date, chemical treatment and biodegradation were used for dealing with these problems. Recently, we proposed an alternative, biotechnological approach based on the use of minced plant materials to mediate the decontamination process (Dec and Bollag 1994). The detoxification effect was due to polymerization of the pollutants through an oxidative coupling reaction that can be mediated by oxidoreductive enzymes, such as horseradish peroxidase or laccases from fungi (Klibanov et al. 1983, Bollag 1992). Numerous plants have been found to possess oxidoreductase activity (Mayer and Harel 1979), and they may be used as enzymatic agents for the oxidative coupling reaction.

In this paper, the usefulness of various plant materials (potato, white radish, and horseradish) for the removal of chlorinated phenols from an industrial wastewater is demonstrated. Horseradish, as the most efficient plant, was subsequently used to optimize the detoxification process. The optimization was performed using defined solutions of 2,4-dichlorophenol as a model pollutant.

EXPERIMENTAL PROCEDURES AND MATERIALS

Chemicals and Enzymatic Materials

The wastewater (pH<1) used in this study was obtained from a company producing 2,4-D. The model compound 2,4-dichlorophenol (2,4-DCP) was purchased from Aldrich-Chemie GmbH (Steinheim/Albuch, Germany). Horseradish peroxidase with specific activity of 102 units/mg of solid was obtained from Sigma Chemical Co. (St. Louis, Missouri). Potato, white radish, and horseradish were bought at a vegetable market and stored in a container with soil. Before the experiments, the respective vegetables were washed with water, then cut into pieces and thoroughly mixed.

Decontamination Reactions

Triplicate samples of the 2,4-D wastewater or model solutions of 2,4-DCP in universal buffer (pH 2 to 11) were amended with horseradish peroxidase or cut plant materials and the reactions were initiated by the addition of H_2O_2 (Dec and Bollag 1994). The pH of the wastewater was adjusted, if required, with 10 M NaOH. Samples without H_2O_2 and with boiled enzyme or plant materials served as controls. After a 2-hour incubation, the reaction mixtures were acidified with acetic acid to pH 1.7 and centrifuged at 6000 × g for 15 min. The supernatants were analyzed by high-performance liquid chromatography (HPLC).

RESULTS

Before plant materials were applied, we tested the decontamination potential of purified horseradish peroxidase. As determined by HPLC analysis, the 2,4-D wastewater contained five phenolic components: phenol, 2-chlorophenol, 4-chlorophenol, 2,4-dichlorophenol, and 2,6-dichlorophenol. Of these, 2,4-DCP was the major pollutant; its concentration in the wastewater was up to 850 mg/L. Horseradish peroxidase (9.5 units/mL + 20.4 mM H_2O_2) proved to be very efficient (Dec and Bollag 1994); over 95% removal of 2,4-DCP was achieved in a remarkably wide pH range (pH 3 to 10). However, application of the purified enzyme on an industrial scale did not appear to be economically feasible.

Table 1 presents the results of our experiments on decontamination of the 2,4-D wastewater using plant materials (1 g plant material cut into 250 pieces). Removal of pollutants was not observed when wastewater was only treated with H_2O_2. The addition of minced plants alone (without H_2O_2) to the wastewater did not result in any transformation of chlorophenols; however, significant physical sorption of pollutants in the plant tissue was observed. In a study with 2,4-DCP, the sorbed pollutant could be recovered almost completely (96.6%) by extraction of the plant material with water and methanol (Roper et al. 1995). In the presence of H_2O_2 (2.65 to 21.2 mM), the reaction mixtures and the surfaces of the dissected plants changed color and a precipitate of oligomerized chlorophenols was formed. The percentage of 2,4-DCP removal increased with increasing amounts of H_2O_2 (Table 1), indicating that transformation of chlorophenols was caused by peroxidase activity. The decontamination effected by plant materials equaled that achieved using purified horseradish peroxidase. After reactions involving H_2O_2, only 1% of 2,4-DCP could be extracted from plant material (Roper et al. 1995).

When minced horseradish roots (0.5 g cut into 128 pieces + 5.3 mM H_2O_2) were incubated for 2 hours with 5 mL of 9 mM 2,4-DCP in universal buffer (pH 2 to 12), large amounts of the substrate (approximately 90%) were removed between pH 3 and 8 (Figure 1). Maximal removal (96.4%) occurred at pH 6. Less removal (68.0% at pH 6) was observed with smaller amounts of horseradish (0.2 g) and H_2O_2 (2.65 mM). For purified horseradish peroxidase (8.3 units/mL + 18.4 mM H_2O_2) the optimal pH was 5, and maximum removal (96 to 99%) was observed at pH 4 to 7, a range narrower than that for the minced horseradish roots.

TABLE 1. Removal of 2,4-dichlorophenol (2,4-DCP) from 2,4-D production wastewater using different plant materials.

Addition of H_2O_2, mM	% Removal[b] of 2,4-DCP using:		
	Potato	White Radish	Horseradish
0	53.3	43.9	78.0
2.65	73.5	70.7	88.8
5.30	90.8	92.2	99.1
10.60	96.5	99.0	100.0
21.20	87.9	95.2	100.0

Source: Dec and Bollag (1994).
(a) 1 g per 5 mL of each reaction mixture at pH 5 (2-h incubation).
(b) The standard deviation ranged betweeen 0.3 and 2.1%.

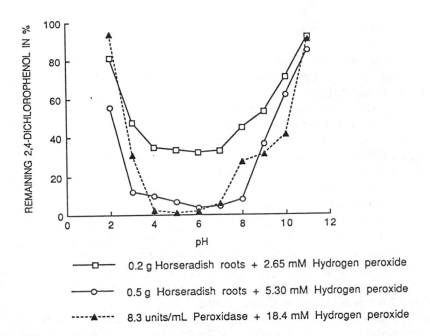

FIGURE 1. **Effect of pH on the transformation of 2,4-dichlorophenol by minced horseradish and horseradish peroxidase in universal buffer during a 2-hour incubation at 25°C (the SD ranged between 0.2 and 3.0%) (Dec and Bollag 1994).**

Reuse of enzymatic preparations makes the procedure more economical. To determine the reusability of horseradish roots, 0.5 g of the plant material cut into 128 pieces was incubated repeatedly (recycled) with 5-mL samples of 0.6 mM and 6.0 mM 2,4-DCP in universal buffer at pH 6; samples were amended each time with H_2O_2 (1.3 mM and 13.0 mM, respectively). The incubated solutions were removed and replaced with fresh portions every 30 minutes. Minced horseradish was capable of repeatedly removing 2,4-DCP from the reaction mixture (Figure 2).

The reusability of the plant material depended on the initial concentration of the pollutant. For 6 mM 2,4-DCP buffer solution, the peroxidase activity in cut horseradish gradually decreased from 100% removal in the two first cycles to 11.6% in the 16th cycle (data not shown). However, with 0.6 mM 2,4-DCP solution, essentially no decrease of horseradish efficiency could be observed during 15 cycles; subsequently a decrease of peroxidase activity was observed, from 100% removal in the 15th cycle to 49.0% in the 30th. If recycling was halted overnight, no change in the efficiency of horseradish was observed. In the first cycle, more than 30% of 2,4-DCP in the 0.6-mM solution and about 40% in the 6.0-mM solution were removed in the absence of H_2O_2 by physical sorption to the plant tissue, and this removal dropped below 5% in subsequent cycles.

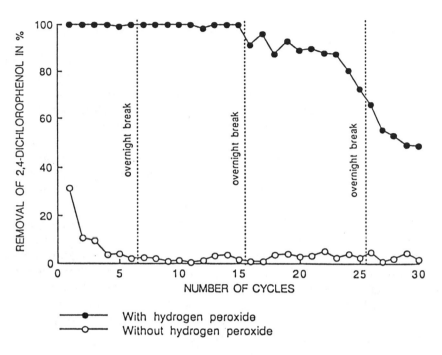

FIGURE 2. **Transformation of 2,4-dichlorophenol (0.6 mM) by minced horse-radish roots during repeated 30-minute incubation cycles with 1.3 mM H_2O_2 (the SD ranged between 0.1 and 4.5%) (Dec and Bollag 1994).**

To examine the stability of peroxidase contained in horseradish roots, 0.5-g amounts of the plant material cut into 128 pieces were stored at room temperature (22°C) for 0, 2, 8, 32, and 128 days and then recycled seven times (every 30 minutes) with 5-mL samples of 9.0 mM 2,4-DCP in universal buffer at pH 6. The analysis for 2,4-DCP removal showed that the cut horseradish gradually lost peroxidase activity during prolonged storage, but retained enough active enzyme to remove 98% of 2.4-DCP in the first cycle after 8 days of storage and 80% after 32 days. The decrease in activity was more evident in subsequent cycles. Nevertheless, horseradish tissue retained some enzyme activity even after 128 days of storage (40% removal in the first cycle).

In a study of substrate specificity, horseradish was able to remove other phenols and anilines from the buffer solutions (Roper et al. 1995). The rate of removal depended on the chemical structure of the pollutant.

DISCUSSION

The results demonstrate that, in addition to simplicity of application, the following properties render plant materials suitable for water pollution control: high decontamination potential in a wide pH range, short treatment time requirement, and stability during prolonged storage. Of special significance, however, is the ability of horseradish to be recycled. This feature also pertains to enzymes immobilized on solid supports, making them valuable for industrial applications (Gianfreda and Scarfi 1991). In fact, the shared properties of minced plant material and immobilized enzymes suggest that the active enzyme is immobilized in the plant tissue. The nature of this immobilization has not been determined. It may be due to the attachment of the enzyme to cell walls or encapsulation in cell particles and organelles.

In addition to horseradish, potato, and white radish, other plant sources should be evaluated for their decontamination potential, especially those containing oxidoreductases that do not require the addition of H_2O_2. The application of plant technology to wastewater treatment can be performed either periodically in reactors with mechanical stirring or continuously through columns packed with the dissected plant material. Plants also can be applied directly to polluted natural aquifers; in the case of polluted groundwater a "pump and treat" approach may be employed. Based on the phenomenon that oxidoreductases mediate the binding of phenols and anilines to humus (Bollag 1992), it may be possible to use dissected plants for decontamination of soils. The used plant material from these treatments may then be disposed of by composting, plowing into soil, or combustion.

REFERENCES

Bollag, J.-M. 1992. "Decontamination of soil with enzymes: An in situ method using phenolic and anilinic compounds." *Environ. Sci. Technol.* 26:1876-1881.

Dec, J. and J.-M. Bollag. 1994. "Use of plant material for the decontamination of water polluted with phenols." *Biotechnol. Bioeng.* 44:1132-1139.

Gianfreda, L., and M.R. Scarfi. 1991. "Enzyme stabilization: State of the art." *Molec. Cell Biochem.* 100:97-128.

Klibanov A.M., T.-M. Tu, and K.P. Scott. 1983. "Peroxidase-catalyzed removal of phenols from coal-conversion wastewaters." *Science* 221:259-261.

Mayer, A.M. and E. Harel. 1979. "Polyphenoloxidases in plants." *Phytochemistry* 18:193-215.

Mills, W.B., D.B. Procella, M.J. Ungs, S.A Gherini, K.V. Summers, L. Mok, G.L. Rupp, and G.L. Bowie. 1985. *Water Quality Assessment: A Screening Procedure for Toxic and Conventional Pollutants in Surface and Groundwater.* EPA/600/6-85/002a.

Roper, J.C., J. Dec, and J.-M. Bollag. 1995. "Using minced horseradish roots for the treatment of polluted waters." (manuscript in preparation).

Treatability Study to Evaluate In Situ Chlorinated Solvent and Pesticide Bioremediation

Joy E. Ligé, Ian D. MacFarlane, and Thomas R. Hundt

ABSTRACT

Exploiting microbial reactions to remediate chlorinated solvents and pesticide contamination appeared to be an attractive remedial alternative for a site located at the Dover Air Force Base in Dover, Delaware. To generate data to evaluate the feasibility of implementing enhanced in situ bioremediation as a remedial technique, a relatively inexpensive and rapid treatability study was designed. Batch microcosm studies were used to mimic in situ redox conditions (oxygenated and unoxygenated) and methanotrophic conditions. No evidence of target compound degradation existed under either the amended or unamended aerobic conditions; however, the activity observed in some samples under anaerobic conditions suggested transformation of chlorinated compounds and pesticides. Results showed that biodegradation may be inducible under certain conditions, but time lags and efficiencies could be expected to vary considerably. A remedial alternative analysis could not be expected to achieve the degree of accuracy and precision necessary without the data resulting from this study.

INTRODUCTION

Enhanced in situ biodegradation is a potentially attractive remedial alternative for some sites, but few standard methods exist to evaluate its feasibility. A treatability study was designed to evaluate the feasibility of implementing in situ bioremediation for a site located on the Dover Air Force Base (DAFB) in Dover, Delaware, which contained elevated concentrations of certain chlorinated solvents and pesticides within the subsurface. This laboratory study provided a relatively inexpensive ($50,000) and rapid (12 weeks) means to generate data for input into the remediation alternative analyses. Specific objectives of this laboratory study were to assess the ability of the indigenous microorganisms to degrade 1,1,1-trichloroethane (TCA), trichloroethene (TCE), tetrachloroethane (PCE), and lindane, and to determine whether biodegradation could be stimulated by the presence of alternative electron acceptors, cosubstrates, and nutrients. This paper presents the methodology used for the treatability study and an overview of the results and application to the feasibility study.

SAMPLING AND ANALYSIS

Soil and Groundwater Collection

The treatability study was designed to mimic oxygenated, unoxygenated, and methanotrophic conditions by collecting acclimated soil samples from areas most representative of aerobic and anaerobic/microaerophilic environments. To identify these zones, two exploratory borings were initially cored using a hollow stem auger equipped with a Hydropunch® sampler in areas known to contain elevated levels of the target compounds. Groundwater samples were collected from two depths within each boring and analyzed in triplicate for redox potential and dissolved oxygen (DO) using a redox meter and DO probe and CHEMetrics® kit, respectively. Based on this analysis, soil in the upper depths (4.6 to 5.2 m) were treated as anaerobic since groundwater samples collected at that zone exhibited less aerobic conditions (DO: 1.5 mg/L and redox: –140 mV) compared to samples collected from the 10.7 to 11.3 m zone (DO: 3 to 5 mg/L and redox: +88 to +195 mV). Subsurface soil samples were then collected from two borings, drilled approximately 1.5 m upgradient from the initial Hydropunch® borings, using a hollow-stem auger equipped with a 7.62-cm inner diameter split spoon. The split spoon was fitted with a sterile butyrate liner to contain the soil cores. Aerobic samples were subsampled aseptically and placed into sterile Whirl-Paks®. Cores treated as anaerobic were subsampled in a portable anaerobic glove bag filled with nitrogen gas and stored in sterile anaerobic Gas-Paks®. All samples were stored at 4°C until processed after 48 hours of collection.

Groundwater was collected from a monitoring well located near the boreholes and refrigerated until use. The groundwater was used as stock solution in the microcosms, so that the natural chemical constituents of the aquifer would be present.

Degradation Assay. Microcosm enrichment studies were conducted to examine the degradation of TCA, TCE, PCE, and lindane over time under various conditions. The microcosms consisted of 40-mL serum vials, containing 8 g of sediment and 31 mL of filtered groundwater. To account for losses of volatile compounds during the mixing of the stock solution, the microcosms were spiked with approximately 1 mg/L each of the target compounds. A nutrient stock, consisting of 20 inorganic compounds, was added to facilitate biomass growth.

Three batches of microcosms were prepared to replicate oxygen-saturated, methanotrophic, and methanogenic conditions. This entailed sparging the ground-water (prior to adding the organic spikes) with pure oxygen for the aerobic cultures and adding a 2:1 combination of oxygen- and methane-sparged groundwater for the methanotrophic cultures. Anaerobic cultures were prepared in an anaerobic glove box, and the groundwater was sparged with a mixture of nitrogen and carbon dioxide to ensure anaerobic conditions. In addition, sodium dithionate was added to the nutrient stock solution as an oxygen scavenger, and sodium lactate was added to stimulate the growth of methogens. The trial microcosms were filled with no headspace (amended with 1 mL of deionized water), sealed with a Teflon™ septum, and stored in an inverted position to minimize volatilization of the chlorinated solvents. A biocide consisting of 0.5 mL each of mercuric chloride and sodium azide was added to control microcosms. The microcosms were incubated in darkness at a constant temperature of 22°C, and the anaerobic cultures were stored in an anaerobic glove box.

Three trial microcosms and two associated controls were sacrificed and analyzed every week (8 weeks) for the aerobic microcosms and every other week, during a 12-week incubation period, for the anaerobic systems. To assess biodegradation, groundwater extracted from each microcosm was analyzed for target compounds, intermediates, and inorganic parameters. Carbon monoxide, carbon dioxide (CO_2), methane (CH_4), ethane, and ethene were measured using a gas chromatograph (GC) equipped with a thermal conductivity detector and a flame ionization detector. The spiking compounds and their potential breakdown products were assayed following EPA Standard Methods 624 and 608.

Microbiological Enumeration. Microbial population density was determined with an acridine orange direct count (AODC) method, modified from Trolldenier (1973) and Wilson et al. (1983). Enumeration was performed on all three batch microcosm systems on the first and last week (week 8 or 12) of the incubation period to measure population growth.

RESULTS AND DISCUSSION

Table 1 presents the mean percent loss (the standard deviation) of TCA, TCE, PCE, and lindane over the incubation periods for the aerobic microcosms relative to controls. The data indicated that there was no significant degradation of target compounds or production of transformation intermediates under either the unamended (oxygen-saturated) or amended (methanotrophic) aerobic conditions. No measurable oxygen consumption or CO_2 production was observed in the unamended and CH_4-amended aerobic microcosms relative to the controls. However, there was some substantial disappearance of CH_4 in sample microcosms relative to the controls (Figure 1). The concentration of CH_4 in samples from Location A became particularly low after 8 weeks. This suggests that the acclimation period for measurable transformation of chlorinated solvents by methanotrophs is significant (> 2 months). After the 8-week incubation period, a significant increase (Table 1) in the number of bacteria was observed in sample microcosms incubated in the presence of CH_4, suggesting that methanotrophic bacteria were consuming some growth substrate; however, there was little to no increase in the indigenous microbial population in microcosms incubated in the absence of CH_4.

Biodegradation of target compounds under enhanced anaerobic methanogenic conditions was significant after 12 weeks, particularly in sediments from

TABLE 1. Summary of degradation and enumeration assay.

Sample Location	Depth (M)	Percent Biotic Degradation (at end of incubation period)				Total Count[a]	Percent Change Relative to Controls (at end of incubation period)		
		1,1,1-TCA	PCE	TCE	Lindane		CO_2 Concentration	Methane Concentration	O_2 Concentration
Oxygen-Saturated Conditions									
A	10.4-11.0	3 (±7)[b]	6 (±4)	6 (±4)	1 (±3)	+14	−54	−8	0
B	10.4-11.0	7 (±3)	0 (±12)	2 (±7)	9 (±8)	+42	−57	−7	0
Methanotrophic Conditions									
A	10.4-11.0	5 (±10)	0 (±2)	0 (±3)	10 (±15)	+859	−13	−96	0
B	10.4-11.0	4 (±4)	4 (±3)	5 (±1)	4 (±10)	+94	−11	−96	0
Methanogenic Conditions									
A	4.6-5.2	23 (±12)	93 (±11)	90 (±15)	42 (±15)	+50	−3	+489	N/A
B	4.6-5.2	17 (±15)	36 (±10)	27 (±7)	8 (±2)	+87	−7	+20	N/A

N/A = Not analyzed.
(a) Percent change relative to initial (week 1) microbial count.
(b) Values in parentheses represent the standard deviations.

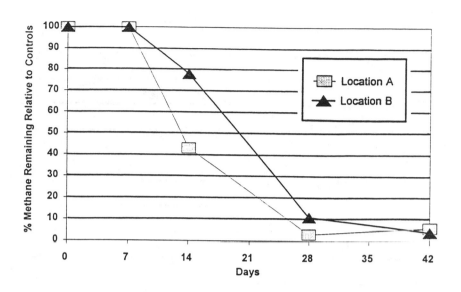

FIGURE 1. Rate of disappearance of methane in methanotrophic microcosms.

Location A (Table 1). Significant levels of 1,1-dichloroethane (1,1-DCA) and *cis*-1,2-DCE in sample microcosms were observed (170 and 1,500 µg/L, respectively), suggesting transformation of parent compounds. Similarly, an increase in CH_4 and a decline in CO_2 (Table 1) suggests that methanogenic bacteria were active in biodegrading the chlorinated parent compounds. Enumeration of bacteria from the anaerobic microcosms indicates that there was an increase in population after the 12-week incubation.

A first-order kinetic model was used to estimate half-lives for those anaerobic samples exhibiting significant biodegradation. The first-order rate model provided an excellent fit for the data collected from Location A, as shown in Figure 2. TCE and PCE half-lives were both 57 days ($k = -0.0122$ and -0.0123, respectively), whereas lindane and TCA had longer half-lives of 133 ($k = -0.0052$) and 267 ($k = -0.0026$) days, respectively. The degradation rates typically were highest after 28 days, indicating a slight lag time. The first-order model was also applied to PCE and TCE data collected from Location B (Figure 2). Degradation of PCE and TCE provided a relatively good fit ($r^2 > 0.70$), although the half-life was significantly longer than Location A samples. This reflects the large variability associated with biokinetics as a result of spatial heterogeneities in distribution and activity of subsurface bacteria.

SUMMARY

The results of the degradation assay suggest that aerobic degradation may possibly be stimulated under methanotrophic conditions, although no significant degradation of the target compounds was measured. In addition, a significant

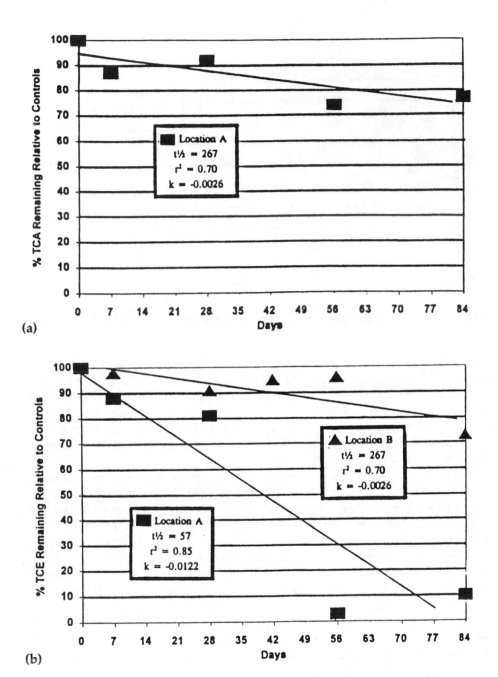

(a)

(b)

NOTE: Kinetic analysis performed only for samples with significant biodegradation and first-order rates. $t_{1/2}$ = half-life, days; r^2 = linear regression correlation coefficient; k = first-order rate constant, days.

FIGURE 2. Rate of disappearance of (a) TCA and (b) TCE in anaerobic
microcosms.

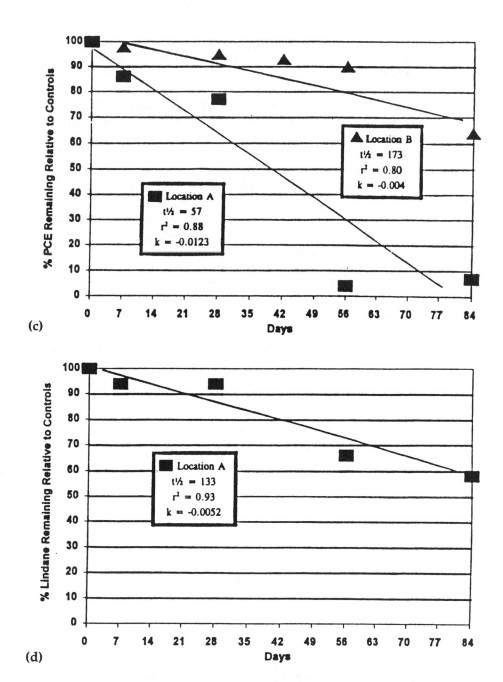

(c)

(d)

NOTE: Kinetic analysis performed only for samples with significant biodegradation and first-order rates. $t_{1/2}$ = half-life, days; r^2 = linear regression correlation coefficient; k = first-order rate constant, days.

FIGURE 2 (continued). Rate of disappearance of (c) PCE, and (d) lindane in anaerobic microcosms.

degradation of the target compounds was measured. In addition, a significant acclimation period was observed, making the field implementation at the site questionable. On the contrary, site microorganisms were capable of degrading TCA, TCE, PCE, and lindane under enhanced anaerobic conditions over a finite time period. However, under these conditions, some chlorinated intermediates (1,1-DCA and *cis*-1,2-DCE) were also identified during the transformation of the parent compounds.

Although not indicative of a full-scale application, the results suggest that implementation of an enhanced in situ bioremediation scheme, particularly under anaerobic conditions, may result in further attenuation of some chlorinated solvents and pesticides. Enhancement of biotransformation could be achieved in the field by the presence of an acclimated microbial population and a biodegradable carbon and energy source. However, if such a system were implemented full scale, time lags and efficiencies could be expected to vary considerably. Successful implementation of this remedial alternative would also be difficult due to the problems of delivering nutrients and maintaining anaerobic conditions throughout the aquifer.

Without the data generated from this study, an accurate assessment could not have been performed on the feasibility of implementing in situ biodegradation as a remedial alternative. Because there was a great deal of uncertainty regarding full-scale bioremediation operations and another alternative (air sparging with vapor extraction) looked more favorable (EA 1994), it was decided not to pursue bioremediation as a remedial technique. However, valuable information was gained from the study providing evidence that natural attenuation may be occurring at the site. This was indicated by measuring low redox conditions in the groundwater, the presence of active and acclimated microbial populations in the field samples, and the positive results for transformation within the soil microcosms.

ACKNOWLEDGMENTS

This study was supported through the U.S. Air Force Environmental Restoration Program. The technical assistance of Liza Wilson and Edward Bouwer of the Johns Hopkins University is gratefully acknowledged.

REFERENCES

EA Engineering, Science, and Technology, Inc. 1994. *Final Feasibility Study Report for Site WP-21, Dover Air Force Base*. September.
Trolldenier, G. 1973. "The Use of Fluorescence Microscopy for Counting Soil Microorganisms." *Bull. Ecol. Res. Comm.-NFR* (Statens Naturventensk Forskningsrad) 17:53-59.
Wilson, J. T., J. F. McNabb, D. L. Balkwill, and W. C. Ghiorse. 1983. "Enumeration of Bacteria Indigenous to a Shallow Water-Table Aquifer." *Ground Water*, 21(2):134-142.

AUTHOR LIST

Adriaens, Peter
University of Michigan
Dept. of Civil & Environ. Engrg.
181 EWRE Building
Ann Arbor, MI 48109-2125 USA

Al-Abed, Souhail
University of Cincinnati
Dept. of Civil & Environ. Engrg.
1275 Section Road
Cincinnati, OH 45237-2615 USA

Alvarez-Cohen, Lisa
University of California, Berkeley
Dept. of Civil Engineering
726 Davis Hall
Berkeley, CA 94720-1710 USA

Arcangeli, Jean-Pierre
Technical University of Denmark
Dept. of Environ. Science & Engrg.
Building 115
DK-2800 Lyngby
DENMARK

Arp, Daniel J.
Oregon State University
Dept. of Botany and Plant
    Pathology
Corvallis, OR 97331 USA

Arvin, Erik
Technical University of Denmark
Dept. of Environmental Engineering
Building 115
DK-2800 Lyngby
DENMARK

Ballapragada, Bhasker S.
University of Washington
Dept. of Civil Engineering FX-10
Seattle, WA 98195 USA

Ballerini, Daniel
Institut Français du Pétrole
Division Biotechnologie et
    Environment
1 et 4 ave de Bois-Préau BP 311
92506 Rueil Malmaison Cedex
FRANCE

Beaulieu, Chantal
National Research Council Canada
Biotechnology Research Institute
6100 Royalmount Avenue
Montréal, Québec H4P 2R2
CANADA

Bielefeldt, Angela R.
University of Washington
Dept. of Civil Engineering FX-10
309 More Hall
Seattle, WA 98195 USA

Bollag, Jean-Marc
The Pennsylvania State University
129 Land and Water Building
University Park, PA 16802-4900 USA

Booth, Steven R.
Los Alamos National Lab
P.O. Box 1663 MS B299
Los Alamos, NM 87545 USA

Brockman, Fred
Battelle Pacific Northwest
Environmental Sciences Dept.
P.O. Box 999, K4-06
Richland, WA 99352 USA

Chang, Chu-Yin
Natl. Chung Hsing University
Environmental Engineering Dept.
Taichung, Taiwan 40227
REP. OF CHINA

Chu, Kung-Hui
University of California, Berkeley
Civil Engineering Dept.
726 Davis Hall
Berkeley, CA 94720-1710 USA

Cirpka, Olaf
Universität Stuttgart
Institute of Hydraulic Engineering
Pfaffenwaldring 61
D-70550 Stuttgart
GERMANY

Corriveau, Alain
National Research Council of Canada
Biotechnology Research Institute
6100 Royalmount
Montréal H4P 2R2
CANADA

Criddle, Craig S.
Michigan State University
Dept. of Civil and Environ. Engrg.
A-126 Research Complex - Engrg.
East Lansing, MI 48824-1326 USA

Davis-Hoover, Wendy J.
U.S. Environ. Protection Agency
National Risk Management Research
     Laboratory
5995 Center Hill Road
Cincinnati, OH 45224 USA

Dec, Jerzy
The Pennsylvania State University
129 Land & Water Building
University Park, PA 16802-4900 USA

Downs, Charles E.
Conoco Inc.
1000 South Pine
P.O. Box 1267
Ponca City, OK 74602-1267 USA

Dybas, Michael J.
Michigan State University
Dept. of Civil & Environ. Engrg.
A-119 Research Complex - Engrg.
East Lansing, MI 48824 USA

Ely, Roger L.
Oregon State University
Dept. of Civil Engineering
2615 NW Mulkey Avenue
Corvallis, OR 97331 USA

Fathepure, Babu Z.
University of Michigan
Dept. of Civil & Environ. Engrg.
181 EWRE Building
Ann Arbor, MI 48109 USA

Fayolle, Françoise
Institut Français du Pétrole
Division Biotechnologies et
     Environnement
1 et 4 Ave De Bois-Préau Bp 311
92506 Rueil Malmaison Cedex
FRANCE

Fennell, Donna E.
Cornell University
School of Civil & Environ. Engrg.
Hollister Hall
Ithaca, NY 14853 USA

Ferguson, John F.
University of Washington
Dept. of Civil Engineering FX-10
Seattle, WA 98195 USA

Field, Jim A.
Wageningen Agricultural University
Dept. of Environmental Technology
Div. of Industrial Microbiology
Biotechnion Bomenweg 2
P.O. Box 8129
6700 EV Wageningen
THE NETHERLANDS

Findlay, Margaret
Bioremediation Consulting, Inc.
55 Halcyon Road
Newton, MA 02159 USA

Fogel, Samuel
Bioremediation Consulting, Inc.
55 Halcyon Road
Newton, MA 02159 USA

Fountain, John
State Univ. of New York at Buffalo
Geology Department
415 Fronczak Hall
Box 601550
Buffalo, NY 14260-1550 USA

Freedman, David L.
University of Illinois at
    Urbana-Champaign
Dept. of Civil Engineering Lab
205 North Mathews Ave. (MC-250)
Urbana, IL 61801-2397 USA

Gao, Jianwei
Battelle Pacific Northwest
P.O. Box 999, MS P7-35
Richland, WA 99352 USA

Gossett, James M.
Cornell University
School of Civil & Environ. Engrg.
Hollister Hall
Ithaca, NY 14853-3501 USA

Gregory III, George E.
DuPont Co.
Environ. Remediation Srvcs.
140 Cypress Station Drive, Suite 140
Houston, TX 77090 USA

Groher, Daniel
ENSR Consulting and Engineering
35 Nagog Park
Acton, MA 01720 USA

Guerin, Turlough F.
Minenco Bioremediation Services
1 Research Avenue
Bundoora, Victoria 3083
AUSTRALIA

Guiot, Serge R.
National Research Council of
    Canada
Biotechnology Research Institute
6100 Royalmount Ave
Montréal, Québec H4P 2R2
CANADA

Hashsham, Syed
University of Illinois at
    Urbana-Champaign
Dept. of Civil Engineering
205 North Mathews Ave. (MC-250)
Urbana, IL 61801-2397 USA

Hawari, Jalal
National Research Council of
    Canada
Biotechnology Research Institute
6100 Royalmount Avenue
Montréal, Québec H4P 2R2
CANADA

Hedrick, David B.
University of Tennessee
Center for Environ. Biotechnology
10515 Research Drive, Suite 200
Knoxville, TN 37932 USA

Herrmann, Jonathan
U.S. Environ. Protection Agency
Office of Research and Development
26 W. Martin Luther King Drive
Cincinnati, OH 45268 USA

Hickey, Robert F.
EFX Systems, Inc.
3900 Collins Road, Suite 1011
Lansing, MI 48910 USA

Hooker, Brian S.
Battelle Pacific Northwest
P.O. Box 999, MS P7-41
Richland, WA 99352 USA

Hundt, Thomas R.
EA Engineering Sci. & Tech., Inc.
Two Oak Way
Berkeley Heights, NJ 07922 USA

Hyman, Michael R.
Oregon State University
Dept. of Botany and Plant Pathology
Cordley Hall 2082
Corvallis, OR 97331-2902 USA

Jain, Mahendra K.
Michigan Biotechnology Institute
3900 Collins Road
P.O. Box 27609
Lansing, MI 48909 USA

Jensen, Hanne Møller
Technical University of Denmark
Inst. of Environ. Science & Engrg.
Building 115
DK-2800 Lyngby
DENMARK

Kitanidis, Peter K.
Stanford University
Civil Engineering Dept.
Terman Engineering Center
Stanford, CA 94305-4020 USA

Kitayama, Atsushi
The University of Tokyo
Dept. of Biochemistry and
    Biotechnology
Faculty of Engineering
Tokyo 113
JAPAN

Koizumi, Jun-ichi
The Natl. University of Yokohama
Division of Bioengineering
Yokohama 240
JAPAN

Kuang, Xin
National Research Council of
    Canada
Biotechnology Research Institute
6100 Royalmount Ave
Montréal, Québec H4P 2R2
CANADA

Lapat-Polasko, Laurie T.
Woodward-Clyde Consultants
426 North 44th Street, Suite 300
Phoenix, AZ 85008 USA

Lasecki, Matthew
HDR Engineering Inc.
4500 SW Kruse Way, Suite 340
Lake Oswego, OR 97035-2564 USA

Lazarr, Natalie C.
Woodward-Clyde Consultants
426 North 44th Street, Suite 300
Phoenix, AZ 85008 USA

Lee, Chi-Mei
Natl. Chung Hsing University
Environmental Engineering Dept.
Taichung, Taiwan 40227
REP. OF CHINA

Lee, Michael D.
DuPont Co.
Environ. Remediation Srvcs.
Glasgow Building 300
P.O. Box 6101
Newark, DE 19714-6101 USA

Le Roux, Françoise
Institut Français du Pétrole
Division Biotechnologies et
    Environnement
1 et 4 Ave de Bois-Préau BP 311
92506 Rueil Malmaison Cedex
FRANCE

Lewis, Ronald F.
U.S. Environ. Protection Agency
Natl. Risk Mgmt. Research Lab
26 W. Martin Luther King Drive
Cincinnati, OH 45268 USA

Ligé, Joy E.
EA Engrg. Science & Tech., Inc.
15 Loveton Circle
Sparks, MD 21152 USA

Lu, Chih-Jen
Natl. Chung Hsing University
Environmental Engineering Dept.
Taichung, Taiwan 40227
REP. OF CHINA

MacDonald, Thomas R.
Stanford University
Civil Engineering Dept.
Stanford, CA 94305-4020 USA

MacFarlane, Ian D.
EA Engrg. Science & Technology
15 Loveton Circle
Sparks, MD 21152 USA

Magar, Victor S.
University of Washington
Dept. of Civil Engineering, FX-10
Seattle, WA 98195 USA

Manley, Scott
Battelle Pacific Northwest
Environmental Sciences Dept.
P.O. Box 999
Richland, WA 99352 USA

Mohn, Henning
University of Washington
Dept. of Civil Engineering, FX-10
Seattle, WA 98195 USA

Murdoch, Lawrence C.
University of Cincinnati
1275 Section Road
Cincinnati, OH 45237-2615 USA

Nagel, Eva
DuPont Co.
301 Glasgow Business Center
P.O. Box 6101
Newark, DE 19714-6101 USA

Nealson, Kenneth H.
The University of Wisconsin-
    Milwaukee
Center for Great Lakes Studies
600 E. Greenfield Avenue
Milwaukee, WI 53204 USA

Nye, Jeffery
Michigan Biotechnology Institute
3900 Collins Road
P.O. Box 27609
Lansing, MI 48909 USA

Odom, J. Martin
DuPont Co.
301 Glasgow Business Center
P.O. Box 6101
Newark, DE 19714-6101 USA

Ogram, Andrew
Washington State University
Dept. Crops and Soils Sciences
Pullman, WA 99164-6420 USA

Palumbo, Anthony V.
Oak Ridge National Laboratory
Environmental Sciences Division
P.O. Box 2008
Oak Ridge, TN 37831-6038 USA

Parkin, Gene F.
University of Iowa
Dept. of Civil & Environ. Engrg.
1136 Engineering Building
Iowa City, IA 52242-1527 USA

Patnaik, Priyamvada
University of Cincinnati
Dept. of Civil and Environ. Engrg.
1275 Section Road
Cincinnati, OH 45237-2615

Payne, William
Battelle Pacific Northwest
Environmental Sciences Dept.
P.O. Box 999
Richland, WA 99352 USA

Peck, Philip C.
Minenco Bioremediation Services
Level 5, 77 Berry Street
N. Sydney, New South Wales 2060
AUSTRALIA

Petersen, James N.
Washington State University
Chemical Engineering Dept.
118 Dana Hall
Pullman, WA 99164-2710 USA

Petrovskis, Erik A.
The University of Michigan
Dept. of Civil & Environ. Engrg.
EWRE Building, Room 108
Ann Arbor, MI 48109-2125 USA

Peyton, Brent M.
Battelle Pacific Northwest
P.O. Box 999, MS P7-41
Richland, WA 99352 USA

Pfiffner, Susan M.
Oak Ridge National Laboratory
Environmental Sciences Division
Oak Ridge, TN 37831-6038 USA

Phelps, Tommy J.
Oak Ridge National Laboratory
Environmental Sciences Division
P.O. Box 2008-6036
Oak Ridge, TN 37831-6036 USA

Puhakka, Jaakko A.
Tampere University of Technology
P.O. Box 527
SF 33101 Tampere
FINLAND

Rhodes, Stuart H.
Minenco Bioremediation Services
Level 5, 77 Berry Street
N. Sydney, New South Wales 2060
AUSTRALIA

Richter, David L.
Conoco, Inc.
124 Environmental Services Div.
P.O. Box 1267
Ponca City, OK 74602-1267 USA

Ringelberg, David B.
University of Tennessee
Center for Environ. Biotechnology
10515 Research Drive, Suite 300
Knoxville, TN 37932-2575 USA

Rock, Jennifer L.
Offutt Air Force Base
55 CES/CEZRMB19
106 Peace Keeper Drive, Suite 2N3
Offutt AFB, NE 68113-4019 USA

Saaty, Ramiz P.
Los Alamos National Laboratory
P.O. Box 1663 M.S. B299
Los Alamos, NM 87545 USA

Saffarini, Daad A.
The University of Wisconsin-
    Milwaukee
Center for Great Lakes Studies
600 East Greenfield Avenue
Milwaukee, WI 53204 USA

Scholze, Richard
U.S. Army Corps of Engineers
Construction Engrg. Research Labs
P.O. Box 9005
2902 Newmark Drive
Champaign, IL 61826-9005 USA

Schraa, Gosse
Wageningen Agricultural University
Dept. of Microbiology
Hesselink van Suchtelenweg 4
6703 CT Wageningen
THE NETHERLANDS

Segar, Jr., Robert L.
University of Missouri - Columbia
Dept. of Civil Engineering
E2509 Engineering Building East
Columbia, MO 65211 USA

Sherwood, Juli L.
Washington State University
Chemical Engineering Dept.
Dana Hall, Room 118
Pullman, WA 99164-2710 USA

Shoemaker, Steven H.
DuPont Co.
140 Cypress Station Dr., Suite 135
Houston, TX 77090 USA

Showalter, W. Eric
Los Alamos National Lab
P.O. Box 1663 MS B299
Los Alamos, NM 87545 USA

Skeen, Rodney S.
Battelle Pacific Northwest
P.O. Box 999, MS P7- 41
Richland, WA 99352 USA

Soong, Adam
Battelle Pacific Northwest
Environmental Sciences Dept.
P.O. Box 999
Richland, WA 99352 USA

Speitel, Jr., Gerald E.
University of Texas at Austin
Dept. of Civil Engineering
Austin, TX 78712-1076 USA

Stams, Alfons J.M.
Wageningen Agricultural University
Dept. of Microbiology
Hesselink van Suchtelenweg 4
6703 CT Wageningen
THE NETHERLANDS

Stensel, H. David
University of Washington
Dept. of Civil Engineering
309 More Hall FX-10
Seattle, WA 98195 USA

Stover, Michael A.
DHHS/PHS/Indian Health Service
New York District Office
223 Genesee Street
Oneida, NY 13421 USA

Strand, Stuart E.
University of Washington
College of Forest Resources AR-10
Seattle, WA 98195 USA

Sun, Wuhua
Washington State University
Dept. Crop & Soil Sciences
Pullman, WA 99164-6420 USA

Tabinowski, JoAnn
DuPont Co.
Glasgow Building 300, Route 896
Newark, DE 19714 USA

Tatara, Gregory M.
Michigan State University
Center for Microbial Ecology
A124 Research Complex-Engineering
East Lansing, MI 48824-1325 USA

Treboul, Chloé
Institut Français du Pétrole
Division Biotechnologies et
    Environnement
1 et 4 Ave De Bois-Préau Bp 311
92506 Rueil Malmaison Cedex
FRANCE

Truex, Michael J.
Battelle Pacific Northwest
P.O. Box 999, MS P7-41
Richland, WA   99352 USA

van Eekert, Miriam H. A.
Agricultural University Wageningen
Dept. of Microbiology
Hesselink van Suchtelenweg 4
6703 CT Wageningen
THE NETHERLANDS

Vane, Leland M.
U.S. Environ. Protection Agency
Natl. Risk Mgmt. Research Lab
26 W. Martin Luther King Drive
Cincinnati, OH   45268 USA

Veiga, Maria C.
Universidade da Coruna
Departamento de Quimica
Fundamental E Industrial
Campus da Zapateira, s/n
15071 A Coruna
SPAIN

Vernalia, Jane
University of California, Berkeley
Civil Engineering Dept.
726 Davis Hall
Berkeley, CA   94720-1710 USA

Vesper, Stephen J.
University of Cincinnati
Dept. of Civil and Environ. Engrg.
1275 Section Road
Cincinnati, OH   45237-2615 USA

Vogel, Timothy M.
Rhône-Poulenc Industrialisation
24 Avenue Juan-Juarès
69153 Décines
FRANCE

Walter, Gunter A.
University of Washington
Dept. of Civil Engrg., FX-10
309 More Hall
Seattle, WA   98195 USA

Weathers, Lenly J.
University of Iowa
Civil & Environmental Engineering
122 Engineering Res. Fac.
Iowa City, IA   52242 USA

White, Darrell G.
Dupont Chemicals
Corpus Christi, TX   USA

Williamson, Kenneth J.
Oregon State University
Dept. of Civil Engineering
Apperson Hall 202
Corvallis, OR   97331 USA

Workman, Darla
Oregon State University
c/o Battelle Pacific Northwest
P.O. Box 999, MS K4- 06
Richland, WA   99352 USA

Wu, Wei-Min
Michigan Biotechnology Institute
3900 Collins Road
P.O. Box 27609
Lansing, MI   48909 USA

Youngers, Greg A.
Conoco Inc.
Environmental Services Division
P.O. Box 1267
Ponca City, OK   74602-1267 USA

Zeikus, J. Gregory
Michigan Biotechnology Institute
3900 Collins Road
P.O. Box 27609
Lansing, MI   48909 USA

Zinder, Stephen H.
Cornell University
Department of Microbiology
W226 Wing Hall
Ithaca, NY   14853-3501 USA

INDEX